SELECTED PRAISE FOR *RISE OF THE MACHINES*

'The major disruptions in our modern society all belong to one big story. A common theme connects war machines, computer networks, social media, ubiquitous surveillance and virtual reality. For 50 years or more the same people and the same ideas weave through these innovations united by the term "cyber", as in cyberspace and cybernetics. Read this amazing history and you'll go: aha!'

KEVIN KELLY, founder of *Wired*

'A fascinating history of cybernetics, and of the visionaries like Norbert Wiener who first imagined the potential – and peril – of machines that would begin to replicate the capabilities of the human mind.'

MARTIN FORD, author of *Rise of the Robots*

'Thomas Rid has provided a gripping account of how after the Second World War, cybernetics, a theory of machines, came to incite anarchy and war half a century later. Thanks to his extensive research we can now read for the first time the real story of Moonlight Maze, the first big state-on-state cyber attack, setting a new narrative standard for historians and journalists alike.'

SIR DAVID OMAND, director of GCHQ during Moonlight Maze

'Everyone I know should read this book. It will be a classic.'

ROBERT LEE, former US Air Force cyber warfare operations officer, author of *SCADA and Me*

'Powerful... Thomas Rid is recent past shapii

Esqui

D1638444

RISE OF THE MACHINES

THOMAS RID is Professor in Security Studies at King's College London. He received his PhD from Humboldt University in Berlin, and worked for ten years in leading think tanks in Berlin, Paris, Washington and Jerusalem. He is the author of four books, including *War and Media Operations* (2007) and *Cyber War Will Not Take Place* (2013). He lives in London.

Follow him at @RIDT

RISE OF THE MACHINES

the lost history of cybernetics

THOMAS RID

SCRIBE

Melbourne • London

Scribe Publications
18–20 Edward St, Brunswick, Victoria 3056, Australia
2 John St, Clerkenwell, London, WC1N 2ES, United Kingdom

Published by Scribe in Australia, New Zealand, and the UK 2016
First published in the United States by W. W. Norton & Company, Inc.
This edition published 2017

Text design by Lovedog Studio
Printed and bound in the UK by CPI Group (UK) Ltd,
Croydon CR0 4YY

9781911344100 (UK edition)
9781925307603 (e-book)

Scribe Publications is committed to the sustainable use of natural resources
and the use of paper products made responsibly from those resources.

CiP records for this title are available from the British Library.

scribepublication.com.au
scribepublications.co.uk

I like to think
    (it has to be!)
of a cybernetic ecology
where we are free of our labors
and joined back to nature,
returned to our mammal
brothers and sisters,
and all watched over
by machines of loving grace.

        —Richard Brautigan, 1967

Contents

Preface

WHAT DOES "CYBER" EVEN MEAN? AND WHERE DOES THE idea come from?

Many people have been asking this question for some time now. My students at King's College London did, as did cyber warfare officers in the US Air Force and Pentagon strategists. Secretive British spies inquired, as did bankers and hackers and scholars. They all struggled with the rise of computer networks, and what that rise means for our security and our liberty. And they all slapped the prefix "cyber" in front of something else, as in "cyberspace" or "cyberwar," to make it sound more techy, more edgy, more timely, more compelling—and sometimes more ironic.

I didn't have a good answer for them, despite having written widely on cybersecurity. This was frustrating. Like so many others, I could only point, vaguely, to a curious origin story: that of a science fiction novel written in the mid-1980s, William Gibson's *Neuromancer*. The indefatigable Gibson plucked cyberspace out of his airy imagination, the story goes, and then punched the future into his olive-green portable typewriter, a Hermes made in 1927. For Gibson, that new electronic space inside the machines, meaningless yet evocative, was the perfect replacement for outer space as a playground for the protagonists of his science fiction novels. He simply wanted an unblemished stage. That origin story, as improbable as it is, has been repeated countless times. Remarkably, one of the biggest challenges to our civil liberties and state sovereignty in the twenty-first century was commonly traced to a fictional story about a drug addict escaping into hallucinatory computer networks.

Could that be right? It was as if history started in 1982 with a fantasy. How did Gibson's story fit into a larger picture, into a cultural and technical trajectory that takes a longer view? How exactly had "cyberspace" made the leap from Gibson's solitary typewriter to the Pentagon's futuristic and lavishly staffed Cyber Command? By 2010, a fast-growing share of the work of the National Security Agency—and of its British counterpart, Government Communications Headquarters (GCHQ)—was somehow "cyber related," as jargon had it.

Then came the great intelligence leaks of 2013. The NSA's and GCHQ's publicly revealed technological capabilities outraged privacy activists and some allies—but humbled many of the world's most fearsome spy agencies; "the leaks demonstrated a capability gap to us," one Chinese intelligence officer told me in Beijing later that year. China had to catch up. Meanwhile, major computer network breaches escalated, with foreign spies and criminals siphoning off vast amounts of intellectual property and sensitive personal information. By 2015, the global cybersecurity market of firms offering a smattering of security solutions had surpassed $75 billion, swelling at double-digit rates. So precarious were the new threats that even in times of economic hardship and austerity, government and military budget lines weren't just safe from cuts, but sloping up fast.

Yet the real story of one of the world's most exciting, most expensive, and most menacing ideas remained an enigma. So where does "cyber" come from? What is the history of this idea? And what does it actually mean?

I started digging. *Rise of the Machines* is what I found.

"Cyber" is a chameleon. For politicians in Washington, the word stands for power outages that could plunge entire cities into chaos at any moment. For spies in Maryland, it stands for conflict and war, and for data being stolen by Russian criminals and Chinese spies. For executives in the City of London, it stands for major security breaches, for banks bleeding money, and for ruined corporate reputations. For inventors in Tel Aviv, it triggers visions of humans merging with machines, of wired-up prostheses with sensitive fingertips, and of silicon chips implanted under tender human skin. For science fiction fans in Tokyo, it stands for an escapist yet retro punk aesthetic, for mirrored shades, leather jackets, and worn-down, dusty gadgets. For romantic internet activists in Boston, it stands for a new realm of freedom, a space beyond the control of oppressive governments and law enforcement agencies. For engineers in Munich, it stands for steely control, and for running chemical plants by computer console. Ageing hippies in San Francisco nostalgically think back to wholeness and psychedelics and "turning on" the brain. And for screen-addicted youth in between, "cyber" means simply sex-by-video-chat. The word refuses to be either noun or prefix. Its meaning is equally evasive, hazy, and uncertain. Whatever it is, it is always stirring, it is always about the future, and it always has been.

One way to clear the fog is to study the history of one of the weightiest and most pivotal ideas of the twentieth century, an idea whose legacy is set to be even more momentous as the twenty-first century moves on: *cybernetics*. Cybernetics was a general theory of machines, a curious postwar scientific discipline that sought to master the swift rise of computerized progress. From the get-go in the early 1940s, it was about computers, control, security, and the ever-evolving interaction between humans and machines.

A crucial moment, it turns out, was World War II—in particular, the air defense problem that emerged as this epic confrontation got under way. To shoot down deadly new bombers, ground-based artillery needed complex ballistic calculations, completed faster and more accurately than human "computers" could perform, or even read off precalculated range tables. Machines needed to be invented for the task. And soon "mechanical brains" started to "think," in the quaint language of the time. The rise of the machines had begun.

In 1940, in the midst of all this, a curious story ran its course at the vast campus of the Massachusetts Institute of Technology. Norbert Wiener, an eccentric mathematician, read about howitzers and artillery shells and was inspired. Shooting at the blue sky, aided by creaking computers, appealed to the roly-poly professor. After the war he took a tangled set of ideas from electrical engineers and weapons designers, straightened it out, refined it, repackaged it, and, with a generous gesture, threw his creation out to an eager public like candy to a throng of hungry children.

The timing was perfect. At the end of the decade, the technological wonders of the military effort were beginning to seep into industry and private households. Somebody needed to explain the new gadgetry and its purpose. Enter cybernetics, the bold theory of future machines and their potential. Wiener and his keen acolytes would enchant the machine; seduced by their own theory, they endowed it with spirit and an appeal that would extend to the cultish. Engineers, military thinkers, politicians, scholars, artists, and activists started projecting their hopes and their fears into the future of thinking machines.

The postwar rise of the machines spans a wide arch: the most crucial anchor point of that arch emerged in the late 1940s with the publication of Wiener's epoch-making book *Cybernetics*. The tweedy scholar with thick, horn-rimmed glasses revealed the magic of feedback loops, of self-stabilizing systems, of machines that could autonomously adapt their behavior and learn. Automata now had a purpose and could even self-reproduce, at least in theory. The machine suddenly seemed lifelike.

The story that ensued is spellbinding: this sizzling-hot new discipline transformed what computers stood for over the next half century—from machines of assured destruction to machines of loving grace. By the late 1940s, the fantastic new way of thinking about cybernetic self-adaptive systems galvanized charlatans and weapons engineers alike, from the Church of Scientology to the Boeing Corporation gearing up for the Cold War. Cybernetics promised to guide stray missiles to targets and lost minds to exaltation. By the 1960s it had given rise to exoskeletons to load nuclear bombers, as well as to deeper connections among wholesome hippies. It improved kill ratios of air force fighters and electronically "turned on" San Francisco's counterculture.

By 1980, a cybernetic space inside the machines had emerged, a mythical place of hope for a freer and better society—and a fierce domain for battle and war. This twisted cybernetic history has a firm grip on what we expect from technology, security, and liberty in the twenty-first century. Ultimately, cybernetics itself—often shortened to "cyber"—would acquire the very features of those mythical machines that it had predicted since midcentury: the idea was self-adapting, ever expanding its scope and reach, unpredictable, and threatening, yet seductive, full of promise and hope, and always escaping into the future.

Rise of the Machines is my attempt to disentangle seven distinct historical cybernetic narratives, each peaking in a particular decade. The book's themes are wide-ranging: they extend from autonomous robots, exoskeletons, and walking trucks to virtual-reality goggles and remailers; sometimes, therefore, a chapter's specific line of inquiry will be abandoned to maintain the book's focus as it moves on to the next decade. The chapters are organized chronologically, but the themes may overlap; in the 1960s, for instance, cybernetics shaped visions of the future of the body, of work, and of community. The narrative at the start of a new chapter may therefore jump back in time to illuminate events that happened simultaneously with events already discussed.

With a focus on the history of cybernetic myths, a word on the mythical is in order. Myths are deeply embedded in our collective memory; they shape our understanding of technology at every turn, even if we aren't aware of this palimpsest, the deep and hidden legacy of the cybernetic history. Contrary to the use of everyday language, a myth—as articulated in a prominent tradition of political theory[1]—doesn't mean that something is factually wrong. Myths don't contradict the facts; they complement the facts. Saying something *works* as a myth doesn't mean that it is "only" a myth. On the contrary. Political and technical myths are very real and work in powerful ways. And they may even align smoothly with the observed, hard facts. Myths are more than real in at least three different ways.

First, myths overcome the limits of fact, experience, and technology at a given moment in time. The rise of the machines was always projected into the future, not into the present or the past. Evidence was always in short supply. The cybernetic promise, of course, was neither wrong nor right. Any vision of the future is neither false nor true until the predicted future, or some pale version of it, actually comes to pass. For their adherents, subscribing to the cybernetic narratives of the future required more than evidence. It required belief. And the myth made it easy to believe. With its inherent uncertainty cloaked in the unflinching language of science and technology, the line between scholarship and worship blurred again and again over the decades. It did so subtly, and seductively.

So powerful was the myth that its own creators kept falling for it. Technology myths have the form of a firm promise: the cyborg will be built; machines that are more intelligent than humans will be invented; the singularity is coming; cyberspace will be free. The myth is underdetermined by fact, yet it purports to be as certain and as hard as empirical evidence can get, shielded from debate and contradiction. Faith dressed as science.

Second, mythologies are remarkable not for their content, but for their form. The basis of the myth appears as fully experienced, innocent, and indisputable reality: computers are becoming ever faster; machines are ever more networked; encryption is getting stronger. But at the same time the myth makes a leap, it adds a peculiar form to the meaning. And this form is always emotional. Myths are convincing because they appeal to deeply held beliefs, to hopes, and often to fears about the future of technology and its impact on society.

These beliefs are informed by visions and projections, by popular culture, by art, by fiction and science fiction, by plays, films, and stories. But the myth often harks back to fiction clandestinely, without making the cultural inheritance explicit. Science fiction novels, for instance, inspired the 1990s national-security debate. And sometimes hard-nosed experts even wrote the fiction, to spell out dystopian visions of future conflict, freed from the unbearable shackles of fact. The crypto activists of the 1990s, a movement of fervent and influential zealots extolling the many blessings of spreading encryption, unabashedly recommended science fiction as the "sources" and the main inspiration for anarchy in "cyberspace."

The third and most crucial feature of cybernetic myths is that they transcend the present. Mythical narratives form a path between the past and the future to keep a community's shared experiences in living memory. For political and historical myths, such as the German raids on the City of London during the Blitz, the more stable anchor point is in the past. The political myth draws a clear line from a past event into the future and sees the present as a dot on this line. It forms the connective tissue of a community's identity over time, evoked again and again by services in St. Paul's and anniversary flyovers by the Royal Air Force.

For cybernetic myths, the reverse is the case: the more stable anchor point is always in the future or, to be more precise, in a shared yet vague imagination of the future—not too close and not too distant. The golden range seems to be about twenty years forward, close enough to extrapolate from the past, yet distant enough to dare brave new ideas of the future. The outcome is equally effective. The technological myth draws a clear line from the future into the past and sees the present as a dot on this line.

The cybernetic myth creates the powerful illusion of being able to predict the future: Trust me, says the mythos; this is how the future will be. This isn't fiction or prediction; this is a hard fact that hasn't happened yet. Keeping a technological myth as an effective and viable path into the future therefore requires permanent use and repetition; it requires restating the mythical promise again and again, so that it becomes and remains gospel. It requires "work on the myth," the German philosopher Hans Blumenberg pointedly observed in *Arbeit am Mythos*.[2]

This is where *Rise of the Machines* comes in. That mythical path into the future can be a clearly marked and straight one, or it can be obscure and looping backward instead of ahead, running into obstacles that had already been overcome. The work on the myth, in short, can either repeat or overcome past misconceptions. It can be obscuring or it can be clarifying, regressive or progressive, a trap or the way out. This book is meant to be *Arbeit am Mythos*.

Rise of the Machines

CHECKING THE LUGGAGE. LONG AIRPORT WALKWAYS. PASSPORT at the ready. Standing in line at security, taking off shoes and belt. The wait at the gate. Finally the ritualized boarding: Group 3, Seat 37B. Carry-on shoved in the overhead bin. The cabin crowded and too hot, the seats too narrow, the entertainment screens too small. As the safety belt's chrome buckle firmly clicks, the spine-tingling thought snaps into your mind: for the next eight hours, as soon as the wheels have lifted off the runway, your life will depend on this machine. On its engines, its fuselage, its rudders and flaps, its instruments, its air supply, its navigation system, its landing gear, its computers and their software, and God knows what else. You tell yourself that, statistically, flying is safer than crossing the street. But that doesn't change the unnerving fact: you just entrusted your life, all you have, to a computerized and engineered black box flying at an altitude of 33,000 feet in air that you couldn't breathe, over a surface of deadly open water 1,800 miles across. And there's nothing you can do about it now.

Over the next eight hours you forget about the unease, numbing your mind with a bad movie. As the plane touches down and starts rumbling along the runway, braking noticeably, you are briefly reminded that, finally, you're almost back in control of what comes next. As the machine taxis to the gate, people are pulling out their phones as if to celebrate the moment. You tap the virtual switch to turn off airplane mode, and wait until the familiar carrier symbol appears. A few notifications pop up as the phone gets a signal. A missed call. A text message from a loved one. E-mails. Swiping away spam. A peek at a social network. Even before you leave the plane, you know that the person picking you up is likely to be a few minutes late, what your friends were up to, and what your colleagues were reading while you were over Greenland.

Machines are about control. Machines give more control to humans: control over their environment, control over their own lives, control over others. But gaining control through machines means also delegating it *to* machines. Using the tool means trusting the tool. And computers, ever more powerful, ever smaller, and ever more net-worked, have given ever more autonomy to our instruments. We rely on the device, plane and phone alike, trusting it with our security and with our privacy. The reward: an apparatus will serve as an extension of our muscles, our eyes, our ears, our voices, and our brains.

Machines are about communication. A pilot needs to communicate with the aircraft to fly it. But the aircraft also needs to communicate with the pilot to be flown. The two form an entity: the pilot can't fly without the plane, and the plane can't fly without the pilot. But these man-machine entities aren't isolated any longer. They're not limited to one man and one machine, with a mechanical interface of yokes, throttles, and gauges. More likely, machines contain a computer, or many, and are connected with other machines in a network. This means many humans interact with and through many machines. The connective tissue of entire communities has become mechanized. Apparatuses aren't simply extensions of our muscles and brains; they are extensions of our relationships to others—family, friends, colleagues, and compatriots. And technology reflects and shapes those relationships.

Control and communication began to shift fundamentally during World War II. It was then that a new set of ideas emerged to capture the change: *cybernetics*. The famously eccentric MIT mathematician Norbert Wiener coined[1] the term, inspired by the Greek verb *kybernan*, which means "to steer, navigate, or govern." *Cybernetics; or, Control and Communication in the Animal and the Machine*, Wiener's pathbreaking book, was published in the fall of 1948. The volume was full of daredevil prophecies about the future: of self-adaptive machines that would think and learn and become cleverer than "man," all made credible by formidable mathematical formulas and imposing engineering jargon.

To the surprise of its publishers, the book became a best seller. "Once in a great while a scientific book is published that sets bells jangling wildly in a dozen different sciences," *Time* wrote in a glowing review in December 1948.[2] The magazine later even did a cover story on the "intensely exciting" new discipline, illustrated with a cartoon of a Mark III computer portrayed as a uniformed naval officer, titled: "The Thinking Machine."

The public celebrated Wiener as the prophet of a second industrial revolution. In the first revolution, engines and production machines had replaced human muscles; now, in the second revolution, control mechanisms would replace human brains, *Time* raved: "They never sleep or get sick or drunk or tired. If such mechanisms are properly designed, they make no mistakes." Wiener would make similarly grand comparisons in his own writings. As men would construct better calculating machines, Wiener explained, and as men would better understand their own brains, the two would become more and more alike. The magazine also eloquently conveyed some of the MIT professor's pessimism about the coming cybernetic age: "Man, he thinks, is recreating himself, monstrously magnified, in his own image," he wrote, in quaint masculine language.

Wiener's was an extraordinary set of ideas, deeply practical and deeply philosophical at the same time. It was not a coincidence that the new discipline emerged during the epic clash of arms of the 1940s: the speed of battle was drastically accelerating, largely as a result of airpower. Shooting down bombers and fighters therefore required faster and more complex ballistic calculations, faster and more precise artillery fire, and faster and more extended communications. The war fathered a range of innovations that forever changed how humans related to machines, especially to computers.

Cybernetics emerged in reaction to these innovations. A small band of scientists soon articulated a *general theory of machines*—not merely a theory of the machines that had been built already, but a theory of all machines, including those that had not been invented yet.

Machines are also about the future. And cybernetics, forged at war, has hence been the vehicle that projected and predicted the future of ever more intelligent automata. Two opposing forces have shaped cybernetic visions of the future. The first was the hope for a better world with less violence, for work to become more humane, for play to become more entertaining, for politics to become more free, for war to become less bloody. Thinking machines brought progress, in that deeply modernist belief.

But an opposing force equally shaped cybernetic visions of imminent technological change: the fear of a world with robots pushing workers into unemployment, of machines harming humans, of critical systems breaking down, of mass surveillance and the loss of privacy, of mechanized regression. Optimism competed against pessimism, liberation against oppression, utopia against dystopia.

This book examines the implications of giving control to machines, and of interacting with and through them. Are machines finally freeing humankind from the need to do dirty and repetitive labor, from the need to sit in mind-numbing traffic jams, generally making work, life, and play more social, more networked, but also safer and more secure? Or are modern societies sleepwalking into a dangerous brave new world that is slowly slipping out of control? Are we inadvertently setting up networked economies, literally reaching into our pockets and handbags, that could grind to a screeching halt at any moment, and possibly blow up at the most critical hubs? How much risk are advanced societies taking by delegating ever more control into the virtual hands of ever more networked and seemingly ever more intelligent machines?

No one can answer these questions about the future. The future hasn't happened yet. From today's vantage point, the future is hazy, dim, and formless. But these questions aren't new. The future of machines has a past. And mastering our future with machines requires mastering our past with machines. Stepping back twenty or forty or even sixty years brings the future into sharper relief, with exaggerated clarity, like a caricature, revealing the most distinct and marked features. And cybernetics was a major force in molding these features.

That cybernetic tension of dystopian and utopian visions dates back many decades. Yet the history of our most potent ideas about the future of technology is often neglected. It doesn't enter archives in the same way that diplomacy and foreign affairs would. For a very long period of time, utopian ideas have dominated; ever since Wiener's death in March 1964, the future of man's love affair with the machine was a starry-eyed view of a better, automated, computerized, borderless, networked, and freer future. Machines, our own cybernetic creations, would be able to overcome the innate weaknesses of our inferior bodies, our fallible minds, and our dirty politics. The myth of the clean, infallible, and superior machines was in overdrive, out of balance.

By the 1990s, dystopia had returned. The ideas of digital war, conflict, abuse, mass surveillance, and the loss of privacy—even if widely exaggerated—can serve as a crucial corrective to the machine's over-whelming utopian appeal. But this is possible only if contradictions are revealed—contradictions covered up and smothered by the cybernetic myth. Enthusiasts, driven by hope and hype, overestimated the power of new and emerging computer technologies to transform society into utopia; skeptics, often fueled by fear and foreboding, overestimated the dystopian effects of these technologies. And sometimes hope and fear joined forces, especially in the shady world of spies and generals. But misguided visions of the future are easily forgotten, discarded into the dustbin of the history of ideas. Still, we ignore them at our own peril. Ignorance risks repeating the same mistakes.

Cybernetics, without doubt, is one of the twentieth century's big-gest ideas, a veritable ideology of machines born during the first truly global industrial war that was itself fueled by ideology. Like most great ideas, cybernetics was nimble and shifted shape several times, adding new layers to its twisted history decade by decade. This book peels back these layers, which were nearly erased and overwritten again and again, like a palimpsest of technologies. This historical depth, although almost lost, is what shines through the ubiquitous use of the small word "cyber" today.

The prequel begins when German bombers took to the sky toward London in the summer of 1940. The first two chapters set the stage. The war, with its gargantuan research effort, forms the crucial back-drop to the vertigo-inducing rise of cybernetics and the hopes and fears it elicited, as outlined in the first chapter, *Control and Communi-cation at War*. The second chapter, *Cybernetics*, charts the early rise of the new discipline. Each of the remaining seven chapters then portrays one distinct theme of the great cybernetic story, roughly structured according to each theme's decade of origin.

The 1950s saw a wide-eyed debate about the coming gloom and boom of *Automation*, with the optimists eventually prevailing over the pessimists. By 1960, another cybernetic myth of mechanized

Organisms inspired the popular view that man was now able to build a superman, that humans were now able to improve on the work of their own creator—and that machines, in turn, could improve on man's creation. By the 1970s, *Counterculture* had discovered the wholesome, cyclical, and spiritually liberating potential of cybernetics—the "machines of loving grace," in the famous phrase of one hippie poet.[3]

This shift of perspective then propelled an even more curious development. The 1980s gave rise to the intoxicating idea that networked computers opened access to a new, uncharted, truly free and lawless virtual *Space*. By the 1990s, with the spread of public cryptography, the myth had turned political: machines, when powered by the right algorithms, could create a superior and truly libertarian political order, crypto *Anarchy*. Then, as the century drew to a close, a curious flip happened. By the 1990s, dystopia was back in full force, with the final cybernetic myth rearing its ugly head, *War*. Malicious code could now bring down entire nations and forever change the nature of military confrontations. The long fall of the machines had begun.

1. CONTROL AND COMMUNICATION AT WAR

THE EVENING SKY OVER LONDON WAS WIDE AND DARK
blue, with only scattered clouds. It was the autumn of 1940. The calm
was deceiving. Suddenly the wail of air-raid sirens ripped through the
twilight. To Londoners, the sound was as familiar as it was unnerving. Night raids had become the norm. Between August 23, 1940,
and New Year's Eve that year, only eight nights were quiet, without
German aerial assaults on Britain. The Germans had begun flying at
night. That made it harder for the RAF's Fighter Command to hunt
the incoming bomber convoys, and it made it harder for London's
defenders to shoot them down. But raiding in the dark was also inaccurate. Pilots were still bombing by sight, often using light incendiary
bombs to mark targets for subsequent heavy bombardment. The Germans focused on sites of military interest, such as centers of industrial
activity and transportation hubs.[1] No light would escape London's
blacked-out windows. On moonless nights the Luftwaffe even relaxed
the rules of engagement.

That night the moon was rising, illuminating the capital's landscape of red and brown rooftops in a soft and silvery sheen. At 85 Fleet
Street, two reporters grabbed their steel helmets and climbed on top

of the *Chicago Tribune*'s fifth-floor office, next to St. Bride's Church, designed by Sir Christopher Wren in 1672. The two American scribes, Joseph Cerutti and Larry Rue, awaited the coming raid.

They looked up: "The sharp beam of searchlights stabbed the sky."[2] Then, to the southeast of London, they saw "a glittering chain of tracer bullets mounting skyward." Next came the crack of antiaircraft batteries, with their shells bursting high in the sky, like shooting stars in reverse. Only then did Cerutti and Rue hear the "remorseless *throb-throb-throb*" of dozens of German bombers, laden with high explosives and incendiary bombs. High over the city the pilots opened the hatches. Their deadly cargo came whistling down, at first invisible, and then the bombs thundered on impact and the incendiaries spread a glaring white flame far and wide. The red glow of fire had become familiar, dimmed at first by clouds of smoke. Flocks of city birds—starlings, sparrows, and pigeons—fluttered into the burning sky. The fire, swiftly taking hold, lit up "in ghastly relief" the huge dome of St. Paul's, London's majestic cathedral church.

Only the approaching dawn brought relief. The slowly returning daylight seemed to repel the relentless waves of bombers. "We're alright now," said Rue. The *Tribune*'s London bureau chief had seen many raids before: "It's dead overhead."[3] Then Cerutti heard the sound of a single plane circling low. A big lone bomb came hurtling down:

> It landed on a solid office block in a neighboring street. I stood rivetted behind a low stone parapet. The bomb exploded with an ear-splitting crash and, in the flash of the explosion, I saw the entire façade of the building rise gently into the air, seemingly intact. Windows and cornices stood out clearly as it mounted, perhaps 50 feet, and then disintegrated, raining debris over a wide area.

The famous air battle of Britain unfolded incrementally, starting in June 1940. On August 1, Adolf Hitler issued Führer Directive number

17. It ordered the Luftwaffe to "overpower the English Air Force with all the forces at its command in the shortest possible time."[4] The bombing raids became more intense during August. After a change of strategy in early September, Hitler chose London as the prime target. On September 15, two hundred German aircraft with a heavy fighter escort took aim at the imperial capital. The pounding continued for months. By day, German bombers and fighters swept across southeast England. By night, they attacked London. The Luftwaffe escalated on the night of October 15–16, sending 235 bombers to the capital. British defenses were dismal: with 8,326 rounds fired, London's defenders managed to destroy only two planes and damage just two others that night.[5] The year ended with a great fire raid on the city, on the night of December 29–30, which famously engulfed St. Paul's Cathedral in flames. Only fourteen enemy aircraft were shot down during all of December.

The Battle of Britain was "truly revolutionary," military historian John Keegan observed.[6] For the first time in history, one state had taken an entire military campaign to the skies to break another state's will to resist. No land or sea forces were attacking Britain; only the mighty German air force. The need for action and improved air defenses was great. It was acutely felt across the Atlantic. In a curious chain of events, the German bombs that were falling out of London's night sky helped trigger a veritable explosion of scientific and industrial research in the United States. Only four years later, before the war in Europe was over, new thinking machines would be deployed to the English Channel—machines that were capable of fighting each other, and of making autonomous decisions of life and death.

Vannevar Bush was one of his generation's gifted visionaries and a prolific inventor. Since 1932, Bush had been vice president of MIT and dean of the School of Engineering. In 1936, the US Army General Staff had cut half of its research-and-development budget, believing that America's weaponry was adequate and that the money would be better spent on maintenance, repair, and more ordnance.[7] After making inquiries, Bush was dismayed to find a military leadership clueless about how science could be useful in war—and scientists clueless about what the military might need in the event of war.

Bush's service on the National Advisory Committee for Aeronautics (NACA), the organization that preceded NASA, gave him unique insights into cutting-edge aeronautical developments: in 1938, he heard a fellow member, Charles Lindbergh, give a talk after his return from a privileged tour of German munitions and aircraft factories. Lindbergh was impressed by the mighty German war machine, especially by the displays of the seemingly invincible Luftwaffe. And few people understood the power of the flying machines as well as Lindbergh did. Eleven years earlier the aviation pioneer had become the first pilot to fly nonstop from New York to Paris. He later compared his plane, the *Spirit of St. Louis*, to a "living creature." High in the air he, Lindbergh, would find unity with the machine, "each feeling beauty, life, and death as keenly, each dependent on the other's loyalty. *We* have made this flight across the ocean, not *I* or *it*."[8] Lindbergh feared that when war came, the unity between man and ever faster, bigger, and more powerful machines would no longer be about beauty and life, but about death from above. America should remain uninvolved in the war, the aviator was convinced.

Bush drew different conclusions. The rugged New Englander had strong views, stamina, and drive.[9] America needed to get ready for war. And that meant that science had to do its bit. In January 1939, Bush, then fifty years old, moved from Boston to Washington to become president of the Carnegie Institution. He was already well connected when he arrived in the District of Columbia that winter: he had chaired a division at the National Research Council, and he had served on NACA. Bush was keen to be involved in the politics of research funding. Carnegie's offices were at the corner of Sixteenth and P Streets, ten blocks north of the White House. With Europe still at peace, in the spring of 1939, Bush became concerned about the "antiaircraft problem," more than a year before the Germans exploited it so devastatingly in the Battle of Britain.

Through his work at NACA, Bush saw that aircraft would grow bigger, faster, and capable of flying at higher altitudes. This evolution, he understood, made it difficult, if not impossible, to bring down the machines with run-of-the-mill gunnery. Hitting the machines directly with artillery shells that exploded on impact was practically impossible. The shells needed to be timed so that they would detonate close enough to the target to bring it down. Yet correctly setting the time fuse became ever harder as speed and distances increased. In October 1939, Bush was elected chairman of NACA, the agency that coordinated research into aeronautics. But NACA's antiagency, an institute coordinating research into air defense, did not exist. Bush proposed to the president that "no similar agency exists for other important fields, notably anti-aircraft devices."[10] On June 27, 1940, Roosevelt established the National Defense Research Committee, better known by the shorthand NDRC.[11] Its purpose was to fund academic research on practical military problems. The NDRC would be spectacularly successful.

Engineers often used duck shooting to explain the challenge of anticipating the position of a target. The experienced hunter sees the flying duck, his eyes send the visual information through nerves to the brain, the hunter's brain computes the appropriate position for the rifle, and his arms adjust the rifle's position, even "leading" the target by predicting the duck's flight path. The split-second process ends with a trigger pull. The shooter's movements mimic an engineered system: the hunter is network, computer, and actuator in one. Replace the bird with a faraway and fast enemy aircraft and the hunter with an antiaircraft battery, and doing the work of eyes, brain, and arms becomes a major engineering challenge.

This engineering challenge would become the foundation of cybernetics. When Norbert Wiener read about the duck-shooting comparison, he instantly fell for it.[12] And he would claim again and again, falsely, that he overcame the related prediction problem. In reality, one of America's most gifted entrepreneurs had already cracked prediction, and built an entire contracting empire in the process. Wiener's successful theory, although the professor never acknowledged it, stood on the shoulders of an engineering giant.

Elmer Ambrose Sperry's business acumen was extraordinary. He founded Sperry Gyroscope in 1910, at 40 Flatbush Avenue Extension in downtown Brooklyn. Sperry's vision was to build a company that provided control as a separate technology: stabilizing ships, guiding airplanes, and directing guns. Sperry products would make these machines perform at a higher level and with greater reliability than human operators could have achieved unaided.

Sperry understood that air defense problems were not limited to the ground. The Flying Fortresses, America's mighty B-17 bombers, were large and vulnerable to fast, small, and swarming fighters. The large aircraft needed novel defenses. Thomas Morgan, Sperry's president in the early 1940s, explained that the firm's primary value of military products would be that "they extend the physical and mental powers of the men in the armed forces enabling them to hit the enemy before and more often than the enemy can hit them."[13]

Sperry turrets were an example. The turret gunners worked individually. Their .50-caliber guns could be operated by line-of-sight, with visual targeting and relatively close range. An airborne fire control mechanism like a gun director was not necessary to operate the hydraulic turrets. The turret's movement was smoothed and stabilized, enabling the gunner to swing around rapidly in pursuit of enemy fighters. But fending off the attacking planes was not automated, although the machine had fail-safes that prevented gunners from hitting parts of their own planes in the stress of battle. Nevertheless, the turrets were taking man-machine interaction to a new level.

Sperry was looking for a new way to depict how soldiers and workers interacted with machines. The firm's engineering graphics department decided to hire an artist with experience in perspective drawing. They settled for Alfred Crimi, a well-known Sicily-born fresco and mural painter from New York City. After an in-depth security vetting, and after getting used to the Italian sporting oriental silk cravats and a goatee, Sperry gave Crimi a private studio. The company didn't know how to take advantage of the artist properly at first, so he had freedom and time to experiment.

Crimi developed a technique of transparent, overlapping drawing. His best-known paintings show gunners in turrets, with their gun sights visible through the body, "as though seen through an X-ray," Crimi explained.[14] He depicted human-machine interaction both at the fighting front and at the home front, detailing the company's assembly lines of navy gun sights, female workers working with great focus at a microscope, giant gyrocompasses at sea, and a high-altitude laboratory that simulated an altitude of 72,000 feet.

One of Crimi's most famous pencil drawings shows a gunner lying in a Sperry ball turret. This tiny spherical cabin, with two guns protruding from it, stuck out from the belly of the B-17 Flying Fortress. The turret was kept tiny to reduce drag on the plane. It housed two light-barrel Browning .50-caliber machine guns, with 250 rounds each. An elaborate chute system at the top of the sphere fed ammo down the outer shell to the guns. The guns extended through the entire contraption, on both sides of the gunner. The turret had several triangular windows, with a large, 13-inch-diameter bull's-eye between the gunner's legs. The turret had no space for a parachute.

The gunner rotated the turret hydraulically, with two hand control grips, similar to joysticks. The ball could move nearly 360 degrees vertically and 90 degrees horizontally. This wide range meant that the gunner could either lie on his back or almost stand on his feet while shooting the twin gun. Each joystick had one button to fire. The gunner's right foot operated a push-to-talk intercom system; his left foot operated a reflector sight that superimposed an illuminated pointer on the target. The gunner, usually the smallest crew member, entered the turret when the plane was on course, after the landing gear had been retrieved. The crew pointed both guns toward the ground, and then the gunner pulled open the door, stepped in, placed his feet in the stirrups and curled down into a fetal position between the two Browning guns. After tightening the straps, he had control of the swiveling weapon.

When the gunner tracked an enemy fighter while attacking from below, "hunched upside-down in his little sphere, he looked like the fetus in the womb," in the words of Randall Jarrell, a celebrated American poet.[15] Jarrell served as an officer in the army air forces during the war. In 1945, he published a powerful five-line poem, "The Death of the Ball Turret Gunner." Jarrell's poem wrestled with the consequences of merging man and machine in industrial warfare. The human operator simply became a cog hunched in the belly of the machine, insignificant and disposable, and eventually torn into pieces by enemy fire and "washed out of the turret with a hose."

The same theme, if not as cruelly drawn, defines Crimi's sketches and drawings for Sperry. In true modernist fashion, Crimi's cutaways visually merged man and machine. The sketches made parts of machine casings invisible to reveal human operators strapped inside the turrets as living parts of the machine. Human bodies, in turn, became transparent to reveal dials and levers, or simply disappeared at the waist. Faces were eerily absent. The drawings were reminiscent of sketches used to teach anatomy to medical students. Crimi and the turrets illustrated how men interacted with machines to increase their muscle power. But the ball turret gunner was still using his own eyes to observe an approaching fighter and his own brain to decide where to aim the guns to destroy it.

Crimi's sketches were intended as "morale builders," for "breaking the monotony endured by the assembly line workers."[16] His drawings were prominently displayed in *Time*, the *Illustrated London News*, *Popular Science*, *Diesel Progress*, and other industry publications.[17] These iconic paintings hit a nerve. The art pieces captured the popular excitement about new forms of human-machine interaction, of "mechanized man"; it was this excitement that Wiener's pathbreaking book tapped into. Crimi, at Sperry, expressed in images what cybernetics would soon express in its own jargon: the relationship between humans and their machine tools was beginning to shift.

Meanwhile, some of the free world's brightest engineers worked on control and communication at war, long before cyberneticists even articulated it as "feedback" loops. Air-to-air combat was difficult enough. But the antiaircraft problem was even more vexing when viewed from the ground. Simply seeing the target could be a challenge. By the time an approaching Junkers Ju 88, a new German bomber used in the Blitz, came into the line of sight, it was probably too late to fire because the aircraft was already too close. Defending against an aircraft from the ground required seeing it before the human eye could. Defense required extending the senses, enhancing perception itself. That trick was accomplished by radar. Existing systems could already be used to guide searchlights in the night sky. But the best radar systems in 1940 were not good enough for automatic fire control against enemy aircraft. The NDRC was determined to overcome this limitation.

The term "radar" originally was an abbreviation, a short form for "radio detection and ranging." The main objective of radar was to determine an object's distance in space from the radar station, its range. By 1940, both the Axis powers and the Allies were starting to use shortwave radar. But neither had yet cracked the design for the much more powerful microwave radar technology. That, however, was about to change. Until the atomic bomb was dropped, the Allies saw microwave radar as the war's most powerful secret weapon, a crucial new technology that stood between survival and defeat at the hands of the Axis powers.[18]

Radar could "see through the heaviest fog and the blackest night," in the words of the *New York Times*. The operating principle was simple, a bit like throwing a stone into a dark hole and waiting to see how long it took to hit the ground: the radar station emitted radio waves, the target reflected some of the waves' energy, and an antenna received the echo. The time it took for the echo to return indicated the target's range. The radar's electromagnetic pulse signal traveled at the speed of light, about 186,000 miles per second. If an object was 15 miles away from the radar, its echo would return 0.00016 second later. The detected range and direction would be displayed to the operators on a "scope," a round screen that resembled the faintly illuminated face of a clock. A number of rings marked the scope, and sometimes a map was superimposed on it. The target would appear on the scope as a small glimmer of light, a "pip." The pip's distance from the scope's center depended on how long the echo took to return.

Radar also measured precise direction—where the target was,

not just how far away it was. The antenna's position indicated the direction: it would rotate and throw out sharply directed pulses, like searchlights of microwaves. The target showed up as a small flickering pip on the operator's round screen when the rotating antenna was pointed straight at the target. The target's altitude was calculated through the antenna's upward-pointing angle. Naturally, radar would pick up noise. Radar manuals in the 1940s often included long sections on "pipology," or "the study and interpretation of all types of contacts seen on radar indicators," as one military handbook defined it.[19] It was an art: operators needed a finely trained eye for the size of the pip, its shape, its bobbing and flickering, its fluctuation in height, its movement in range and bearing. Their task was daunting: mistaking noise for a signal could mean firing at a rock, or firing at friendly forces instead of the enemy.

The US Army's first radar system, the SCR-268, was designed in 1937. It was clumsy. The 268 had vast antennas, about 40 feet wide and 10 feet high. It was also inaccurate. The 268's problem was its long wavelength: 5 feet. Using this radar was a bit like using a map in bird's-eye view, without the ability to zoom in and see details. The solution was theoretically simple but practically hard: shorter wavelengths, or *microwaves*. Shorter waves with a higher frequency had a critical advantage. The shorter the wave, the more accurate the beams, and therefore the higher the resolution of the picture that operators could see. Using the new radar, in theory, enabled them to zoom into the map at high resolution. It would be an incredibly powerful tool. But there was a catch: physicists knew microwaves existed, but nobody had figured out a way to generate and emit microwaves in sufficient power for a useful radar set.[20] German engineers did not even consider microwave radar to be technically possible.[21]

War brought an answer to MIT. It wasn't without irony: by attacking England, Germany helped create one of the most potent weapons that would ultimately bring it down. The fierce German aerial attacks on London and southern England meant that Britain had to focus all her energies on immediate war production. It became harder to continue basic research. So Sir Henry Tizard, then the chair of Britain's Aeronautical Research Committee, set out on a mission to enable US applied research to take advantage of some of England's most prized top secret experiments on microwave technology. Researchers at the University of Birmingham had made a sensational discovery in late 1939 and named it the "cavity magnetron."[22]

The tiny contraption was remarkable: it could produce the coveted shortwaves, below 10 centimeters, even down to 3 centimeters. Even better, aircraft and boats could carry the magnetron's much smaller antennas. The potential was tantalizing for US planners: not only would they be able to see the enemy at high resolution at all times while he was unable to see them, but the technology held even more promise, in that radar could become mobile, enabling aircraft to fly in the blackest of nights and ships to maneuver in the densest of fogs. That still wasn't all: 10-centimeter and 3-centimeter radar sets were much harder to jam than radar with longer wavelength. This meant that the Allies could jam the enemy, blinding him, while enhancing their own perception.

America's radar program changed radically on August 28, 1940. A fierce tropical storm lashed the mid-Atlantic states that Wednesday. Vannevar Bush had dinner with Tizard at the Cosmos Club in Washington, DC. The two got along well and discovered a shared passion for applied civilian research. The dinner set in motion a series of events that led to Bush's NDRC taking control of microwave research. The army and the navy had terminated their own microwave research in 1937 and didn't object. With the magnetron, Bush recalled, "we ran away with the ball."[23]

In October 1940, MIT's Radiation Laboratory was established, initially with a few dozen researchers and just a few rooms. Over the next months, the lab made breathtaking progress. The MIT engineers made another brilliant discovery: they realized that if the reflected radar pulse could be amplified, through feedback, to control the servo-mechanisms of the radar's antenna, then the same faint signal—now vastly more precise, thanks to the smaller microwaves—could also control a howitzer. If radar could automatically track a target, then entire guns could automatically track their targets as well.

At the end of May 1941, the Rad Lab demonstrated its experimental automatic angle-tracking radar system at MIT. The engineers hauled a .50-caliber machine gun turret originally designed for the B-29 bomber to the roof of an MIT building and then hooked it up to one of their experimental radar stations. They set up the system so that the gun would automatically point at a tracked aircraft flying by, even when it was behind cloud cover. "It was very impressive," recalled Ivan Getting, who led the demonstration:

> You could look through the telescope mounted on the radar mount, and the airplane would go behind a cloud, and you wouldn't see anything but a cloud. When the airplane emerged from behind the cloud, there was the airplane right on the cross hair. It was just like magic.[24]

The engineers knew what to do: take the roof-mounted contraption, redesign it, and mount the automatic antiaircraft gun on a truck. In early December 1941, the Rad Lab took its experimental truck, the XT-1, to the US Army's Signal Corps at Fort Hancock, New Jersey, for demonstration. On Friday evening, December 5, the engineers celebrated the success of their new machine with generous amounts of beer at the fort. Two days later, on Sunday morning, Japan attacked Pearl Harbor.

Over the course of the next four war years, the Rad Lab swelled into a research giant that took over most US radar work, with a $4 million monthly budget and a staff of about four thousand that included one-fifth of all of the nation's best physicists.[25] The Rad Lab operated its own manufacturing plant, ran its own airport at Bedford, Massachusetts, and had its own field radar stations across the United States and around the world. The lab became the NDRC's largest project and one of the most celebrated scientific institutions of the war. By May 1945, less than five years after the Tizard mission, the army and navy had contracted $2.7 billion of MIT-inspired radar equipment. This remarkable investment laid the foundation for America's mighty postwar electronics industry.

The lab's most notable achievement was the truck-mounted microwave radar, the XT-1. The army renamed it SCR-584. The machine's name stood for "Signal Corps Radio 584." It was a formidable device. The 584 made nearly all earlier radar systems obsolete. The machine was precise enough to display on its scope the trajectory of a 155-millimeter artillery shell as it approached its target. When the small pip and the large pip converged on screen, both simply disappeared.

Enhancing muscles through gun-pointing hydraulics was impressive. Enhancing perception through radar was even more impressive. Both advances weren't enough, though. Hitting a German bomber from far away required more than seeing the plane ahead of time and being physically able to point a powerful gun at it. Hitting the enemy bomber required *knowing* where to aim the gun before firing. The shell didn't travel at light speed like the radar pulse from the roof of MIT did: a 155-millimeter shell could be in the air for up to twenty seconds before it reached the German Junkers bombers en route to London, and the targeted plane could fly up to 2½ miles between the moment the enemy antiaircraft gunners fired the shell and the moment the shell hit the plane. Like the hunter shooting ducks on the wing, the gunner had to predict and aim at a point in the future. A separate mechanical brain was needed to make this prediction.

Military units in charge of shooting big guns are called "batteries."

Fire control—accurately aiming the complex artillery guns—was hard. In the early days the different elements of an antiaircraft battery could be several hundred feet apart, depending on terrain and tactics. The battery's independent components were linked by telephone lines. To hit a target, an observer had to relay data to an officer by telephone. The officer would then input the data into a primitive computer and obtain the output variables. He would then telephone the gun installations and read the targeting data to them. The gunners used the data to set the shell's fuse and aim the gun, and then they fired. Lines of communication were half the task. So perhaps it isn't too surprising that a crucial role in the history of fire control fell to a telephone company: Bell Telephone Laboratories, a mighty research institute founded by AT&T and Western Electric based in Manhattan.

Accurately firing a gun at a moving target required two separate calculations: *ballistics* and *prediction*. The ballistic solution is more straightforward: how to fire a shell so that it explodes at a specific point in space and time. To accomplish that task, a gunner needed to provide three values to the gun: azimuth and elevation to determine the direction of fire, and timing for the shell's fuse setting to determine when it would explode. The traditional, nonautomated method for artillery crews was to look these values up in a firing table. These tables had long columns for elevation, azimuth, fuse settings, time of flight, and drift.

As gunnery evolved, the range correction factors became more elaborate: muzzle velocity, headwind, tailwind, air temperature, air pressure, and more. Studying booklets under fire became impractical. In response, mechanical gun directors automated the tables by converting them into strangely shaped metal cones dotted with pins, a bit like the revolving cylinders of an old-fashioned music box that would play a particular melody. These cylinders, so-called Sperry cams, looked like twisted and curved tree trunks. Yet they were manufactured with precision. The tables-turned-cones were the read-only memory—later called ROM—of what, in effect, was a primitive mechanical computer. The machine was able to look up and combine precalculated values.

Prediction, the second computing task, was more challenging. Calculating how to fire a shell so that it would hit a specific point in space and time was one problem. A harder problem was calculating where that specific point in space and time would be in relation to a fast-flying aircraft. To simplify the situation, engineers made an assumption: the targeted enemy aircraft was flying straight and level, not up and down and curving to evade fire. The gun-directing machine assumed a constant trajectory on a horizontal plane. That assumption was unrealistic, but it wasn't so unrealistic that the prediction became useless.

To reproduce that straight line, Sperry's state-of-the-art gun directors at the start of World War II physically represented the behavior of the approaching bomber in both horizontal and vertical dimensions: "The actual movement of the target is mechanically reproduced on a small scale within the computer," a defense journal reported in 1931. "The desired angles or speeds can be measured directly from the movements of these elements."[26] By 1940, Sperry had been at the bleeding edge of control system engineering for nearly thirty years and was perhaps better equipped than any other company to meet the complex challenge of mechanically predicting a flight path. Sperry's mechanical computer, the M-7, had eleven thousand parts and weighed 850 pounds.

This was the situation before Bell Labs entered the fray. Bell Labs' pitch on gun control started with a dream. In May and June 1940, one of the lab's physicists, David Parkinson, worked on a small project, the automatic "level recorder." Parkinson tried to plot rapidly varying voltage on strip-chart paper. To that end, he simply linked an instrument that measured voltage—a potentiometer—to a pair of magnetic grasps that held a pen. The voltage thus led the pen, drawing curves on paper. When the voltage dropped, the potentiometer dropped the grasps with the pen, so that the curve on the paper dropped.

While Parkinson was working on the level recorder, the Battle of Dunkirk shook Europe. Between May 26 and June 4, 1940, Nazi Germany routed the French, British, and Belgian defenders and forced them to evacuate. The attacks by Stuka dive-bombers were widely reported in the US press and radio. Twenty-nine-year-old Parkinson, troubled by these events, had "the most vivid and peculiar dream" one night.[27] He later recorded his dream in a diary:

> I found myself in a gun pit or revetment with an anti-aircraft gun crew. . . . There was a gun there . . . it was firing occasionally, and the impressive thing was that *every shot brought down an airplane*! After three or four shots one of the men in the crew smiled at me and beckoned me to come closer to the gun. When I drew near he pointed to the exposed end of the left trunnion. Mounted there was the control potentiometer of my level recorder![28]

As he woke up the next morning, Parkinson didn't find it difficult to understand his odd dream: the pen was a gun! If a potentiometer could control motions of a pen fast and precisely, then it could also control the motions of a gun fast and precisely. The signal simply needed to be amplified.

When he arrived at work that day, Parkinson pitched the idea to his boss at Bell, Clarence Lovell. Lovell instantly saw the idea's potential: the Bell machine's core would be a computer. But not a clumsy, creaking mechanical lookup mechanism that didn't actually compute. Bell's electrical computer would really compute, not just look up and combine precalculated values. Lovell and Parkinson's "range computer," as they called their invention, eliminated the trunk-shaped mechanized cones at the heart of Sperry's M-7. Calculating the timing for the fuse required determining the distance from the point of observation to the target by radar. The dream machine represented that distance "in the form of an electrical difference of potential."[29]

Coming up with the idea for electrical calculation, and implementing it in practice, required a range of skills that went beyond what a manufacturing firm had to offer, even one like Sperry. A telecommunications company had what was needed: experience in communications engineering, such as filter design, smoothing and equalization techniques, manufacture of potentiometers, resistors, capacitors, and feedback amplifiers. And the nation's leading telecom lab in 1940 was Bell Labs.

The founder and onetime president of Bell Labs was Frank Jewett. The former instructor in electrical engineering at MIT held a holistic view of communication. In 1935 he had challenged conventional wisdom on electrical signals at a lecture to the National Academy of Sciences: "We are prone to think and, what is worse, to act in terms of telegraphy, telephony, radio broadcasting, telephotography, or television, as though they were things apart."[30]

For Jewett, the electrical signal was the common, universal element. Bush had put him in charge of Division C—communications and transportation—of the newly founded National Defense Research Council. Warren Weaver, a science administrator formerly of the Rockefeller Foundation, led a wide range of the NDRC's projects on automatic controls, including gun directors and radar devices, under the title D-2. Jewett at Bell was keenly aware of the urgency of the fire control project, and inclined to see it as a communication problem. Weaver agreed: "There are surprisingly close and valid analogies between the fire control prediction problem and certain basic problems in communications engineering," he wrote later. Bell Labs got Weaver's second contract. On November 6, 1940, with support from the army's Signal Corps, Weaver's new D-2 shop and Bell Labs signed the contract for "Project 2."[31]

Weaver appreciated that the Bell group had deep experience with electronics. On paper, the planned equipment looked too good to be true when compared with existing mechanical gun directors: electrical gun directors required less skill, time, and cost in production—while in operation they afforded higher accuracy, speed, and flexibility. For the first time, the computer (the M-9) would place mathematics inside the feedback loop. Bell's computer enabled the gun director to calculate simple mathematical functions, such as sine and cosine, through resistors, potentiometers, servomotors, and wipers. The math, amplified, would drive a heavy 90-millimeter antiaircraft gun.

But state-of-the-art gun directors were limited, even when coupled with automated radar tracking. Once the time-fused shell left the gun's muzzle, it would either hit or miss. Since shells as well as planes were flying faster and higher, setting the fuse precisely enough became ever more difficult, even if done automatically by the gun, not by the gunner by hand. Targeting was open loop: there was no feedback to the shell after it was fired. If only there was a way to tell the shell to explode a little later or earlier than timed, depending on the actual situation up there at an altitude of 10,000 feet.

Johns Hopkins University, also NDRC funded, would come up with an ingenious way to close that feedback loop: the proximity fuse, also known as the "variable-time fuse" or simply "VT fuse." The shell would be smart, able to sense when it was close to the German bomber and only then explode. The difference was subtle but crucial. Timed fuses were set before being fired; the detonation of proximity fuses was determined by information gathered in flight. The fuse mechanism had to be sensitive yet rugged enough to withstand the shock of being fired by a mighty, 5.8-ton M-114 howitzer. A force twenty thousand times that of gravity would impact the shell in the gun. Worse, the projectile would spin at high speeds in flight. And it had to be safe and not blow up as it left the muzzle.

The new American fuse was a miniature radio station—with a sender, antenna, and receiver—all within the small nose of an artillery shell. When a 155-millimeter shell left the howitzer gun at almost double the speed of sound, its tiny radio station would switch itself on and start emitting a continuous wave. As the projectile approached the German bomber or cruise missile high in the sky, the radio waves would be reflected back by the target, like light in a mirror. The shell would sense the reflection, amplify it, and pass it on to a tiny gas-filled thyratron tube, which would serve as a switch to detonate the charge. The engineering challenge was dizzying. For a decade already, the best German minds had worked in vain on proximity fuses that could be used for air defense.[32]

Closing the final antiaircraft feedback loop required several inventions. The first problem was building tiny glass vacuum tubes, of the type that had been used in hearing aids. The fragile glass needed to withstand the howitzer. Testing was tough: the Johns Hopkins academics first fortified the hearing-aid tubes with methods they had gleaned from bridge and skyscraper design. Then they dropped them in steel containers, slammed them against lead blocks, whirled them, and fired them with a homemade smooth-bore gun. They found that the glass tube needed to be cushioned in rubber cups and a wax compound. After painstaking testing, they succeeded: the miniature glass tube would survive being fired by the big guns.

The tiny radio station also needed a tiny power plant. This battery, of course, had to pass similar stress tests—but the shock of firing and the spinning in flight could also be opportunities. The Johns Hopkins engineers developed a liquid battery, with two electrolytes separated by a glass ampul. When the gun was fired, the glass broke, effectively switching on the battery. Safety required a short delay, however, so that the shell's radio fuse, upon leaving the muzzle, wouldn't mistake its own artillery battery for the target. The ingenious idea was to utilize the projectile's spin in a mercury switch: when the shell left the muzzle, it took a short moment of spinning around its own axis in order to force the mercury through a porous diaphragm out of the contact chamber, turning the switch on. By the time the mercury was tumbled out, the artillery shell was on its way to the intended target. Now the radio shell was whizzing through the air, armed, on its own, waiting to sense feedback from unsuspecting enemy aircraft.

The radio shell was a breakthrough. "As a secret weapon, it ran second in importance to the atomic bomb only," the *Baltimore Sun* breathlessly reported after the war.[33] The Nazis, unable to design the devices themselves, coveted the highly prized fuses. In June 1942, the FBI captured eight German spies who had been tasked with finding out more about the project. Merle Tuve headed the NDRC's special Division T, which funded the development of the revolutionary device. The shell's mass production was kept secret even from the ten thousand factory workers who built 130 million miniature vacuum tubes over the course of four years. Approved assembly line production started in September 1942. By the end of 1944, 118 plants managed by 87 companies were producing more than forty thousand fuses daily.

Only top managers at about half of these companies knew what they were actually producing. The workers who produced the vacuum tubes were told they would be manufacturing hearing aids. One factory dubbed the device "Madame X." Monthly production eventually reached a peak of two million. So secret was the fuse that its use was permitted only over open water, at least until late 1944. The sky above the sea was safer for two reasons: it was harder for the enemy to observe antiaircraft firing closely, and it was impossible to pick up a dud from the ground in order to figure out how it worked. The ocean swallowed the secret.

Commanders who used the new technology were ecstatic. George Patton, commander of the Third United States Army, reportedly was so impressed by the device that he expected the very nature of war to change: "I think that when all the armies get this shell we would have to devise some new method of warfare."[34]

III

The NDRC's fire control division under Warren Weaver funded eighty projects over five years, from 1940 to 1945. The contracts practically mapped the world of control systems at the time, as the MIT historian David Mindell pointed out.[35] D-2 awarded fifty-one contracts to companies and laboratories in the private sector, and twenty-nine to academic research institutions. More than sixty projects tackled problems of land-based antiaircraft fire. The average funding volume was $145,000. Weaver's largest and possibly most successful contract, at $1.5 million, was the Bell gun director that resulted in the M-9. D-2's smallest and possibly most inconsequential contract, at just over two thousand dollars, went to Norbert Wiener, to explore how to predict flight patterns.[36]

As early as February 1940, five months after Nazi Germany had invaded Poland, Wiener joined a subcommittee under the direction of the Princeton mathematician Marston Morse. The scientists discussed how the American Mathematical Society could help during a national emergency that "we hope will never arise."[37] In July, the Battle of Britain got under way. By late July, shortly after Hitler ordered the invasion of Britain, Wiener learned that the armed forces had received his suggestions on the use of incendiary bombs, and he reiterated his desire to participate in the war effort.[38]

Later, on September 11, Wiener attended a meeting of the American Mathematical Society at Dartmouth College that made computing history. Bell Laboratories had started operating the "Complex Computer" at their old headquarters at 463 West Street in New York a few months earlier. The machine had 450 relays and 10 crossbar switches, as well as remote terminals, each with a keyboard for input and a teletypewriter for output. One was in Dartmouth. Bell's George Stibitz was familiar with Wiener's work and invited the participants to challenge the computer with problems involving the addition, subtraction, multiplication, or division of complex numbers. Wiener stepped up to the keyboard, trying to stump the machine. But the New York–based computer outsmarted him. The teletypewriter magically hacked out the correct number. September 1940 was thus Wiener's first encounter with a thinking machine.[39]

Meanwhile, the Germans pounded London. The Battle of Britain was intensifying, with the Royal Air Force bombing launch points for what the British feared would be an imminent German invasion. On September 20, 1940, Wiener wrote to Vannevar Bush, "I . . . hope you can find some corner of activity in which I may be of use during the emergency."[40]

Section D-2 of the newly established NDRC was holding its first meeting also in September. Warren Weaver concluded that the most pressing issue was improving antiaircraft fire control for the army. Two months later, on November 22, Wiener submitted a four-page memorandum for the antiaircraft predictor to Bush's research committee, "to explore the purely mathematical possibility of prediction by an apparatus" and then "to construct the apparatus."[41] Shortly before Christmas 1940, the project was approved: the NDRC awarded $2,325 to the MIT professor.

Wiener hired a twenty-seven-year-old MIT graduate in electrical engineering and mathematics, Julian Bigelow, as the project's chief engineer. Bigelow was ambitious and valued precision. Always immaculate in suit and tie, Bigelow also was an active amateur aviator, which gave him a skill set that would be useful for his new project. The two academics knew that they were tackling one of the most difficult problems of fire control. The Blitz was at its worst. Just days after Wiener's project got under way, on the evening of December 29, the Luftwaffe strafed the City of London harder than ever. In three hours, 120 tons of explosives and twenty-two thousand incendiaries were dropped. That night, Herbert Mason took the iconic photograph of St. Paul's dome towering over the smoke of London's raging fires.

Never had the antiaircraft problem been more urgent. When under fire, Wiener believed, pilots "will probably zigzag, stunt, or in some other way take evasive action."[42] To illustrate this back in a classroom at MIT, the professor drew a sharp zigzag line on a blackboard. Bigelow pointed out that the pilot's behavior was constrained by the aircraft itself.[43] The pilot did not have complete freedom to maneuver in a fast-moving, inert plane. Zigzag wasn't so easy. Wiener began to understand that human psychological stress and the plane's physical constraints made the man-machine system more predictable. This realization made it easier to calculate a plane's future flight path from its past behavior. He erased the zigzag line and instead drew a line of smooth curves.

Wiener and Bigelow occupied a former math classroom in MIT Building 2, Room 244, and turned it into a "little laboratory," as they called it. There they experimented with an improvised apparatus. They faced one problem at the outset: they didn't have any data on actual pilot behavior, so they needed to simulate flight under fire. This was a challenge. To replicate the random curves that German pilots were flying over London and elsewhere in wartime Europe, Bigelow installed a motor-driven white spotlight that projected a smooth, circular, but nonuniform flight pattern onto the wall of their makeshift lab. It took about fifteen seconds to traverse the wall.[44]

This was the ideal flight path. To simulate the actual flight path of a stressed pilot, the two researchers installed a second, red spotlight. A mock pilot then had to follow the white curves by pointing this red light at it, chasing the spot on the wall. It was a difficult task, deliberately so. Hunting the light required controlling a "sluggish contrivance," as Wiener called it, which was designed to be complicated and to feel "completely wrong."

The resulting erratic behavior, Wiener believed, would simulate the constrained flight patterns of enemy pilots under stress. Now the two set out to mathematically model pilot-aircraft behavior. Meanwhile, other NDRC projects were making breathtaking progress. By the end of May, the Rad Lab had successfully tested its automated B-17 turret on one of MIT's roofs. Wiener wasn't aware of the Rad Lab research. Days later, on June 4, 1941, Weaver arranged for Wiener and Bigelow to visit a Bell facility in Whippany, New Jersey. On observing Bell's experimental design, Bigelow recalled his surprise at the simplistic assumptions of the engineers: "They had no random variables in them at all and took no account of evasive action or even the natural curvature of the plane's flight path."[45] But Bell had no interest in the abstract math that Wiener presented.

On February 1, 1942, Wiener submitted a lengthy report to Weaver's D-2 section at the NDRC. It had a mouthful of a title: "Interpolation, Extrapolation, and Smoothing of Stationary Time Series." Wiener made an academic contribution to a range of arcane theoretical debates in abstract mathematics. But this work was useless for the war effort; Pearl Harbor had happened only two months earlier, and US auto factories were just switching from commercial production to war production.

Wiener's 124-page paper lacked any sense of urgency. It didn't mention the unsuccessful experiments in the little MIT lab or any mechanical implementation of his theory. The paper mentioned the antiaircraft problem only two times, buried in a forest of mathematical formulas on page 76. Neither the title nor the introduction or index contained a single reference to the problem that had motivated the project's funding. Instead, Wiener offered a mind-numbing alphabet soup of abstruse mathematics: Brownian motion, Cesàro partial sum, Fourier integral, Hermitian form, Lebesgue measure, Parseval's theorem, Poisson distribution, Schwarz inequality, Stieltjes integral, Weyl's lemma, and many more. When Weaver received the paper, he had it classified and bound in an orange cover. Engineers nicknamed the document "yellow peril," a joking reference to the paper's impenetrable theory and lack of practical relevance.

Five months later, on June 10, 1942, Wiener and Bigelow submitted a brief interim report on their attempts to build the promised apparatus. After playing with lights in darkened Room 244 for several months, they had developed some of the core ideas that would later shape Wiener's cybernetic worldview. They understood that man and machine were forming an entity, a system, a joint mechanism. That combined mechanism would, in effect, behave like a servo, they argued, a device built to autocorrect its performance in response to error:

> We realized that the "randomness" or irregularity of an airplane's path is introduced by the pilot; that in attempting to force his dynamic craft to execute a useful manoeuver, such as straight-line flight, or 180 degree turn, the pilot behaves like a servo-mechanism, attempting to overcome the intrinsic lag due to the dynamics of his plane as a physical system, in response to a stimulus which increases in intensity with the degree to which he has failed to accomplish his task.[46]

These observations, however, were based on only their simulation. "Little information on the nature of the paths of combatant aircraft was accessible either in terms of actual space trajectories, or observational data from the tracking apparatus," they pointed out.[47]

A little more than two weeks later, on July 1, 1942, Stibitz and Weaver visited Wiener's improvised little lab. The visit happened eighteen months into the two-year project. The Germans had just conquered Sebastopol on Crimea, and Rommel's Afrika Korps had invaded Egypt. The United States was about to launch its air offensive against Nazi Germany. The war was even inching closer to New England: just three days before Wiener's meeting, the eight German spies seeking information about the proximity fuse were captured from a boat off Long Island.

The professor boldly told his government sponsors that the equipment he had built at MIT was "one of the closest mechanical approaches ever made to physiological behavior."[48] His theoretical predictor was based on what his sponsors believed were "good behavioristic ideas." Wiener explained the more general ambitions of his project: he would try to predict future actions of an "organism" by studying the past behavior of that organism, not by studying the structure of that organism.

Stibitz was deeply impressed by Wiener and Bigelow's presentation of the predictor. "It simply must be agreed that, taking into account the character of the input data, their statistical predictor accomplishes miracles," he observed. "The behavior of their instrument is positively uncanny," Stibitz wrote in a report. But Weaver was more skeptical. He wasn't yet convinced whether Wiener's machine was "a useful miracle or a useless miracle." Weaver jokingly told Wiener they would "bring along a hack saw on the next visit and cut through the legs of the table to see if they do not have some hidden wires somewhere."[49]

Joking aside, the short prediction time, at most about one second, was a very serious limitation. Moreover, the NDRC funders questioned one of the very basic assumptions underpinning Wiener's model: that enemy pilots under stress would behave in a consistent way. The pilot might not see bursting shells. Individuals might react differently to the same events. Bigelow himself became skeptical. He told Weaver that Wiener's statistical method had "no practical application to fire control at this time."[50]

Yet the proud professor could not bring himself to "make this statement," Bigelow told Weaver. Admitting failure was hard for the former child prodigy. Wiener's ambition and pride convinced him of the feasibility of statistical prediction. On August 20, 1942, Wiener was still convinced his work would be of the highest significance. The Institute of Mathematical Statistics planned a public workshop that fall. When Wiener heard about the conference program, he sent an urgent note to the organizers: even the titles of presentations could be "a tip-off" to the enemy, he warned, darkly. His work on statistical prediction was "vital and secret in more ways, and in vastly more important ways, than I have been able to tell you," Wiener told one of the organizers.[51]

Ten days later, Weaver lost patience. On September 1, 1942, Weaver vented in a memo that he was "highly skeptical" of Wiener's business. In the search for data that they could use for their white-versus-red light experiment on the wall in their improvised lab, Wiener and Bigelow had taken the initiative to visit several army installations, including the Ballistic Research Laboratory of the Army Ordnance Department at Aberdeen Proving Ground, Maryland; the Frankford Arsenal in Philadelphia; and the Antiaircraft Artillery Board at Camp Davis, North Carolina. They also went to Fort Monroe in Virginia. The trip didn't go well. In Weaver's words:

> [Wiener and Bigelow] have gaily started out on a series of visits to military establishments, without itinerary, without any authorizations, and without any knowledge as to whether the people they want to see (in case they know whom they want to see) are or are not available.[52]

The army's ordnance officers had other priorities. Germany was pushing into Stalingrad and looked invincible. The American war machine had shifted into high gear. The NDRC also was busy: revolutionary VT proximity fuses started coming off the assembly lines by the tens of thousands. Yet Wiener and Bigelow made little progress and instead appeared to waste others' precious time: "Inside of twenty four hours my office begins to receive telegrams wanting to know where these two infants are," Weaver fumed. "This item should be filed under 'innocents abroad.'"[53] Wiener's $2,325 contract was terminated after less than two years, in late 1942.

The MIT professor was disappointed. "I still wish that I had been able to produce something to kill a few of the enemy instead merely of showing how not to try to kill them," he cryptically wrote to Weaver on January 28, 1943.[54] As an engineer, Wiener had failed. His antiaircraft predictor never worked as intended, and it was never even close to improving antiaircraft fire. Five years later, Wiener cryptically mentioned the unpleasant episode in the introduction to his landmark book, *Cybernetics*: "It was found that the conditions of anti-aircraft fire did not justify the design of [a] special apparatus for curvilinear prediction."[55] Reality had failed, he believed, not his idea.

Nevertheless, *Cybernetics* centered on the antiaircraft problem, on man-machine interaction, control, and feedback. The hapless professor later implied again and again that he had invented these concepts during the war, single-handedly improving antiaircraft fire. Wiener's work itself represents a fascinating case study of human-machine interaction under stress: the mechanical antiaircraft problem gave Wiener the urgency, the passion, the inspiration, the language, and— most important—a powerful metaphor to articulate cybernetics.

Meanwhile, Sperry was already selling feedback loops with spectacular success. By 1942, the company had carved out a highly profitable approach to selling control systems. Sperry had created "a whole new field of scientific accessories to extend the functions and the skill of the operator far beyond his own strength, endurance and abilities," the company's history recounts.[56] With the war effort accelerating, demand for its products was exploding. Sperry and its subsidiaries had worked for two decades on product development, and now they were unable to handle the requested staggering quantities: in 1942, Sperry was contracted to produce over a billion dollars worth of control systems.

If anything, this number was understating the significance of Sperry's contribution to the war, its executives pointed out: "This billion dollars' worth of technical equipment will fill the vital gap between the one hundred billion dollars' worth of weapons and the thousands of men who must operate them." World War II was industrial war at its height, fought with mechanical beasts of steel and iron, clashing on land, at sea, and in the air. Man, unaided, was unfit for war in the machine age. The Sperry Corporation wrote:

> His airplanes have become so big and fly so far that he must have automatic pilots instead of flying by hand. The machine gun turrets must be moved by hydraulic controls. The targets of his antiaircraft guns now move so fast in three dimensions that he can no longer calculate his problems and aim his gun. It must all be done automatically else he would never make a hit.[57]

At the beginning of World War I, airplanes flew at altitudes under 2,000 feet, and at a speed of about 70 miles per hour. Antiaircraft weapons were simple: "Those guns that could be trained at high angles were fired the same manner that a hunter would fire at a duck in flight," wrote Preston Bassett, then president of the Sperry Gyroscope Company, again using the preferred hunting comparison.[58] But the engineers at Sperry understood that shooting down ever-faster objects required ever more automation:

> The time factor has now become so small for all operators involved that the human link, which appears to be the only unchangeable

factor in the whole problem and which is a very erratic one, has become more and more the weakest link in the chain of operations until it is quite apparent that the human link must be dropped out of the sequence.[59]

During World War I, an air defense battery had about twenty operators who cooperated in the full sequence from searching for the target to firing at it. By 1935, that number was down to eighteen at night and twelve by day. By the beginning of World War II, the number of humans was down to ten. When the war ended, only three to four operators were left. "All that can be said now is that those few remaining must also go," Bassett observed after the war. "Time has now run out for the slow reaction of a human operator."[60] Shooting down autonomous machines required autonomous machines.

After Pearl Harbor, "American industry really took its coat off and went to work." The number of Sperry employees reached its peak of thirty-two thousand in 1943. Twenty-two subcontractors were building Sperry products during the war; the Ford Motor Company, for instance, produced antiaircraft directors, and Chrysler manufactured gyrocompasses to navigate ships. The total number of employees producing Sperry equipment would swell to well over a hundred thousand during the war.[61] Between 1942 and 1945, the value of shipped Sperry products amounted to over $1.3 billion.[62] This was at a time when a flight from New York to Chicago with Capital Airlines cost $18.[63]

Without Sperry's feedback gear, neither machines nor men could operate under fire. Sperry provided the unity, the connective tissue, the man-machine interface. "Without this equipment, neither men nor weapons would be effective," the company knew.[64]

In the early hours of June 13, 1944, General Sir Frederick Pile was awakened "by an unusual uncertainty" on the part of London's air-raid siren.[65] The "Alert" sound went off, that creepy up-and-down howling siren, followed almost immediately by the constant howl of the "All Clear" sound—only to revert back to "Alert" seconds later.[66] It was completely dark in the general's bedroom. Pile was the commander of Britain's Anti-Aircraft Command. A few moments later his telephone rang. The duty intelligence officer reported that "the Diver" had finally arrived. "Diver" was the British code name for the V-1, a terrifying new German weapon: an entire unmanned aircraft that would dive into its target, not simply drop a bomb.

The so-called buzz bomb was the world's first cruise missile, a self-propelled single-use vehicle flying on a nonballistic trajectory at relatively low altitudes. The second Battle of London was about to begin. Seven flying bombs were spotted in this night's opening salvo. One of them was headed straight to East London. It crashed into an elevated railway viaduct at Bethnal Green, blocking all the train lines headed out of Liverpool Street, one of the capital's largest train stations. The public was kept in the dark about the nature of the attack—and about the nature of the defense. The breathtaking engineering achievement that would now creak into action was among the Allies' best-kept secrets. This secret would take many years to come out in clarity. When the first cyberneticists articulated their theories about feedback loops and adaptive machines in the coming years, they had no idea that a spectacular version of this future had already arrived high over the English Channel.

When the German V-1s were finally lobbed from their rail launchers on the Continent toward London, a shrewd, high-tech system lay in wait on the other side of the English Channel, ready to intercept the robot intruders. As the low-flying buzz bombs cruised over the rough waves of the Atlantic toward the coast, invisible and much faster pulses of microwaves gently touched each drone's skin, 1,707 times per second. These microwaves set in motion a complex feedback loop that would rip many of the approaching unmanned flying vehicles out of the sky.

A tiny amount of the energy waves was reflected by the V-1's skin back to its emitting source, a parabolic antenna on top of an MIT-designed mobile radar station. The antenna would catch the reflection, turn it into a signal, and pass the signal on via cables and vacuum tubes and a series of filters. The signal would then be amplified and fed back to the motors that controlled the antenna. But in reverse: when the signal weakened, the antenna would turn toward the stronger signal, in effect tracing the V-1. The system locked on to what would now become a target. More tiny microwaves bouncing off a drone's skin would enable the watchful system to measure, precisely, the range, bearing, and elevation of the novel German weapon system.[67]

By mid-1943, Allied intelligence services knew from intercepts, photographs, and informants that Germany would soon launch robot bombs against southern England.[68] Pile, an artillery gunner by training, understood that a state-of-the-art gun predictor would make all the difference. "We badly wanted the SCR 584," he recounted.[69] The Air Defense Command also used "radio shells" against the buzz bombs. The new "variable time" proximity fuse autonomously tripped a trigger that detonated the TNT in the shell's body. The fuse had the size of a pint milk bottle. Once it came within 70 feet of the targeted buzz bomb—it blew up, knocking everything within the blast zone out of the sky.

One week after D-day, the Germans started sending robots across the English Channel. The first wave of V-1s was ineffective. Few made it across the sea. The German launch crews, trained as artillerymen, didn't know how to handle this odd and novel "weapon of vengeance," *Vergeltungswaffe*, or simply "V-weapon." They learned fast. Soon the V-1s were being hurled against the capital of the British Empire from about fifty launch sites. More than seven thousand pilotless bombs were fired against London in total. The V-1 was immediately dubbed the "robot bomb" by the press. "Even though Allied troops have cleared the Calais coast of Nazis and their bomb launching platforms, the robot bomb threat is not ended by any means," the *Christian Science Monitor* reported, proudly announcing that Ford would now be building "robombs" for the United States.[70]

By the summer of 1944, about five hundred guns, many of them equipped with the new radio fuses, were in place to counter the incoming buzz bombs. "The result of the advent of VT fuses was truly sensational increases in 'kills,'" an official army report said, in a statement that was widely reported in newspapers across the United States after the war ended.[71] In the fourth week of July 1944, 79 percent of V-1s flying across the channel were downed. By August, on the last day V-1s were launched in quantity against Britain, the air defenses worked particularly well, Pile recalled: "On a Sunday in August the Germans sent 105 buzz bombs across the Channel but only three arrived."[72] The new antiaircraft artillery brought down sixty-eight V-1s; the rest were downed by fighters or malfunctioned.

The confrontation in the sky over the English Channel during the late summer of 1944 was remarkable: never before had one autonomous weapon clashed with another autonomous weapon with so little human interference.[73] The future of war had arrived. "Now we saw the beginning of the first battle of the robots," Pile observed at the time: "Human error was being gradually eliminated from the contest: in the future, the machines would fight it out."[74] It was automaton against automaton, VT robot against V-1 robot.

The V-1 was both hard and easy to hit at the same time. At first glance, the German cruise missile's strengths were its size, speed, and altitude. It was small, fast, and low: it was shorter and slimmer than a fighter plane; at 400 miles per hour it was significantly faster than a Spitfire; and it cruised at 2,000–3,000 feet. Worse, the weapon didn't have a "one-shot stop," a single point of failure that would bring it down when hit: it didn't have a pilot, and its engine was highly resilient to damage. But the new weapon had one crucial weakness. The V-1, ironically, lacked precisely the problem that Wiener had tried to solve: an erratic flight path. The *Vergeltungswaffe* flew straight and level. Its flight path was predictable. Any pilot under stress would do the opposite: try to evade enemy fire.

2. CYBERNETICS

THE US MILITARY WAS DEEPLY IMPRESSED WITH GERMANY'S accomplishments during World War II. By the end of the war, the Luftwaffe had fielded the world's first ballistic missile, the V-2. The rocket, in contrast to the cruising V-1, was thrust high into the atmosphere by its initial phase of flight, followed by a steep and fast descent into its target. London and the rest of England were utterly defenseless against a projectile that would strike like a lightning bolt, three times faster than the speed of sound.

Engineers in the Third Reich had been working on a high-powered rocket that would have been able to strike across the Atlantic, but they did not finish it in time. When the US Army invaded and occupied Germany in 1945, it took back stateside some of the Reich's brightest engineers, as well as cutting-edge technology. The most highly prized individual was Wernher von Braun, the Reich's young and dashing missile engineer, and his vast team from Peenemünde by the Baltic Sea in Germany's north.

Between 1945 and 1952, the US Army continued working on the V-2, building and launching sixty-seven missiles from White Sands

Proving Ground, the site of the world's first nuclear explosion and the largest military installation in the United States.[1] US-modified missiles that had originally been used to attack London also took the first images of Earth from space and, in February 1947, carried the first animals into space (fruit flies).[2]

Sometime in late 1946, a missile researcher from the Boeing Airplane Company in Seattle reached out to Norbert Wiener. In 1946, Wiener's NDRC-funded yellow-peril study on predicting flight patterns was still classified. The executive told Wiener in a letter that his company was working on guided missiles. Would the professor be willing to share a copy of his classified manuscript? It was a cheeky request. Wiener, of course, wasn't supposed to possess, let alone share, classified material. But Wiener saw this letter from the Boeing executive as an opportunity. An opportunity to voice his frustrations with Pentagon-funded work—and to do so in a highly public way.

Wiener jotted down a response to Boeing, rejecting any cooperation in the sharpest of tones. He invoked Hiroshima and Nagasaki. He said that science had lost its innocence. He wrote that the exchange of ideas must stop "when the scientist becomes an arbiter of life and death." Indeed, Wiener was so angry that he forwarded the letter to the editor of the *Atlantic*. The monthly magazine published it in full in January 1947, under the title "A Scientist Rebels."

"The potential use of guided missiles can only be to kill foreign civilians indiscriminately, and it furnishes no protection whatsoever to civilians in this country," Wiener argued in his letter to the editor. He saw an opportunity to take a public stance against militarism more generally, and against guided missiles specifically:

> I cannot conceive a situation in which such weapons can produce
> any effect other than extending the kamikaze way of fighting to

whole nations. Their possession can do nothing but endanger us by encouraging the tragic insolence of the military mind.[3]

The *Bulletin of the Atomic Scientists* republished the letter the same month. Even Albert Einstein commended Wiener for his courage. Wiener's eloquent line on the "tragic insolence of the military mind" powerfully resonated with a rising number of postwar critics and pacifists.

Yet these were surprisingly naïve statements, coming from a man who had conducted a two-year research project on the antiaircraft problem. Guided missiles, of course, could be used as interceptors against incoming bombers; Wiener's own research (albeit unsuccessful), the very research that Boeing had requested, was about improving the hit-to-miss ratio against enemy aircraft. The exact opposite of what he wrote was true, and he knew it: the potential use of guided missiles could very well be to down foreign bombers highly selectively, not to kill civilians "indiscriminately."

But the world had changed in the meantime, between 1943 and 1946. American scientists had helped put the most devastating weapon ever invented into the hands of the US government: the atomic bomb. The government had decided to use it, without any debate or consultation with the inventors, let alone the public. Hundreds of thousands of innocent civilians were indeed killed indiscriminately when Paul Tibbets, then a colonel, pushed the button in his Boeing B-29 Superfortress bomber to release the bomb over Hiroshima.

For many a scientist, this historic and horrific event meant that uncomfortable questions about consequences could no longer be ignored. On the very day of Hiroshima, Wiener had pondered the momentous implications of nuclear weapons. "I had no share in the atomic bomb itself," he recounted later. "I was nevertheless led into a very deep searching of soul."[4] He didn't know what to do.

Already by August 1945, Wiener had been convinced that his new ideas about machines were outright dangerous. He had not yet found a name for those ideas. The inspiration for his book title, *Cybernetics*, would come to him only one year later, in the summer of 1946 in Mexico, at the Instituto Nacional de Cardiología, where Wiener worked with his friend Arturo Rosenblueth. Wiener's ideas on control and communication had already been formed when the war came to such an abrupt end that August. "Cybernetics," Wiener wrote in his memoir, "was not as revolutionary as the atomic bomb." Nevertheless, he felt he had given the world a scientific concept that might well be misused; the "possibilities for good and for evil were enormous," he wrote.[5]

Worse, Wiener felt strangely betrayed, although he couldn't quite articulate his feeling. It wasn't that he had given a sharp tool to the army to play with. After all, his antiaircraft research had failed. The military didn't change anything as a result of Wiener's scientific work. The defense establishment had even rejected him. Even more disconcerting for the soul-searching scientist, the reverse was true: the army, in a way, had given *him* a sharp tool to play with, cybernetics. And he himself embraced military examples in his own thinking (and later writings). Wiener didn't change what the military did; the military changed what Wiener did.

The father of the emerging new scientific discipline wondered whether he should censor himself—keep his cybernetic insights secret—so that the dangerous ideas would cause no harm. But he also understood that his ideas were already out there. Cybernetics couldn't be undone. Not even he could stop it. The genie was out of the bottle. This could mean only one thing: somebody needed to warn the world about the dangers of the coming rise of the machines.

"I thus decided that I would have to turn from a position of the greatest secrecy to a position of the greatest publicity," Wiener wrote later.[6] When the letter from Boeing landed on his desk in late 1946, he knew what to do.

In the winter of 1947, Wiener decided to hold an interdisciplinary seminar. The idea was to bring together scientists and practitioners on what he called communication. "He was launching his vision of cybernetics in which he regarded signals in any medium, living or artificial," recalled Jerome Wiesner, then a professor of electrical engineering at MIT, and later MIT's president. In the spring of 1948, Wiener eventually started what became a series of weekly gatherings. Every Tuesday evening, philosophers, engineers, psychologists, mathematicians, and experts in such fields as acoustics and neurophysiology met for dinner. One would present an ongoing research project and discuss it with the group. The first session, Wiesner recalled, reminded him of "the tower of Babel," so unintelligible were the language and jargon of scientists in other disciplines. That changed over the coming months. "As time went on, we came to understand each other's lingo and to understand, and even believe, in Wiener's view of the universal role of communications in the universe."[7]

For most participants, these dinners were formative experiences, where new ideas and new collaborations were forged. Some even joked about the "Wiener circle," in reference to the Vienna circle that had connected such towering figures of analytical philosophy as Karl Popper and Rudolf Carnap ("Wiener" is German for "Viennese"). One of the regular participants in this circle to discuss communications was J. C. R. Licklider, a man who would later play a key role in laying the groundwork for the internet.

Three ideas were at the core of this novel approach to thinking about automation and human-machine interaction. The first core idea of cybernetics was control. The very purpose of machines and living beings is to control their environment—not merely to observe it, but to master it. Control is fundamental. The concept of entropy illustrates just how fundamental. Entropy is a measure of disorder, of uncertainty, degradation, and loss of information.[8] Nature has a tendency to increase entropy, to gradually decline into disorder: cold things warm up; hot things cool down; information gets lost in noise; disorganization gradually takes over. Halting or reversing this trend toward disorder requires control. Control means that a system can interact with its environment and shape it, at least to a degree. Environmental data are fed into a system through *input*, and the system affects its environment through *output*. For Wiener, this was the essence of the cybernetic worldview:

> It is my thesis that the physical functioning of the living individual and the operation of some of the newer communication machines are precisely parallel in their analogous attempts to control entropy through feedback.[9]

This quote introduces the second core concept of cybernetics: feedback. For Wiener, "feedback" described the ability of any mechanism to use sensors to receive information about actual performance, as opposed to expected performance. An elevator is an example. Feedback will tell the door-opening mechanism whether the elevator has actually arrived behind the sliding doors and only then open the doors for waiting passengers; not using feedback would increase the risk of error, possibly allowing unsuspecting people to step into an empty elevator shaft.

Another of Wiener's examples is, unsurprisingly, an artillery gun that uses feedback to make sure the muzzle is actually pointing at the target. The mechanism that controls the turning of the turret, for instance, requires feedback. The actual performance of the turret-turning mechanism varies: extreme cold thickens the grease in the bearings and makes turning harder; sand and dirt might affect turning even more. So verifying output through feedback is critical.

Feedback tends to oppose what a system is already doing—for instance, stopping a motor from turning a turret or telling a thermostat to switch off a heater. Such feedback is called "negative feedback," and it generally serves the purpose of stabilizing a desired state. Feedback, then, is "the property of being able to adjust future conduct by past performance," wrote Wiener.[10] The mechanism itself may be very simple, like turning a turret, or it may be complex, like regulating the human body's temperature. In Wiener's mind, feedback would provide a kinesthetic sense to machines, akin to the human awareness of the position or movement of arms and legs.[11] This analogy led him to another core idea.

The third core idea of cybernetics described a tight relationship between humans and machines. In the antiaircraft predictor project, the enemy pilot was recognized as effectively forming a single entity with the bomber, behaving "like a servo-mechanism," as the team at MIT had noted already in early 1941. It wasn't just the plane and the pilot that formed a servomechanism. Even the antiaircraft gun behaved as a "dynamic system," composed of multiple human operators and complex mechanics joined in the struggle against ever-increasing entropy. But there are other aspects to the human-machine parallel. Wiener tended to anthropomorphize machines: switches corresponded to synapses, wires to nerves, networks to nervous systems,[12] sensors to eyes and ears, actuators to muscles. Conversely, he also mechanized man by using machine comparisons to understand human physiology.

The first two cybernetic concepts—control and feedback—could be abstract, technical, and hard to grasp. But the third idea—the merging of human and machine—stirred the imagination.

In one of Wiener's most brilliant passages, he illustrated his cybernetic argument by considering the way an artificial prosthesis interacts with the human body. A wooden leg, the MIT professor wrote, is a mechanical replacement of a lost limb of flesh and blood. Thus, a man with a wooden leg is "a system composed both of mechanical and human parts."[13] The most primitive form of prosthesis is the peg leg, pirate-style. Such a limb-shaped wooden leg, he thought, would be not that interesting. But in Russia, Wiener would later recount after a visit to Moscow, work was done on artificial limbs that makes use of cybernetic ideas. The idea excited him. Wiener embarked on a fascinating thought experiment.

Suppose a man loses a hand at the wrist. The hand itself is cut off, with its skin, finger bones, tendons, and muscles. But the stump of the forearm still has the strong muscles that used to extend into the hand. These muscles can still contract. Such contraction won't move the lost hand and fingers, but the muscle nerves will produce electrical signals, so-called action potentials. Electrodes can pick up these signals, amplify them, and feed them to electrical motors in an artificial hand. That way the amputee is able to control the movement of his new, battery-powered hand. So far, so good; such prostheses had already been produced at the time.

Yet the feedback loop was missing. Artificial hands couldn't feel, couldn't touch. The signal, however, could be transmitted from the mechanical limb back to the amputee's brain. Pressure gauges in the artificial fingers, Wiener theorized, could report back a vibrating sensation to the skin in the stump. Over time, the amputee would learn to replace the missing natural tactile sensation with the vicarious sensation. This cybernetic limb story only prepared the ground for Wiener's thought experiment.

Engineering moving and feeling prostheses was possible. The task would require the construction of systems "of a mixed nature," systems that included human as well as mechanical parts. But why should this kind of design be limited to replacing lost body parts? Why not add entirely new artificial limbs and more to the human body? "There is a prosthesis of parts which we do not have and which we never have had," Wiener reasoned.[14] In his mind, these were scenarios from a not-too-distant future.

In some ways that future already existed. A ship's propeller and its obstacle-avoiding, depth-sounding apparatus extend the captain's and the crew's bodies. And in an aircraft, radar extends the pilot's eyes and the plane's wings and jet engines actuate instead of muscled limbs, "while the nervous system that combines them is eked out by the automatic pilot and other such navigation devices."[15] Human-mechanical systems, Wiener concluded, were relevant and useful in a "large practical field." In some situations, such cybernetic devices were indispensable.

The founder of cybernetics would stretch the concept so wide that cars and ordinary telephones qualified as indispensable cybernetic devices—undoubtedly a large practical field in 1950s America, enchanted by the wonders of modern transportation and telecommunication technologies. Cybernetics, of course, could easily be applied beyond one man and one machine. Collective entities—say, firms or entire societies—formed systems that lent themselves to cybernetic analysis. Social science could study control, communication, feedback, adaptive behavior, and organizational learning, and many scholars would do so. Wiener had seen that potential already in the late 1940s: "Properly speaking, the community extends only so far as there extends an effectual transmission of information," he wrote in *Cybernetics*. That perspective would offer "a basis for political thinking," Wiener told the *New York Times* a few months after his book was published.[16]

The *Times* didn't just review Wiener's book once. The nation's leading newspaper covered the book three times. Just before Christmas 1948, the *Times* introduced its readers to the new science. Cybernetics, the *Times* quoted Wiener, "combines under one heading the study of what in a human context is sometimes loosely described as thinking and in engineering is known as control and communication."[17] Cybernetics was about finding the features that are common to automatic machines and the human nervous system. Brain and machine overlapped in a wide area, the *Times* wrote, and the brain itself behaves very much like a machine. Constructing and operating more and more complex machines would help scientists understand how the brain itself operated. And there was no reason why the theory could not be applied to all complex systems.

Many leading minds in engineering, mathematics, biology, and psychology, but also sociology, philosophy, anthropology, and political science, would initially be drawn to the new thinking of adaptive systems. The best-known early cyberneticists were the mathematician John von Neumann, a fellow polyglot, computer pioneer, and prominent professor at the Institute for Advanced Study in Princeton still in his early forties, nine years younger than Wiener; the American neurophysiologist and neural networks pioneer Warren McCulloch; the Austrian American physicist Heinz von Foerster; and the Mexican physician Arturo Rosenblueth, one of Wiener's closest friends and collaborators. Among the best-known thinkers who were subsequently influenced by cybernetic thought, and contributed to it, are the British neurophysiologist W. Grey Walter and management theorist Stafford Beer; the Austrian-born biologist Ludwig von Bertalanffy; the Chilean philosopher Humberto Maturana; the German-speaking political scientist Karl Deutsch; and the American sociologist Talcott Parsons. Two of the most visionary and pivotal cyberneticists, as will become clear shortly, were Ross Ashby, an English researcher and inventor, and Gregory Bateson, a British-born American anthropologist and social critic.

The new discipline itself had fuzzy borders. Its proponents were united in the belief that cybernetic-systems theory would lead to a revolution in our understanding of the human mind and human behavior, "both normal and abnormal," as Wiener had hoped. The extent of that coming conceptual revolution, at least in the mind of the field's ambitious founding father, would be comparable to the scientific revolution brought about by relativity theory or quantum mechanics.

The press coverage of the budding new science frequently repeated the misleading story that Wiener had successfully built gun directors during the war. The behavior of any particular pilot was impossible to predict, the *Times* explained to its readers, but the evasive actions of many thousands of pilots could be analyzed, thus revealing the most probable escape tactic. "This was actually done during the war, and knowledge of the most frequent dodging techniques was built into anti-aircraft gun directors," the paper of record reported.[18] Except it wasn't. This was Wiener's theory. But he never had the data of "thousands of pilots." It didn't work in practice. No such machine was ever built.

II

The *New York Times*, in its ecstatic review of *Cybernetics* one winter Sunday, the twenty-third of January 1949, recounted Wiener's daring predictions of a future with actual thinking machines. That future came rather quickly: *Time* magazine arrived at newsstands that very evening and declared that the future was already here. The world's first "thinking machine" had actually been built.[19] Not at MIT or at Bell Labs, but in a mental hospital in Barnwood, a village near Gloucester, in England's West Country.

Memories of the war were fresh in England in 1948. Traumatized military officers and clergy preferred Barnwood House, a sanatorium surrounded by the peaceful rolling hills and grazing sheep of the Cotswolds in England's most picturesque countryside. Ross Ashby was the forty-five-year-old chief of research at Barnwood House and a major in the Royal Army Medical Corps.[20] It was at this remote hospital that Ashby invented the "homeostat," a quirky machine inspired by his work with mentally disturbed patients. Ashby's curious gadget soon enthralled scientists and quickly became a worldwide media sensation. In January 1949, the gentleman from the quiet English countryside confidently told *Time* magazine from New York City that his machine was "the closest thing to a synthetic human brain so far designed by man."[21]

It had taken Ashby fifteen years to design his protobrain, and another two years to build it. It cost him £50.[22] The contraption looked like four old-fashioned car batteries set in a square on a large metal foundation. It was too heavy for one person to carry. Ashby and his assistant, Denis Bannister, used magnetically driven potentiometers, electrical wiring, valves, switches, and small water troughs to build the machine.

The mechanical brain's gray matter was coated in black and consisted of surplus control devices used by the Royal Air Force in the Second World War. "It has four ex RAF bomb control switch gear kits as its base, with four cubical aluminium boxes," recorded Ashby in his notebook. The only visibly moving parts were four small magnets, swinging like compass needles inside one of four small water troughs that were installed on top of each box. Each of the four boxes had fifteen crude switches to change various parameters. At first glance, the boxes appeared to be physically unconnected. But the units were designed to interact in arcane ways. When the machine was switched on, the magnets in one unit would be moved by the electrical currents from the other units. The magnets' movements, in turn, altered the currents, which then changed the movements again, and so on. The setup was dynamic and fragile. Or so it seemed.

The machine was designed to keep its four electromagnets in a stable position, with the needle above each box centered in the middle of a trough. This was the homeostat's normal, "comfortable" position. The experiment was to make the machine "uncomfortable" and then see what it would do. The inventive Dr. Ashby found a number of ways to make his creation uncomfortable: reversing the polarity of a connection; reversing the polarity of a trough; changing some of the machine's feedback; reversing a magnet; restricting the movement of a magnet on one side; joining magnets together with a bar. The idea was to disturb the machine's equilibrium, indicated by the compass needles, and see how the homeostat would react. Whatever the doctor did to his machine, it soon found a way to adapt to the new conditions, swiftly re-centering its magnets' compass needles. Ashby's gadget, he believed, "actively" resisted any attempt to disturb it, producing "coordinated activity" to restore balance.[23] It was a very British machine indeed.

Ashby was convinced that his contraption would actually "think." After all, it was able to choose, independently, the proper way to become "comfortable" again, with all four magnets centered. The machine would be "deciding" which of its 390,625 theoretically possible ways to act would be best to overcome the problem at hand, Ashby told the reporters who flocked to Barnwood to write about the curious invention. Ashby exuded the authority and credibility of a doctor, psychiatrist, and military officer. The homeostat, an appropriately sturdy and steady device, would click gently while doing its thinking work. A reporter from the *Daily Herald*, one of Britain's largest newspapers, was stunned: "These clicks are 'thoughts,'" he wrote. "The machine always thinks out its problem and puts itself right again."[24]

On December 13, 1948, the *Herald* carried a front-page article titled "The Clicking Brain Is Cleverer Than Man's." The machine's inventor, the paper reported, was confident that the machine would one day be developed into an artificial brain "more powerful than any human intellect," capable of finally tackling and solving the world's intractable political and economic problems.

Like Wiener, Ashby was inspired by the "goal-seeking" properties of killing machines. He also used the example of antiaircraft gunnery during the war. Thanks to innovative feedback-based technologies—radar tracking, prediction, and variable-time fuses—air defense had become refined enough to show such goal-seeking behavior. An undefined property such as life or a mind did not enable behavior toward a predefined goal; negative feedback did.

"Any machine, however inanimate, which has negative feedback will show this feature," Ashby was convinced.[25] And the homeostat proved it. The contraption embodied the then predominant scholarly fixation with behavior and function. Ashby's thinking machine was celebrated as "epoch-making" and written up in popular magazines from England to Australia to the United States.

But fellow scientists were more skeptical than hacks. Ashby was invited to present his ideas on homeostasis on Thursday and Friday, March 20–21, 1952, at a Macy conference in New York City. Twenty-one regular participants and eleven guests met at the Beekman Hotel, two blocks east of Central Park. The Josiah Macy Jr. Foundation, a philanthropic outfit founded in 1930, funded a series of ten meetings, with transportation, meals, and cocktails included. Many distinguished cybernetics-inspired scientists were in the audience, among them Gregory Bateson, his then wife Margaret Mead, Warren McCulloch from MIT's Laboratory of Electronics, Julian Bigelow, who had worked with Wiener on the antiaircraft predictor, and Arturo Rosenblueth from Mexico City, to whom Wiener had dedicated his foundational book. Wiener himself was not present.

The Macy sessions were thrilling, at least for scholars. Mead, a famous anthropologist, attended the first meeting and later recalled the atmosphere: "That first small conference was so exciting that I did not notice that I had broken one of my teeth until the conference was over," she wrote in a memoir.[26]

Ashby's presentation on the homeostat was sandwiched between a talk on emotions in the feedback loop and a talk on learning octopuses. He had carried his thinking machine across the Atlantic and was ready to demonstrate its awesome capabilities.

Ashby's goal was to explain how organisms achieve homeostasis, how animals and plants maintain the balance of life. He viewed an organism as a "mechanism" that faces a hostile and dangerous world. The main job of this mechanism is "keeping itself alive."[27] That could mean keeping its body temperature in a certain range, keeping its blood sugar level under control, keeping enough water in its tissue, or maintaining any other equilibrium—homeostasis, in short. "If the organism does not behave suitably, it will pay the penalty of its inefficiency by being killed," Ashby told the audience of interdisciplinary scientists.[28]

Ashby had been invited as a guest, and the program listed him as "Department of Research, Barnwood House, Gloucester, England." Most of the American participants had no idea that the gentleman from the English countryside worked at a mental institution. Now, in his lecture, the doctor was especially focused on learning, on learning organisms; he didn't want to limit himself to just machines. Ashby was fascinated by the idea that an organism could "reorganize its neuronic equipment" in response to changes in its environment. But his ideas on the body in its environment turned out to be rather different from those of most other early cyberneticists. Most took it for granted that there was a fundamental distinction between system and environment. Ashby didn't. So he moved carefully. He tried to outline how a body and its environment related to each other: was water, for instance, part of an organism or part of the environment? The distinction between system and environment wasn't as clear as many thought, Ashby implied.

"May I interrupt for a moment?" one participant asked. The American cyberneticists were baffled. Julian Bigelow wanted to make sure he understood this odd idea correctly. "You started out by saying that the environment consisted of everything under the sun," he said.[29]

Ashby tried to carry on with his talk. He introduced his homeo-stat—a mechanism that, he thought, fulfilled all the criteria: it faced a hostile world, the environment was part of its mechanism, and it behaved rather suitably if it had to, learning to maintain a stable bal-ance in response to being disturbed, so that it would not be killed, Ashby explained. His £50 contraption built from RAF switches, mag-nets, and troughs, he told the cream of American postwar interdisci-plinary science gathered at one of Manhattan's finest hotels—that machine—was itself alive.

A heated discussion ensued. For more than an hour, Bigelow and other skeptics disputed that the homeostat would be alive, or that the machine would indeed learn or adapt in the first place. They didn't even buy Ashby's ideas about the fluid environment.

Some scientists in the audience cautiously sided with Ashby. One fellow psychiatrist, from the University of Chicago, asked Ashby if his machine could develop maladaptive behavior; "in other words, it gets a neurosis?" he asked. "What I have proposed so far does not contain sufficient complexity to allow the system to develop a neurosis," Ashby responded.[30]

Gregory Bateson was intrigued. He quizzed the homeostat's inven-tor: "If it should evolve a recipe about how to try, it would then have a neurosis, wouldn't it?"

"It could, yes," said Ashby.

"How is your brain specified?" asked Jerome Wiesner, the brilliant electrical engineer who had joined MIT's Radiation Laboratory in 1942. Wiesner was referring to Ashby's mechanical brain in front of him, not the natural brain in his skull. But the question was phrased somewhat ambiguously, Ashby thought. "The brain—natural or mechanical—is specified only in that it shall be filled indiscriminately with switch gear," he said, obscurely.[31]

Ashby then explained his ideas of organism-environment interac-tion. Delay and latency were problems. If a creature was rewarded or punished, say, five minutes after it had done something, "it is notorious that this makes it very difficult for the living organism to adapt," Ashby said. "At the moment that sort of delay doesn't happen," he added. "Action by the organism is followed almost at once by the signal back from the environment."

Bigelow was confused. Ashby increasingly annoyed him. "Dr. Ashby, is this an existing model?" he interrupted, his tone impatient. "Yes," Ashby said, "I'm describing the homeostat."

Ashby's response only added to Bigelow's confusion. The box had to represent the organism *or* the environment. It couldn't be both. "Well, which object is the homeostat?" Bigelow barked.

"The homeostat is the whole thing, organism *and* environment." Ashby's was a fundamental insight, a historical insight even. But it didn't conform to early cybernetic theory, which insisted on clear separation. The homeostat was both system and environment at once. The electrical magnets were both a disturbance and a means to overcome the disturbance.

Ashby tried to explain, again:

> You can arrange it, if you please, so that one unit tries to control three units, a small brain trying to control a large environment, or so that three units try to control one, a large brain controlling a small environment.[32]

Ashby was convinced: a brain must be given an actual environment to adapt to. In his machine, any of the four units could be regarded as the environment to which the rest must adapt. His mechanical brain had its environment already built in, much like some of the patients he dealt with at the Cotswold sanatorium.

The other thing that annoyed Bigelow was that Ashby kept talking about his black contraption as if it were a living, learning being. "Sir," Bigelow said sharply, "in what way do you think of the random discovery of an equilibrium by this machine as comparable to a learning process?"[33]

"I don't think it matters," Ashby responded. "Your opinion is as good as mine."

Bigelow persisted: "But I am asking, if you place an animal in a maze, he does something which we call learning. Now, in what way does that machine do something like that?"

Ashby kept insisting his machine was learning, because it evidently changed its behavior, just as an animal would. "The dog jumps on the chair; you beat it three times in succession and then it doesn't jump on the chair any more," he replied.

Bigelow agreed on the dog: "It has learned," he granted, adding, "This machine does not do that." Ashby wasn't having any of it, so he turned the question back to Wiener's former assistant: "Would you agree that after an animal has learned something, it behaves differently?"

"Yes," said Bigelow.

"Well, the homeostat behaves differently," Ashby contended, adding that his machine would respond to "punishment," just like a dog. Bigelow's cross-examination continued for a while, but the disagreement between the two cyberneticists only deepened. "It may be a beautiful replica of something, but heaven only knows what," the engineer said in frustration.[34]

Meanwhile, Gregory Bateson was listening intently. He was intrigued by this fiery discussion of living organisms and the environment. The social scientist and anthropologist was putting one and one together, and started thinking of biological environments, of an "environment that consists largely of organisms," like forests, animals, and tribes. Nature, Bateson thought, also didn't distinguish between organism and environment, yet it learned.

The anthropologist asked the prominent ecologist in the room, G. Evelyn Hutchinson of Yale University, whether the machine could be compared to nature itself: "Is not the learning characteristic of Ashby's machine approximately the same sort of learning as that which is shown by the ecological system?"[35]

"Yes, definitely it is," Hutchinson agreed.

For Bateson the conclusion was obvious. Survival wasn't a problem for just the organism or the mechanism. Survival was also a problem for the environment—especially "these wibbly-wobbly environments," he said, vaguely. Bateson's thoughts galloped. This discussion with Ashby would mark one of two defining moments in Bateson's life, as he realized only later. Over the next two decades the anthropologist would explore the wholesome power of cybernetics and spiritually rearticulate the idea for the next generation.

The homeostat had another feature that seemed only too human: it exhibited goal-seeking behavior, but it didn't have a *goal*.[36] The homeostat's objective, seemingly, was to do nothing, and it wasn't possible to assign it a goal. One influential robotics engineer and cybernetics pioneer, W. Grey Walter, compared the homeostat to "a fireside dog." Like the dog, comfortable on a pillow in front of the fire, the homeostat "only stirs when disturbed, and then methodically finds a comfortable position and goes to sleep again."[37] Tongue in cheek, he called it *machina sopora*, the "sleep machine." Ashby's ideas went a step too far for most of the modernist cyberneticists of the early 1950s. The engineers in this progress-obsessed decade were focused on behavior, purpose, and shedding light into black boxes. They did not appreciate the unwieldy, seemingly useless and obscure gadget from Gloucester.

Norbert Wiener, remarkably, was an exception. He couldn't make it to New York that week in March 1952, but he heard about the English machine, and about the heated discussion that ensued after its presentation. Ashby's celebrated book *Design for a Brain* came out the same year. Wiener read it immediately. He was impressed. So impressed that he updated the second edition of one of his best-selling books, *The Human Use of Human Beings*, first published in 1950. He needed to reference Ashby's work. Ashby had not simply invented an experimental machine. He had invented a learning machine. Even more significantly, Ashby made a trailblazing philosophical statement in his Manhattan presentation on the homeostat and in his subsequent writings. It was, Wiener thought, "extremely inspiring":[38]

I believe that Ashby's brilliant idea of the unpurposeful random mechanism which seeks for its own purpose through a process of learning is not only one of the great philosophical contributions of the present day, but will lead to highly useful technical developments in the task of automatization.[39]

Ashby indeed had a different take on cybernetics. He wasn't a mathematician. But in the '50s, the new discipline had a reputation of requiring advanced mathematics and knowledge of electrical engineering. Eight years into the discipline, it was time for a fresh approach with a new introduction. Cybernetics is a "theory of machines," Ashby wrote in his best-selling textbook, *An Introduction to Cybernetics*, published in 1956.[40] That wasn't a new concept. Engineers and mechanics had been building complex machines for centuries. Theorizing about machines and about control had been done before.

But cybernetics wasn't about levers and cogs and revolving axles and potentiometers. The question it posed wasn't "What is this thing?" but "What does it do?" as Ashby put it. Cybernetics wasn't about mechanics; it was about behavior. The cyberneticists' view of the machines reflected the midcentury modernist Zeitgeist. In architecture, functionalists designed machines to live in or to sit in—such as Le Corbusier's efficient modernist concrete buildings and chrome-and-leather lounge chairs. In psychology, behaviorists measured and experimented with the mechanics of the mind; Ivan Pavlov's pioneering research on conditioned reflexes is an example. Cybernetics, Ashby suggested, was "essentially functional and behaviouristic."[41]

To Ashby, the popular imagination had it wrong. This wasn't about thinking mechanical devices. The critical indicator of whether a machine qualifies as a brain wasn't a machine's ability to think. More important was whether the machine was doing something. "The brain is not a thinking machine, it is an *acting* machine," he wrote in December 1948. "It gets information and then it does something about it."[42] This was an engineer's perspective of the brain. He considered it simply an input-output device.

Like every other organ in a living body, the brain is a means to survival. And if it doesn't produce action, it doesn't aid survival. The British scientist had long held the view that the nervous system is a physical machine, "a physico-chemical system."[43] The main function of this machine wasn't to sit in isolation in a skull. It constantly worked to bring the organism "into adaptation with the environment." Ashby tried to define what he meant by environment, as precisely as possible. It actually was simple: variables that affect the organism when changed, as well as variables that are, in turn, affected by the organism's behavior. Both were intricately linked: "The free-living organism and its environment, taken together, form an absolute system," he wrote in 1952, just like his homeostat.[44]

III

The world affects the organism. And the organism affects the world around it. That's how Ashby saw it. The world is quite different from experimental settings in laboratories. Scientists deliberately isolate the rat or the student they happen to be working on from the feedback that their subjects experience in the wild. Saying that an experiment is controlled is really saying that the environment is controlled.

"Organism and its environment are to be treated as a single system,"

Ashby was convinced. He glimpsed the consequences of this idea: "The dividing line between 'organism' and 'environment' becomes ... arbitrary."[45]

Ashby knew that this sounded strange, even unreasonable. How could the dividing line between a human body and, say, a tree, become arbitrary? Of course, he added, a man's arm was separate from a hammer, and a hand wasn't a chisel. One is alive, made of bones, blood, muscles, and skin; the other is made of inanimate steel and wood. Humans' tools aren't part of their anatomy. But these facts would be misleading, Ashby argued. "Purely anatomical" features should not obscure *function*. And from a *functional* point of view, the division of the system into "organism" and "environment" was blurred.

What if a mechanic with an artificial arm is trying to repair an engine? Is the arm part of the organism that is fiddling with the car engine? Or is the prosthesis part of the machinery that the man is trying to control? The answer, Ashby thought, is both at the same time: "The arm may be regarded either as part of the organism that is struggling with the engine or as part of the machinery with which the man is struggling."[46]

Something similar applies to the chisel in a sculptor's hand. The chisel is either part of the "biophysical mechanism" shaping the wood, or the tool is part of the material that the sculptor's nervous system is trying to control. But what about the bones in the sculptor's arm? They could be either part of the organism or part of the environment of the sculptor's central nervous system.

This wasn't simply dry theory. Bleeding-edge neurological research would illustrate the potential of the machine. The nervous system in the animal had a remarkable ability to adapt to changes in the environment. But what if the body was part of the brain's environment?

Changing the body and then observing the brain's reaction had long been an exciting field of neurological research. Already in the late 1890s, Alessandro Marina had embarked on a remarkable experiment that would later influence Ashby. Marina knew that the brain could handle some damaged body functions. The neuroscientist from northern Italy wanted to see how good the brain was at self-correcting. He had the idea to modify the muscles that control a monkey's eye movements, the so-called extraocular muscles. Each eye has only six of these muscles. Marina's nineteenth-century experiment was cruel: the scientist surgically opened one of the monkey's eyes and then cut the eye's extraocular muscles and reattached them in a crossed position. The other eye would be left intact.

Marina expected that the monkey's eyes would be literally crossed: contracting the two similar muscles would move the normal eye to the left and the modified eye to the right. But after the wound healed, Marina made a surprise discovery: the monkey's eyes were still moving together, preserving binocular vision. The brain had simply adapted to the surgical change. Forty years later, cyberneticists were intrigued.

In late 1947, when Norbert Wiener was launching his informal discussion circle in Boston, Roger Wolcott Sperry from the University of Chicago published the results of a similar experiment. It was no less cruel: First Sperry put several red spider monkeys to sleep. He cut open the elbows of the anesthetized animals. Then he looked for the muscles that flexed and extended the monkeys' arms, for instance, when the monkeys wanted to pick up food. He then crossed the nerve fibers controlling the muscles.[47]

If one of the monkeys wanted to bend its arm to bring a piece of banana to its mouth, the arm would straighten instead, and vice versa. It was like using a computer mouse upside down: any move would be the opposite of the intended move. At first the monkeys' arm movements were confused and chaotic. But after some "reeducation," all monkeys started picking up food again, adjusting the performance of their rewired arms. Sperry later won a Nobel Prize for his work on the brain. In the 1940s his experiment became a powerful illustration of cybernetic principles.

But this wasn't just about a monkey's arm. Ashby used this example in his influential introductory textbook: Suppose the red spider monkey needs to push a lever to get food from a box, as when opening a door handle to a pantry. Crossing the nerves in the monkey's elbow is just like replacing the lever on the box with another one that needs to be pulled up instead of pushed down. It makes little difference for the brain, Ashby reasoned. For the brain, reversing the mechanics of a wooden handle is just like reversing the mechanics of nerves in an elbow or eye.

This was a challenging insight for man and machine. Something that was outside the body (a lever) and something that was part of the body (the elbow) had similar relationships to the brain. Again, this comparison raised a hairy question: What is the difference between the environment and the system itself? The answer, it seemed to Ashby, is arbitrary. The arm is as much a part of the brain's environment as the door handle is. "Spinal cord, peripheral nerves, muscles, bones, lever, and box—all are 'environment' to the cerebral cortex," the doctor from Barnwood House explained. The brain, like his homeostat, would simply use negative feedback to adapt to the disturbance at hand.

Cyberneticists moved elegantly between electronic engineering and the life sciences, blurring the line between living and nonliving systems. Ashby had another powerful trick up his sleeve. This time it came from engineering, not biomedical research: the black box. Black boxes were invented to help engineers interact with machines. The black box isn't an actual black box, of course. It's an idea, a concept that is used to shed light into the dark.

The homeostat was a black box. Taken literally, it was four black boxes, made of aluminum. But the idea of a "black box" also popped up in the 1940s as a term of art. Engineers and cyberneticists used it first. Calling something a black box was an elegant way to describe something they couldn't describe, to understand a device that wasn't properly understood. The internal workings of the "box" remained in the dark, unseen and "opaque." A machine would produce output from input. What happened in between was unclear.

The gun director was a black box for its operators. They didn't know what exactly happened inside, but they knew how to fire the gun with the output values. The proximity fuse was a black box: most military officers had no idea how the radio mechanism inside the shell worked, but they used it to good effect. Office workers did not know how their brand-new IBM thinking machines worked, but they knew how to input data and how to read the computer's output and act on it.

Ashby didn't suggest that black boxes behave like real objects. He was more radical. Ashby suggested that real objects "are in fact all black boxes."[48] For those with a cybernetic mind, it didn't matter what was inside the black box: switches, tubes, and wires—or blood and gray matter. What mattered was input and output. In this view, the body itself is a prototype of human-machine interaction. Man himself is a black box.

Ashby's textbook came out in 1956. At the time, the general debate about cybernetics was becoming more and more influenced by the rise of one particular type of black box: electronic computers. These computers were gigantic, expensive, industrial calculating machines that instantly gave rise to similarly outsize expectations. The popular press often referred to these new machines as "big brains."

IV

By 1960, approximately thirty-five hundred electronic computers were crunching numbers in the United States. Renting a computer, depending on its capabilities, cost between $1,600 and $300,000 a month; buying one, between $60,000 and $13 million.[49] The big brains were housed in special climate-controlled rooms, closely guarded by a priesthood of technicians. Only these technicians knew how to handle reels of magnetic tape, disks, drums, and decks of punch cards. The machines were used to issue paychecks or to calculate financial reports. Computer science as a separate field of study was only beginning to emerge. Stanford University opened a division of computer science within the mathematics department in 1961, and a full department only four years later. And, of course, the thinking machines were already put to good use in the military, to calculate ballistic trajectories and to deal with an ever-more-challenging air defense problem in the early years of the Cold War. These new and seemingly magic machines required interpretation.

Meanwhile, the cybernetic research program appeared serious, even threatening to some. Cybernetics was coherent and convincing enough to be an emerging field, and it was incoherent enough to attract a wide range of scholars.[50] Mathematicians, physicists, and biologists, as well as philosophers and sociologists, seized the cybernetic vision and produced thousands of books and research articles between the late 1950s and early 1970s. New journals sprang up. International conferences were held. New graduate degrees on cybernetics emerged. The neologism found its way into the world's languages and dictionaries. Early cybernetic work sought to articulate and apply the principles by which systems could be controlled. Later cybernetic scholarship became more abstract: cyberneticists asked how systems would describe themselves, how they organize themselves, and how they control themselves.

The leap from science to myth was small, smaller than many nonscientists would imagine. The mathematical mind is trained for abstraction. Algebra isn't limited by the availability of fresh apples to count or to multiply. Arithmetic is dealing with abstract entities. The ambitions of the emerging discipline were equally expansive. The "real machine" could be electronic, mechanical, neural, social, or economic. This alone meant that the realm of cybernetics was vast.

This ambitious vision of cybernetics is best expressed through an analogy: cybernetics relates to the machine as geometry relates to the object in space. Ashby's machines-and-math comparison was an inspiration, a stroke of genius. Nature provides a range of geometrical objects in space: stones, apples, snakes, horses, or something more complex, like trees or mountains. Geometry contains these objects and can very well calculate the surface of an apple or the volume of a stone. But real, existing objects don't limit geometry. Geometry is far more expansive—a framework to grasp all possible forms and shapes and sizes. It is just the same with cybernetics and machines: "It takes as its subject-matter the domain of 'all possible machines,'" Ashby argued in 1956. It is of secondary interest whether some of these machines "have not yet been made, either by Man or by Nature."[51]

No mathematician would ever feel constrained by five apples on the kitchen counter. Why should a cyberneticist be constrained by vacuum tubes?

One of the common criticisms of cybernetics, even in the striving 1950s, was that its vision was fanciful, that many of the technologies had never been built and would not become reality anytime soon. However, theorizing about nonexistent machines was not a bug, Ashby argued, but rather a feature. Physics, the master discipline, was leading the way. Physics is also an important and highly successful scientific discipline that studies nonexistent systems: springs without mass, particles with mass but no volume, gases that behave perfectly. Such things don't exist. Yet understanding them in pure theoretical form is crucial for understanding even simple things like a watch. Ashby took this abstraction-first approach to its logical conclusion: the cyberneticist would survey the possible relations between man and machine first "in the general theory." Only then would he get his hands dirty by looking more closely at "terrestrial machines" that are found in some scientific or industrial installations.

Ashby's ploy was brilliant: the absence of evidence wasn't a problem. That the theoretically—and cybernetically—predicted future had not arrived yet didn't mean it would not arrive soon. It was the same as with quantum theory. Cybernetics was liberated from the imperfect devices of the present. Ashby's powerful ruse would propel the new idea of cybernetics forward for many decades, leading it into uncharted territory. Some of the participants of the early Macy conferences—Gregory Bateson more than anybody else—would later help lift cybernetics to a higher level.

By the early 1960s it became ever clearer that something big and fundamental was happening. Computers were becoming more capable. Automation in factories was marching on. And the new discipline of cybernetics opened new avenues of study: boundaries were beginning to crumble—between system and environment, between brain and body, between machine and worker. All this created a degree of confusion and uncertainty.

Wiener himself was torn. His overall belief that the theory of machines would change everything was as unshakable as his belief in the antiaircraft predictor had been twenty years earlier. Humankind wanted its tools to be intelligent, to be able to handle more and more complex tasks nimbly at higher speeds. The possibilities were enormous. Machines, there was no doubt in Wiener's mind, would simply be superior to their human creators:

A digital computing machine can accomplish in a day a body of work that would take the full efforts of a team of computers for a year, and it will accomplish this work with a minimum of blots and blunders.[52]

Yet with increased machine intelligence, the risks increased as well. In 1963, Wiener commented that the world of the future would be an increasingly harder struggle against the limits of the human brain, "not a comfortable hammock in which we can lie down to be waited upon by our robot slaves."[53] By now, the MIT professor commanded veritable star power, and his comments would be reported by the nation's leading newspapers. The father of cybernetics wasn't sure whether the machines would turn out to be a force for good or evil, whether the robots would serve their human creators or perhaps rise up and revolt. It was the duty of any self-respecting scientist, he believed, to reflect on the possibilities for abuse, and to warn the world of the coming danger, just as responsible scientists should have warned of the coming atomic bomb. One of the scenarios that most concerned Wiener was computers simulating war, and potentially making automatic decisions about when and how to use force.

This was not a far-fetched scenario. The US Air Force had started investing vast sums of money in automated air defense systems. Delegating decisions to machines courted disaster. "There is nothing more dangerous to contemplate than World War III," Wiener told the *Chicago Tribune* in 1961. "It is worth considering whether part of the danger may not be intrinsic in the unguarded use of learning machines." As machines increase the speed of automated decisions in war, Wiener surmised, man would find it harder to stop them. "To turn a machine off effectively, we must be in possession of information as to whether the danger point has come."[54]

On December 27, 1959, at sixty-five years old, the left-leaning mathematician held a high-powered press conference at the 126th meeting of the American Association for the Advancement of Science in the Morrison Hotel in Chicago. He painted a dire picture of the future of humankind. "It is quite in the cards that learning machines will be used to program the pushing of a button in a new push button war," he told the journalists.[55]

If the rules of victory in a war game do not correspond to what we actually wish for our country, it is more than likely that such a machine may produce a policy which will win a nominal victory on points, at the cost of every interest we have at heart, even that of national survival.[56]

Such escalation could be an accident, or, indeed, ever-more-intelligent mechanized servants could deliberately turn on their human creators. "We wish a slave to be intelligent," Wiener told the press. "However, we also wish him to be subservient." And therein was the contradiction, said Wiener: "Complete subservience and complete intelligence do not go together."[57] The father of cybernetics had no doubt that the rise of the machines was only a question of time:

If the machines become more and more efficient and operate at a higher and higher psychological level, the catastrophe of the dominance of the machine comes nearer and nearer.[58]

3.   AUTOMATION

THE IDEA OF AUTOMATED WARFARE EMERGED FROM THE still-smoking rubble of London and Hiroshima. One city had been pummeled by ballistic missiles; the other, shattered by a single nuclear bomb. Some air force officers understood immediately that war had changed forever. One of the most eloquent and influential visionaries was Henry "Hap" Arnold, commander of the Army Air Forces during World War II, soon to become the US Air Force. To the five-star general, the future of war seemed obvious less than three weeks after Japan's unconditional surrender. In mid-August 1945, he held a press conference in Washington.

"It was terrible—and it was simple," *Newsweek* reported, under the heading "Push-Button Warfare." "Only two new weapons were needed to make it a reality": the American atomic bomb and the German V-2. The two terrifying technologies were practically made to be combined. Soon the range, speed, and destructive power of the new machines of war would exceed anything seen in the most devastating clash of arms that was World War II.

The lethal mix would be enhanced by improved communication and control. "Perfect communications systems between air and ground" will enable the most intricate maneuvers of pilotless aircraft and missiles, the general told the assembled press corps. "Aerial combat as we know it," Arnold predicted correctly, "will disappear." This meant that the defense had to be reorganized. It was nearly impossible to defend against such automated annihilation, as the V-2 attacks on London had demonstrated. The awesome speeds of the new weapons would further remove human operators from air defense. Now machines were the strongest defense. Arnold placed his hopes in rockets that would be "automatically seeking out those planes and missiles."[1] For fighter pilots like Arnold, the progression was intuitive: take out humans, increase speed and reach, and the defense becomes harder—while the offense dominates.[2]

Arnold spelled out his vision three months later in classified reports—and in the *New York Times*. The future machines of war, he wrote, could see terrain by way of radio pulses "fed into mechanical brains" inside the automatic flying weapons. Arnold, combining the lessons of the war against Germany and Japan, foresaw advanced spaceships, carrying nuclear bombs, hurled at their targets through the ionosphere at more than 70 miles altitude and speeds faster than 3,000 miles per hour, powered by solid fuel. These terrifying machines, five-star Arnold wrote in November 1945 in the *Times*, "are forerunners to a 'push-button' warfare in which their derivatives may flash at tremendous speeds and altitudes to perfect strikes hundreds or thousands of miles away."[3]

The consequences were stark. "It must be made perfectly clear to an aggressor," Arnold wrote, that pushing this button would be answered by "a devastating air-atomic counter-attack on him." The same month, he urged the government's civilian leaders to take action. "War may descend upon us by thousands of robots passing unannounced across our shorelines—unless we act now to prevent them," he told Secretary of War Henry Stimson in November 1945.[4] Acting now meant building superior automated machines of war. Only a certain and certainly devastating counterattack would convince the enemies of the United States that peace was in their own interest. Any attack should trigger a counterattack, with automated certainty. The new machines of war brought with them the new logic of deterrence. "If war comes, air power has in a sense failed," the future commander of the mighty US Air Force concluded, in somber tone.

For Norbert Wiener, the idea of automated military confrontation was folly. "Behind all this I sensed the desire of the gadgeteer to see the wheels go round," he observed with a sneer in the *Bulletin of the Atomic Scientists*. "The whole idea of push-button warfare has an enormous temptation for those who are confident of their power of invention and have a deep distrust of human beings." Wiener knew such people. Some of his own colleagues at MIT seemed to trust machines more than they trusted men. But he resented the war, the way it was won, and its innovations in weaponry. And the new competition for military might brought more such "gadgeteers" into positions of power. Worse, Wiener knew one thing for certain: that more such deceptive innovations were imminent.[5]

On September 23, 1949, President Harry Truman announced that the Soviet Union had detonated a nuclear device. The American public was shocked. German weapons engineering had a towering reputation for deadly creativity. The Russians, by contrast, were victorious because of vast lands, vast numbers of soldiers, and devastating winters, not ingenuity or engineering skill. In the summer of 1949, therefore, no one expected any potentially hostile country to develop the atomic bomb for many years to come. The United States felt safe.

Cocksure and arrogant, most American scientists—and most military officials—had been fooled.[6]

From a defensive point of view, the implications were devastating: the antiaircraft problem that had jolted Vannevar Bush into action in 1939 would soon look benign. Again it was a scientist at MIT who took the lead on a vast air defense project. And again it wasn't Wiener, but one of the derided gadgeteers. George Valley had been a professor of physics at MIT and a member of the Air Force Scientific Advisory Board since 1946, a body that preceded the creation of the air force as a separate service. Valley, who held a PhD in nuclear physics, instantly understood how inadequate US air defenses were. Though he was initially skeptical that an engineering solution was the right approach, the new threat convinced him: "I realized that my almost-completed new house was vulnerable to the blast wave of the first bomb to hit Boston."[7]

Automation was the only plausible response. The basic ingredients of the updated air defense problem of the 1950s were similar to those of the Blitz in London a decade earlier, it seemed to Valley: detecting incoming enemy bombers, tracking them, computing target coordinates, and hitting them at the right moment. But a few things were different. The bombs were far more devastating, the planes faster, and the distances wider. Any solution to the new bomber problem would have to be scaled up as well. The entire country, in effect, would become an air defense battery. The problems of command and communication that had previously vexed antiaircraft crews were thus magnified to continental scale. The problem was mind-boggling.

So was the response: the Semi-Automatic Ground Environment. SAGE was a bold idea, first articulated by the Air Defense Systems Engineering Committee in 1950, which became known as the Valley Committee, a body set up by the air force's Scientific Advisory Board. A proof-of-concept site was built on Cape Cod, Massachusetts, in 1951, complete with its own airfield and dozens of radar stations. The

Lincoln Laboratory at MIT, founded in 1952, then led the development of the vast air defense project. The outlines of the SAGE network had taken shape by 1954, when IBM was awarded the first production contract.

SAGE combined a continental network of radar stations—many of them north of the Arctic Circle, and some as far out as 200 miles at sea—twenty-three warehouse-sized supercomputers all over the United States, custom developed by IBM, each connected to more than a hundred field sites by AT&T's nationwide telephone lines. The first gigantic control center became operational at McGuire Air Force Base in New Jersey on July 1, 1958. NORAD, the North American Air Defense Command, had a staff of approximately two hundred thousand at the time. When the system came online, Earle Partridge, NORAD's commanding general, estimated the costs of SAGE alone at $61 billion over a fifteen-year period (more than $500 billion in 2015 money), with operating costs of $8–$10 billion per year.[8]

The system was monumentally ambitious. SAGE logged the course, speed, altitude, and location of all aircraft flying over North America at any given moment, watching friend and foe alike. Each direction center received data from a range of sources: long-range radar, sweeping the skies along America's vast frontiers; airborne early-warning planes, flown by the air force and navy, patrolling the coasts hundreds of miles offshore; radar picket ships, always on watch; gap-filler radar, filling empty spaces between the long-range radar stations; the so-called Texas Towers, radar sentries on stilts along the East Coast. Up-to-the-minute commercial and noncommercial flight data were transmitted to the SAGE central computers, via a dedicated data link on the telephone network. The machines then combined the data with known flight plans of airliners, live weather information from field stations, and the ground situation on airfields in North America. All these information streams fed into the largest computer ever built. The machines then crunched the numbers.

When the early-warning radar detected an enemy bomber forma-
tion, the system would flag it as hostile. An officer would then select
a weapons assigner to attack the enemy. The giant SAGE mainframes
would match the incoming bombers with the available defensive weapons: the computer displayed each incoming plane on a round, 30-inch
cathode-ray tube as a glowing track with a number. The track's length
and direction indicated speed and course.

Operators used odd devices, so-called light guns, to associate a
track with other information, such as identification number, altitude,
or armament. The operator would point a stubby light gun with a
cable connection on the gun's handle right at the glass front of the
bull's-eye-shaped screen, aim the gadget, and then pull the trigger, as
if shooting into the computer. The gun's spotlight would illuminate
the screen at the position of interest and send a signal to the computer,
prompting the machine to update the deflection register's contents—
in effect, telling the system to associate a track with another piece of
information, such as speed or identification or armament (the gun did
what a mouse would do decades later).[9]

The computer would then superimpose a map showing geographical
features, available antiaircraft fire, and airfields that had fighters ready
to scramble. Small squares indicated the points where US fighters
would, it was hoped, take out Russian bombers, with numbers indicating the time to intercept. SAGE displayed all this information in less
than sixty seconds after spotting incoming bombers. If the operators of
the SAGE supercomputers decided an intrusion was real, a semiauto-
mated command would go out to interceptor aircraft that were always
fueled and ready for takeoff, as well as to Nike batteries, America's first
operational antiaircraft missile system, ready for instant action.[10]

Turning the entire North American continent into an air defense
battery meant that radar stations, computers, and interceptors needed
to be linked in real time and on a massive scale. Already in September
1950, MIT had succeeded in sending digitized data from a radar sta-

tion at Hanscom Field in Bedford, Massachusetts, to a computer in Cambridge, via 13 miles of commercial telephone line. The system would later transmit digital data at a speed of 1,300 baud, or symbols per second, on voice-grade telephone lines. The phenomenally large air defense system had a most surprising and underappreciated impact: it helped lay the foundation for networking computers, and ultimately for the internet.

Valley understood that commercial telephone lines were the least expensive means of reliable communication, even for the most expensive continental air defense system in the history of the United States. Valley's plan was to build a system that would rely on rented AT&T telephone lines for air defense "during battle," while two superpowers were engaging in nuclear exchange.[11] But the MIT professor encountered significant cultural resistance from the military.

During Valley's first visit to a dilapidated radar station from the war, he noticed that the officers were using clumsy and unreliable field radios. He asked why they didn't use the telephone. A grizzled officer launched into a "sermon," Valley recalled later, citing an Egyptian pharaoh, Darius of Persia, the Battle of Marathon, the Fall of Rome, Napoléon, and the Civil War: never, never must an army entrust its lines of communication to civilians. It thus became one of Valley's more delicate tasks to sell the AT&T-enabled defensive system to the lieutenant general at the helm of the Air Defense Command, Ennis Whitehead.

The three-star general, tall and bald, was known for his erratic temper, his ruthless persona, and his tough-guy arrogance. Valley didn't like the assignment and was scared of the general. Reluctantly, he made the trip to Long Island, where Whitehead's command was temporarily housed before moving into a permanent headquarters. The general and his staff politely listened to Valley's presentation about telephone lines and air defense, asking only a few softball questions. They then treated the doctor to a sumptuous luncheon. Over appetizers and cocktails, Whitehead related that he also had done a research project. "Darkter," he told Valley in a cigar-ravaged, gravelly voice, "my research is on the subject of blood."[12]

The MIT professor assumed that Whitehead was referring to research on controlling the blood pressure of interceptor pilots at high altitude, or something like that. Medical research in the air force was not unusual. Valley smiled and nodded, avoiding a martini refill, and took another fresh shrimp. "Darkter," the general continued, "my research tells me that when you have bled a nation white, you have it at your mercy."

The general proceeded to cite statistics from various historical campaigns, again Napoléon and the Civil War. Once 10 percent of a population was dead, 20 percent of all men killed, and twice that many wounded, the fight would be over. The doctor signaled that he needed a martini refill. "Yes, Darkter," Whitehead continued, "all they have left are old men and boys, and they have to give up." Valley glanced around the table and noticed he was being hazed. He cut up his roast beef, put too much of it in his mouth at once, let a little juice dribble down his chin, and responded, "General, that is the best piece of military research that has been done since Clausewitz." Whitehead looked him in the eye, then at the chin, grunted, and left. The air force was signed on to AT&T.

Earth's curvature had two surprising consequences: it necessitated more computing and more networking. By the 1950s, aircraft were flying even higher and faster than they had during World War II, and most aviators were excited about this trend. The improved flying ability presented an unexpected problem to military planners: spotting a high-flying plane on radar is simple; spotting a low-flying plane is hard. The newly established US Air Force tended to overlook one particular problem: if a plane was flying at an altitude as low as 500 feet, hills, mountains, and other topographical features swamped any image with "ground clutter." This noise was thousands or even millions of times stronger than the plane's signal. Existing ground-control radar stations worked well against high-flying aircraft, and at long ranges. But long-range radar's field of vision was shaped like a cone on its head: narrow at the ground and wide at the top. The lower an aircraft flew, the bigger the gaps between the cones.

The easiest way to spot and track aircraft flying low and under the radar was to increase the density of radar stations on the ground along the system's three early-warning lines that spanned North America. Earth's curvature meant that hundreds of radar installations were needed to create a network dense enough to thwart the low-altitude threat, including many unmanned so-called gap-filler radar stations that formed a "microwave fence" along the outermost Distant Early Warning Line, north of the Arctic Circle. When the DEW Line was put into service in 1957, the US Air Defense Command operated 182 radar stations, with more to come online.[13]

This vast grid needed connectivity. The conventional telegraph rate was 75 baud. Bell Laboratories saw no need to speed up data communications until Valley asked for the improvement.[14] But data transmission on a telephone line proved far harder than voice transmission. What telephone engineers called "hits" or "bats" of noise (very minimal delays) did not affect voice quality, but SAGE data links turned out to be vulnerable to hits and bats. AT&T solved this problem by dedicating privileged private lines to the air force.

Bell also developed modulator-demodulator terminals—simply called "modems" later. These terminals converted digital data to and from analogue waves, which meant that data could be sent over voice-band telephone lines.[15] By 1960, Bell was managing data transmission speeds of up to 2,000 bits per second on the switched network, and up to 2,400 baud on private lines.[16] These communications were vital. So the air force specified that AT&T had to use two independent, geographically separate trunk routes between the direction center and many of the radar sites.

SAGE became the largest data transmission network the world had ever seen.[17] The capabilities of the final system were extraordinary by any standard: SAGE could remotely control manned interceptors in flight, relaying real-time computer-generated directions by radio to the lead plane's autopilot.[18] When the fighter was in range to engage the enemy bombers, airborne equipment took over from the remote computer. "The pilot may have to use judgment if something goes wrong," the trade journal *Electrical Engineering* reported. "Otherwise, he has only to get the plane in the air and land it after the action."[19] Again, air defense was leading the way. "America is now armed with instant electronic reflexes," IBM boasted in a promotional film in 1960.[20]

At least, that was the theory. The nuclear attack never came. And SAGE was never deployed against incoming bombers as advertised. Yet the system, with its continental scale and astronomical budget, embodied the Cold War planetary threat of nuclear annihilation. By 1960, the vast North American air defense system was the most advanced automated computer system ever built. And yet again, the challenge of antiaircraft fire control would inspire and terrify technology pioneers and intellectuals in surprising ways.

Back in December 1949, when MIT's George Valley was recruited to conceive of the world's most advanced automated system, MIT's Norbert Wiener had already been theorizing about the spiritual consequences of such automation.

"The first industrial revolution," Norbert Wiener wrote in the preface of *Cybernetics*, "was the devaluation of the human arm by the competition of the machinery." Now, one and a half centuries after those dark, satanic mills, the world was again in the throngs of a second industrial upheaval. "The second industrial revolution," Wiener was convinced, "is similarly bound to devalue the human brain, at least in its simpler and more routine decisions."[21] Automation, the argument went, would require more high-skilled scientists and administrators. Less-skilled workers would be crowded out of the labor market and pushed into unemployment.

"Let us remember that the automatic machine," Wiener wrote in 1950, "is the precise economic equivalent of slave labor and must accept the economic condition of slave labor." Still jarred by his participation in war-related research, Wiener warned the public of the impending dangers: "It is perfectly clear that this will produce an unemployment situation in comparison with which . . . the depression of the thirties will seem a pleasant joke."[22]

On Friday, May 5, 1950, the Dramashop at MIT revived a famous science fiction play in the Peabody Playhouse, on Charles Street in Boston's West End. The play, originally written in Czech, had given most languages a new word: "robot." It was Karel Čapek's *R.U.R.*, short for *Rossumovi univerzální roboti*, or "Rossum's Universal Robots." The play tells the story of a factory that makes artificial workers, called *roboti*. The robots rebel against their creators and end up wiping out humankind. The story is full of biblical references. The play, written in 1920, was a runaway success. Three years after publication, the text had been translated into thirty languages. The robots in the story weren't simply unwieldy chunks of metal. They were made of organic matter as well as steel, to the delight of theater costume designers.

As the actors in the Peabody were getting ready, fitting into their clunky metallic costumes, Wiener took to the stage at the West End amateur theater. By now, the best-selling professor was used to being in the limelight. He had given countless interviews, and he had seen his face in magazines and newspapers nationwide. Wiener was a celebrity scholar, and he enjoyed it. He stepped forward to address the audience in the Peabody: "When the play was written," Wiener said, "the automatic machine was still in its infancy, or perhaps it is even better to say it was still in its gestation." He pointed out that the world has seen not just a succession of automatic machines, "but a philosophy of automatic machinery itself." He implied that he was the master philosopher and engineer of tomorrow. Therefore, he knew that Čapek's play wasn't science fiction. *R.U.R.* predicted the immediate future. "Machines demand to be understood, or they will take the bread from the mouths of our workers," he said, eyeing the student audience through his thick-rimmed glasses. "Not only that, they demand that we understand man as man, or we shall become their slaves and not they ours."[23] The prologue certainly sounded convincing.

"And now I would like to show you one of these robots," Wiener said, and with that, he turned toward one wing of the tiny stage, clapped, and commanded, as if calling a dog, "Come here, Palomilla!" The curtain ruffled and moved. A young man appeared, holding a flashlight that he pointed at a small contraption on wheels, about 18 inches long. The machine scuttled toward Wiener, whirring.

Palomilla was a tricycle cart, with two plate-sized wheels in the back and one tiny wheel in the front. The cart had a rectangular metal base. At the two front corners, Palomilla had two photocells each, eye-like sensors that reacted to light. It looked a bit like a dachshund on wheels. The output of the light-sensitive cells was amplified and fed into a tiller that controlled the tiny front steering wheel. The setup meant that Palomilla was drawn to the light, moving toward the torch like a moth—or, if the direction of the photocells' output voltage was reversed, away from the light, more like a bedbug. That's how the MIT researchers referred to the contraption, as moth or bedbug, depending on its settings.

But Palomilla, the clumsy robot on the amateur stage, wasn't meant to be cute. Wiener had designed it specifically to simulate two neural conditions: Parkinson's disease and intention tremors. And he succeeded. He simulated complex human behavior with the simplest technical means. This was a disquieting scenario. The trembling mechanical moth soon drew the attention of the US Army Medical Corps. The medical officers contacted MIT and photographed Palomilla's neural conditions, as Wiener presented them, to compare them with photographs of actual cases of nervous tremors in humans in order to assist the work of army neurologists.[24] But what really made people tremble was not the prospect of machines acting like the sick. The scary scenario was machines acting like the healthiest and the most capable people—even surpassing them as better workers, custom built, never calling in sick.

Wiener's vision hit a nerve in the early 1950s. Bertrand Russell, who had just won the Nobel Prize in Literature, thought that *The Human Use of Human Beings* was "a book of enormous importance," as he wrote in a review in the glossy British magazine *Everybody's* in September 1951. The famous English philosopher, then seventy-nine years old, foresaw existential perils to democracy. Substituting robots for human beings would give power to a small elite. The future governing clique would—"of course"—

> make machines that can utter extraordinarily eloquent political speeches with all the appropriate gestures, and the machines can be taken from meeting to meeting without ever getting hoarse, as human politicians do. Nor can they be intimidated by having rotten tomatoes thrown at them, or even rotten eggs.[25]

Russell's piece was titled "Are Human Beings Necessary?"

Three hours west of Boston, at a General Electric factory in Schenectady, New York, one of the employees was especially taken with this idea. The twenty-seven-year-old war veteran saw how GE's new milling machines cut rotors on jet engines and turbines faster and more precisely than any worker could. And what he saw scared him.[26]

He had read of Wiener's dark predictions, and he decided to write a novel about the coming automated future.

Kurt Vonnegut's first book, *Player Piano*, was published in 1952. The Schenectady factory is the thinly concealed backdrop of the novel's futuristic story: Ten years after a third world war, automated factories have replaced almost all workers with machines. American society is divided between managers and engineers on the one side and everybody else on the other side. The displaced workers lead meaningless and undignified lives in mass-produced houses. The main protagonist is Paul Proteus, one of the privileged engineers, who becomes disillusioned and leads a failed revolt against the mechanized system.

At one point in the story, Paul walks past his secretary's desk on his way into the office. She holds the manuscript of a freshly typed speech, "That's very good, what you said about the Second Industrial Revolution," she tells her boss.

"Old, old stuff," Paul says.

"It seemed very fresh to me—I mean that part where you say how the First Industrial Revolution devalued muscle work, then the second one devalued routine mental work. I was fascinated."

"Norbert Wiener, a mathematician, said all that way back in the nineteen-forties," Paul tells his secretary. He and his secretary are incredulous that people would ever waste their time on mindless, repetitive work. The idea that hangovers, family squabbles, resentments against the boss, debts, "the war," or indeed happiness and holidays would show up in a product makes no sense to them. Eventually, a new machine is invented that makes Paul's own white-collar job obsolete.

"*Player Piano* was my response to the implications of having everything run by little boxes," Vonnegut later told *Playboy* magazine. "The idea of doing that, you know, made sense, perfect sense."[27]

The book's title was a reference to pianos that are modified to be played automatically, as if by a ghost, controlled by a pattern of holes in a scroll. Its subtitle promised a glimpse into the near future: *America in the Coming Age of Electronics*. The book's publisher had sent an advance copy to Norbert Wiener for comment. But a character in the story, one of the rebels, was a "Ludwig" von Neumann. This was a thinly disguised reference to John von Neumann, the computer pioneer and early cyberneticist. Wiener didn't think it was okay to use the names of living people in a trendy science fiction novel. He wrote a terse letter to Vonnegut's editor, insisting that her rookie author could not "play fast and loose with the names of living people." But he could, and he did. Wiener was probably miffed that the final book mentioned von Neumann, nine years his junior, more than twenty times, while he was mentioned only once in passing.

By 1955, science fiction was an up-and-coming literary genre. Authors were hungry for ideas, and cybernetics inspired them. One story from 1955 was "The Cyber and Justice Holmes." The text was the first solo story of Frank Riley (born Rhylick), an American fiction author well known in the 1950s. Riley also took the idea of the second industrial revolution and ran with it. His plot is about a future court system in which intelligent and highly efficient machines have replaced human judges. Riley calls these fictional judgment devices "cybers." Naturally, such a significant innovation came with enthusiastic supporters, as well as skeptical opponents. The story quotes from a fax that praises the qualities of automated decision-making:

> We've seen what other counties have done with Cyber judges. We've witnessed the effectiveness of cybernetic units in our own Appellate Division.... And I can promise you twice as many prosecutions at half the cost to the taxpayers ... with modern, streamlined Cyber justice![28]

Riley's hero is Walhfred Anderson, a bow-tied and conservative-minded judge about to retire. Anderson remains doubtful about the impending changes in the country's criminal justice system. The flesh-and-bone judge is nearly swayed by the advantages that the steel-and-diodes judges have brought to the high table. The reader learns of the judge's self-doubt:

> The Cybers were fast. They ruled swiftly and surely on points of law. They separated fact from fallacy. They were not led down side avenues of justice by human frailty. Their vision was not blurred by emotion.[29]

But perfectly rational behavior wasn't everything, even for a judge. Emotions and empathy mattered as well. The story ends with the latest model of the mighty, seemingly all-knowing computing machine crashing when asked to calculate the "magnitudes of a dream." Riley's short story perfectly captures that typically mid-twentieth-century blend of a firm belief in progress and subdued, lingering fear.

In the 1950s, "automatic" was a generic term of technological progress. Engineers understood that computing machines would be used to control factories. "Calculating machines are, in fact, closely analogous in operation to factories," one 1956 engineering book pointed out, "for their function is the processing of information, just as that of the factory is to process metal, food or chemicals."[30] Yet throughout the 1950s and early 1960s, computers remained so rare that the collective imagination wasn't yet bogged down by buggy software and slow hardware. Most people never experienced the technology simply not working.

By mid-1961, only 5,371 computers of all types had been installed in the United States. Military and governmental applications made up 40 percent of the market, with most machines deployed for the SAGE and later BMEWS, the Ballistic Missile Early Warning System, an installation that extended North America's air defenses beyond nuclear bombers to intercontinental missiles. That year saw the delivery of more than $1 billion of computer equipment to military and commercial users.[31] Technological progress would come fast and in very large strides. But facts and figures were in short supply, and many technology writers felt inclined to contemplate the future rather than the present.

One of the most appealing and exciting ideas at the time was that of self-replicating machines. By the summer of 1961, Wiener was pondering the possibilities of machines that could reproduce themselves, like biological life-forms such as plants or animals or even humans. In an interview with the *Christian Science Monitor* in Boston, Wiener paraphrased the Bible to hint at the theological implication of scientific progress: "The machine created a machine in its own image," he suggested. Wiener, curiously, told the journalist not to quote him by name. Such a request was unusual for the publicity-hungry professor. The newspaperman suspected that the respected mathematician was so coy "perhaps partially because in jazz lingo this field is 'way out.'"[32]

Two years later, Wiener nevertheless decided to attach his good name to a remarkable little book, *God and Golem, Inc.*—a comment on certain points where cybernetics impinges on religion, as the book's subtitle added obliquely. The book indeed had something spiritual.

III

Religion required taboos. The very essence of worship was comprehending the incomprehensible, seeing what cannot be seen. Wiener invoked an image from his childhood: a lonely New England farmhouse, the front parlor closed, blinds drawn, wax flowers under a bell jar on the mantelpiece, gilded bulrushes set around the portrait of his deceased grandfather on an old wooden easel, and a harmonium in black walnut that was never played except at weddings and funerals.

The mute instrument stood for the enforced silence, for the taboo: "We must avoid discussing God and man in the same breath—that is blasphemy," he wrote, his sarcasm thinly veiled.[33]

The religious taboo reached far beyond the confines of New England's red-painted wooden farmhouses on lush green lawns and serene churches overlooking windswept fields. "Even in the field of science," a sixty-seven-year-old Wiener reflected, "it is perilous to run counter to the accepted tables of precedence." The taboo wasn't just about equating God and man; equating living beings and machines in the same breath also was blasphemy. Living beings would be living beings in all their parts, and machines were made of lifeless metal and plastic and glass in all their parts. The two could not even be compared. Automata, unlike animals, had no mysterious fine structure that gave them purpose. And for cyberneticists, purpose was the key feature of all systems driven by negative feedback. "Physics—or so it is generally supposed—takes no account of purpose; and the emergence of life is something totally new."[34]

This was the taboo. And in Wiener's view, this forbidden fruit stood in the way of the advance of knowledge. Wiener set out to attack the taboo in his final book, re-examining several themes that have been discussed in sacred texts—but that have a close cybernetic analogy. Three aspects of cybernetics had religious implications, Wiener wrote: that machines could learn, that they could self-reproduce, and that they illuminated magic and legend. Science, as he saw it, was invading the territory of religion, pushing God out. Cybernetics would make the irrational rational.

At first glance, machines that are able to learn don't seem to have religious significance. But Wiener, the former child prodigy, was versed in the classics. He knew better. Learning machines are linked to one of the deepest and most controversial theological problems: how to explain the suffering of humans in the here and now, how to explain God's tolerance of evil, a problem that theologians know as "theodicy." It was the main question of the book of Job, a key text in the Hebrew Bible and the Christian Old Testament, and of *Paradise Lost*, John Milton's epic seventeenth-century poem. Justifying evil was akin to the question of whether God could play a game with a creature he had made: in both of these religious texts the devil, as Wiener saw it, was playing a game with God, either for the soul of Job or, in *Paradise Lost*, for the souls of all of humankind. Both works also saw the devil as a creature of God; this was a necessary assumption of monotheism. So the crucial question was, "Can God play a significant game with his own creature?"

For Wiener, this question was mimicked in the relationship between humans and their own creatures, machines. In general terms, the question could be rephrased like this: "Can *any* creator, even a limited one, play a significant game with his own creature?" Wiener himself was a limited creator of machines. If he could play a significant game with a machine, that would answer the question. The answer of cybernetics was clear: yes. Yes because even a mechanical creation can win against its human creator. It had already been done, with checkers. Building a mechanism that would beat even the best players of the Russian school at chess was only a question of time, Wiener correctly foresaw. Man's creation could outsmart man, so God's creation could surely outsmart God.

Mechanical power, the MIT professor implied, showed the limits of divine power.

There was another, even more sacrosanct taboo ripe for mechanical reprocessing: that only God can create life. Only God is the creator, and there can be only one God. The hallmark of living beings is that they can remake themselves in their own image, because they were created by God in His image. Siring and creating new life in God's image was the cause of divine nature; creating new life-forms in somebody else's image was against the natural order of things. It was blasphemous: "In our desire to glorify God with respect to man and

Man with respect to matter, it is thus natural to assume that machines cannot make other machines in their own image."[35]

But that is exactly what automation promised: humans were soon able to make machines that could, in turn, make other machines *in their own image*.

If a machine is supposed to create another machine in its own image, Wiener suggested, understanding the notion of an image becomes necessary. What, then, was the image of a machine? "The machine may generate the message, and the message may generate another machine," the professor suggested, somewhat cryptically.

Passing a design on is relatively simple in the case of software. A "soft" machine—a program, even a complex program—can simply be copied, transferred to a different computer, and installed there. The message—the disk image—re-creates the virtual machine. In the case of simple hardware, the machine's image would be its design plan. And that plan could be sent over a telegraph line as a message that would then be used to generate a copy of that machine at the other end.

But cybernetics wouldn't have to stop here. After all, life could be compared to machines now. Organically implemented feedback devices—humans, for instance—were just more complex machines, with negative feedback loops stabilizing body temperature instead of room temperature and blood pressure instead of tank pressure. So in principle, Wiener was convinced, that complex machine made up of molecular structures could be turned into an image, transferred, and reassembled elsewhere, just as simple machines could be disassembled and reassembled with the manual's design blueprint: "It is conceptually possible for a human being to be sent over a telegraph line," he wrote in *God and Golem*. The professor hastened to add that the practical difficulties of sending a person by telephone far exceeded even his own ingenuity.

By the late 1950s, automation had acquired a "mystic" aura, as one author opened a paper on the social and economic implications of this new phenomenon, read at a conference of the British Electrical Development Association.[36] A joke that was making the rounds in the early 1960s captures this mystique: A technician fiddling with a giant computer, impressed by the contraption's growing prowess, asks the machine, "Since you know so much, tell me—is there a God?" Back comes the answer: "There is *now*." Alistair Cooke recounted the story on the BBC in an episode of his famous *Letters from America* on the new "big brains" in January 1962.[37]

Automation and large machines were depicted as autonomous agents. Computers were electronic brains. Robots were portrayed as humanoids in cartoons and films. Extreme and often dark prophecies dominated the popular-press coverage of new contraptions. Modern cybernetics was greeted with the same reprobation that had been attached to the sin of sorcery in former ages, at least according to Wiener, as he laid out in his final book on God and the machines, which he finished in the summer of 1963. Surely, he argued, if a scholar two centuries earlier, in the 1760s, had claimed to make mechanical creatures that could learn to play games, or even to proliferate, such a man would have to wear the sanbenito, a yellow coat embroidered with images of monks, dragons, and demons that meant its wearer was a heretic condemned to burn at the stake. For Wiener, modern humans had just gotten over the insult of being compared to apes by Darwin. Being compared to machines was the next affront in line.

But Wiener had the opposite objective: he didn't want to see the *machine as magical*; he wanted to do just the reverse—to show that the *magical was mechanical*.

To illustrate the mechanics of magic, Wiener used fables and short stories deeply embedded in the collective mind. One of his oft-used stories was Goethe's famous 1797 poem "The Sorcerer's Apprentice." In the poem the master sorcerer briefly leaves his workshop to run a few errands. He tells his apprentice to fill the bath with water in the meantime. Bored, the apprentice tries a magic phrase on a broom, turning the wooden stick into a water-fetching slave. After the initial excitement, the apprentice notices that he has forgotten the magic command and now can't stop the broom's tireless work. The entire workshop is filling up with water. When the apprentice tries to break the broom with an axe, two halves continue working, bringing two times as much water. Finally, the old sorcerer returns, speaks the magic lines, and the broom returns to its corner.

For Wiener, the story followed an archetypical pattern that was reflected in countless fables and horror tales and religious texts: man, eager and ambitious, would summon the help of magic, or of a "Jinni," only to discover that he had lost control of his own creation. But even in these situations of hubris, the *deus ex machina* would still avert catastrophe.

Wiener's favorite tale—"The Monkey's Paw," a classic horror story from 1902 written by W. W. Jacobs, an English author of short stories—carries this principle to a ghastly extreme. A British sergeant-major brings back from India a talisman, a dried monkey's paw. Visiting his friend Herbert, the sergeant-major tells the friend and his parents that the monkey's paw has the power to grant its owner three wishes, and that his comrade used the third wish for death. He throws it in the fire, but Herbert's father retrieves the magic paw. The man wishes for £200. The next day his son leaves for work at a factory. Later that day, a factory supervisor brings the news that Herbert was killed in a machinery accident, offering £200 to the father as compensation. A week after the funeral the mother, consumed by grief, talks the father into using the second wish of the monkey's paw to get her son back. One evening the father utters the wish. Soon after comes slow and heavy knocking on the door. Excited, the mother rushes to unlock it. The father, realizing that a hideous, undead creature is outside, uses the third wish, and the knocking stops.

Cybernetics finally made sense of these stories. The robots in Čapek's *R.U.R.* were fictional machines, so they obviously symbolized machines escaping their human master. But so did the magic broom in Goethe's "Sorcerer's Apprentice," or indeed the monkey's paw in Jacobs's horror tale. In Wiener's mind, the broom and the paw stood for the machine, for man's imperfect creation of magic. And these magic devices obeyed the rules in a literal way, as single-minded and stubborn as the ladder logic of a control device programmed in binary machine language. The broom executed the user's command to the letter, like binary code, until somebody entered the magic command to stop it. The paw executed the user's wish like command-line entry, until a third wish undid the grisly error of the second.

In Wiener's view, the machine wasn't magic. Rather, the reverse was the case: these stories and poems were the precybernetic way of articulating the logic of machines. Of course, Wiener had an all-encompassing notion of communication and control at that point. But somewhere along the way the professor had lost his orientation. He wanted to understand religion through the lens of cybernetics—but he ended up doing the reverse. Wiener inadvertently understood cybernetics through the lens of religion. The machine was meant to dominate the mythical, but the mythical dominated the machine instead.

Thirteen years after *Cybernetics* was published, Wiener decided to take stock and reflect on what had happened in the meantime. The vast amounts of resources spent on SAGE were foremost in Wiener's mind, and he was very much concerned about push-button war and the dark side of automation. In a widely read article published in *Science* in May 1960, he again invoked the stubborn and cruel logic of "The Sorcerer's Apprentice," "The Monkey's Paw," and the jinni: "The magic of automation," the professor concluded, "may be expected to be similarly literal-minded"—and therefore, he implied, similarly horrific and dangerous. By now, Wiener believed that whenever two agencies that were foreign to each other—a human and a computer—were coupled to achieve a common purpose, calamity beckoned: "Disastrous results are to be expected not merely in the world of fairy tales but in the real world."[38]

On his mind was modern war, fought with computer-controlled intercontinental ballistic missiles tipped with devastatingly powerful nuclear weapons. Not just two of them, but thousands of them. If a giant computer with perfect rationality was programmed to win according to rigid rules, the machine would not entertain any additional considerations except winning according to those rules, however narrowly defined. The obvious example was a war game, a nuclear war game. A machine playing the war game needed a preprogrammed notion of victory. If that interpretation of victory was simple and conventional, then victory would be the machine's goal at any cost, "even that of the extermination of your own side."

Meanwhile, push-button war had moved from design to reality. IBM had actually built and installed twenty-four colossal command-and-control computers. One single AN/FSQ-7 machine weighed 250 tons, had 49,000 vacuum tubes, and required a power supply of 3,000 kilowatts. The massive SAGE installations needed their own power stations with generators and diesel engines, complete with a cooling tower. The machines were installed in a duplex system. Two computers were running at each facility, one in active control and one on standby to take over the operational load, should the other computer go down. Both systems were always online, for decades, with an average maintenance downtime of 226 minutes per year.[39] The IBM computers were so big that they were housed on an entire floor of the vast command complex, separate from the displays and the operators, who were two floors up from the actual machines.

By the early 1960s, the debate on computers and automation was changing. The hype in the '50s had come in two phases: first the doomsters dominated; then the dreamers surged. When the automation of air defense was announced in newspapers across the land, the debate flipped from excessive pessimism to excessive optimism.

One authority who shaped the discussion of automation through-out the 1950s and '60s was John Diebold. He was a successful entre-preneur, consultant, public intellectual, and editor of *Automatic Control*, a journal exploring the technological cutting edge of the 1950s. Diebold also worked in an influential research group on auto-matic control mechanisms at Harvard Business School. He helped coin the term "automation" in the late 1940s, when he was in his midtwenties. The trained engineer was more conservative in his pre-dictions. "Writers such as Norbert Wiener," he wrote in a withering comment, by comparing "automatic control systems and the nervous system of humans and animals, have made the world of science fic-tion seem indeed to be upon us, with a race of human-like robots already in the making."

Diebold referred to the perennially popular articulations of these fears, Frankenstein, *R.U.R.*, and the golem. But the facts were dif-ferent. Machines didn't think, Diebold insisted already in his first landmark book, *Automation*, published in 1952. The problem was one of semantics, one of metaphors, one of applying humanistic concepts to mechanistic entities. "No interpretation of the facts could be more perverse—or disturbing."[40]

Diebold had followed the automation of air defense closely, and by the early 1960s he had turned to the Pentagon for the most powerful example available at the time: "The greatest progress in devising auto-matic control systems has been made by the military," he observed.[41] SAGE illustrated both advantages and limitations of centralized, real-time, simulation-based systems, he argued. SAGE maintained an up-to-date picture of air traffic in all of North America, encompass-ing more than fifty thousand flights daily. The system was set up to "control modern air defense weapons rapidly and accurately," Diebold recounted from the official manual. To the automation pioneer, the military's vast air defense system had made a substantial contribution to the private sector: it showed that an online and real-time computer operation was possible. It showed that computerized control could be flexible and deal with a wide range of options, such as many different aircraft, their routing, and various defensive weapons. Finally, the military's system demonstrated that it could filter, and spot, single bad events in a large field of data. These lessons could be applied in running large private-sector companies as well.

The limitations of the system were even more revealing. Diebold was concerned that the human decision-maker still had an important role to play in SAGE—the *S* in the name, after all, stood for *semi-*automated ground environment; it wasn't *fully* automated. Strategy remained a human task. Machines executed strategy only by directing weapons to their targets. To Diebold, that was a problem. The lead time between attack alert and response was becoming smaller. Meanwhile, the number of available weapons was growing bigger. Both made the allocation of weapon systems harder—too hard for inefficient humans with their slow and fallible brains. "Dependence on a human decision maker in our present military environment is hazardous, to say the least," wrote the man who had coined the word "automation."[42] Like gadgeteers before and after him, Diebold trusted the machine more than its designers.

Yes, military strategy was complex, Diebold acknowledged. But business strategy was even more complex. There were simply too many unknown variables in the private sector. Therefore, it was not surprising that the future had arrived in uniform first. At the time, to be sure, some oil refineries were controlled electronically, as were the crude-oil pipelines of Texas. Big New York banks cleared their checks automatically.[43] Insurance companies sent out premium notices to millions of customers without human interference. But industry was still trying to come up with the first corporate *information* system. According to Diebold, however, information systems were a long way from actual control systems that could run industrial production plants with many moving parts. And superimposing a control system on top of such an information system was an even longer way off. "There is *no* control system comparable to SAGE operating in the industrial environment," Diebold emphasized in 1964.[44]

What Diebold missed entirely was that SAGE, the automated system that impressed him so much, had been rendered obsolete already. By the late 1950s, the air defense nightmare predicted by Arnold directly after Hiroshima and the V-2 had become reality. SAGE could deal with incoming bomber formations, but it was useless against incoming ballistic missiles. The supersonic rocket flight from the Soviet Union to the United States would take merely thirty minutes. An updated automated system was needed.

Intercontinental ballistic missiles were so fast that it wasn't enough to spot when they were incoming already. The early-warning radar needed to pick up such an event in the launch phase. NORAD announced a network of seven new radar stations, the Ballistic Missile Early Warning System, or BMEWS. The chain of radar stations had a 3,000-mile line of sight, designed to spot Russian intercontinental ballistic missiles five minutes after launch.[45]

The US Air Force recommended building three forward-deployed early-warning sites—in Greenland, Alaska, and the United Kingdom. Construction started in the summer of 1958. Thule, Greenland, came online in December 1960; Clear, Alaska, six months later; and Fylingdales Moor, in Yorkshire, England, became operational in September 1963. In an emergency, BMEWS offered a decision window of less than fifteen minutes. An entire human chain of command had only a precious moment to receive authorization to respond from the president of the United States.

IV

The president then was John F. Kennedy. On Wednesday, February 14, 1962, Kennedy held a press conference in the State Department. One journalist asked the youthful president about the impact of computers and automation on employment. A few defense intellectuals and officers were concerned about automation on the battlefield at the time; the American public was more concerned about automation in the workplace.

"Mr. President, our Labor Department estimates that approximately 1.8 million persons holding jobs are replaced every year by machines. How urgent do you view this problem—automation?"

"Well, it is a fact that we have to find, over a ten-year period, 25,000 new jobs every week to take care of those who are displaced by machines," Kennedy responded, firmly. "I regard it as the major domestic challenge, really, of the '60s, to maintain full employment at a time when automation, of course, is replacing men."[46]

The president's statements were grounded in fear, not fact. The automated future hadn't quite arrived yet. But that didn't mean it could not be predicted. "Cybernated systems perform with a precision and a rapidity that is unmatched in humans," one influential report proclaimed in January 1962. "They also perform in ways that would be impractical or impossible for humans to duplicate."[47]

The author was Donald Michael, formerly of the Brookings Institution, a respected Washington think tank. Sponsored by the Center for the Study of Democratic Institutions, then an influential think tank in Santa Barbara, the report was darkly titled *Cybernation: The Silent Conquest*. During the early 1960s, "cybernation" was a popular and widely used word to describe computerized automation in industry. Michael described the machines in apocalyptic terms:

They can make judgments on the basis of instructions programmed into them. They can remember and search their memories for appropriate data, which either has been programmed into
them along with their instructions or has been acquired in the
process of manipulating new data. Thus, they can learn on the
basis of past experience with their environment. They can receive
information in more codes and sensory modes than men can.
They are beginning to perceive and to recognize.[48]

For Michael, it was "no fantasy" to be concerned with the implications of "the thinking machines."[49] Within the next two decades,
research laboratories would churn out machines that would be capable
of original thinking. Worse, there was no basis for knowing how and
where this process of machines overtaking humans would stop. "The
capabilities and potentialities of these devices are unlimited," Michael
wrote in the widely circulated report, voicing grave concern. Cybernated systems would have "extraordinary implications for the emancipation and enslavement of mankind."

The report's final chapter was ominously titled "After the Take-over."
Twenty years from 1962, most people "will have had to recognize,"
Michael predicted confidently, "that machines by and large can think
better than they." Thinking computers would run large parts of the
economy. The resulting system would be so complex and hard to understand that it would "be beyond the ken even of our college graduates."[50]
Those having the talent to work with the machines would have to be
taught from childhood and trained as intently as the classical ballerina.

"There will be a small, almost separate, society of people in rapport
with the advanced computers." This was Vonnegut's vision from Schenectady, without the fiction. These privileged cyberneticists, Michael
foresaw, would have established a special relationship with their
machines that could not be shared with the average man. But what
about the rest of society?

The report was front-page news for the *New York Times* and was widely covered in the country's other top newspapers.[51] America's paper of record pointed out that the economic advantages of automation made both cybernation and its consequences inevitable—in the Soviet Union as well as in the United States. Michael told the *Times* reporter that the ultimate effects of this development "certainly would not be conducive to maintaining the spirit of a capitalistic economy."[52]

But for Michael, and for many others, this was a good thing.

The idea of cyberculture was first articulated by the mathematician Alice Mary Hilton in 1963. Her point of departure was unique and rather surprising. Automation was radical, even revolutionary. Calling the effects of automation simply a second nineteenth-century industrial revolution would be "far too narrow," Hilton believed. Automation was bigger: it wasn't merely more industrial mechanization, not just an extension of man's physical power. The nineteenth century, she believed, only completed a prehistoric development: agriculture was enabled because men invented tools as an extension of their physical prowess; automated cybernetic systems were simply going one step further by extending humans' mental prowess. The twentieth century would hence unleash a genuinely new development: humankind's emancipation from repetitive tasks. Creative minds would truly be free to think. Everything that human beings might need or want, Hilton foresaw, would very soon be produced by machines—"solely by machines without any human intervention or labor."[53]

The industrial revolution of the nineteenth century wasn't big enough as a comparison, Hilton argued. The agricultural revolution was the better analogy. The ability to *cultivate* crops and livestock turned food gatherers and primitive hunters into food growers and organized communities. Cultivating plants and domesticating animals freed "some" of their energy to create civilization. Automating production would have a similar effect yet again: now "all of human energy" could be freed from the task of providing for survival. Instead of an *agri*cultural revolution, humankind would now face a *cyber*cultural revolution.

Hilton, an exceptionally eloquent mathematician, was certainly the most potent cheerleader of that revolution. She tried to come to terms with the consequences of it in a book published in 1963: *Logic, Computing Machines, and Automation*. In it, she explored the effect that the widely predicted revolution would have on human nature:

> Could human beings become truly civilized if we could live in a world free of human drudgery, a world where the jungle no longer threatens to swallow us, where we need no longer be afraid that the slightest slip will send us tumbling into slimy swamps, where no-one needs to swing a whip over the backs of others to keep them pulling the plow because he is afraid he might have to pull the plow himself, if he relaxes his watchfulness. In an era of cyberculture, all the plows pull themselves and the fried chickens fly right onto our plates.[54]

Hilton's optimism was unbridled and ambitious. Keen to help shape the future, she had reached out to Norbert Wiener, seeking his counsel and cooperation. But Wiener, as usual, remained skeptical. On March 5, 1963, he wrote a short note to Hilton expressing his disapproval: "I don't like the name 'cyberculture,'" he wrote. "This initial jargon is, I think, one of the curses of modern life," he volunteered. "These portmanteau words rub me the wrong way and they sound to me like a streetcar making a turn on rusty nails."[55] Yet despite his skepticism, Wiener agreed to serve on the editorial advisory board of Hilton's planned book series, *The Age of Cyberculture*.[56]

Hilton was a tireless organizer and advocate, teaching and networking labor activists across the United States and beyond. For her and her fellow community organizers, Norbert Wiener had become a pop star, an idol, and an inspiration, as she made clear in several letters to him and others. Just as Hilton was preparing to launch a major report in a veritable media blitz in late March 1964, she woke up to sad news: "Dr. Norbert Wiener Dead at 69; Known as Father of Automation," read the *New York Times* obituary, on the front page, with a typical image showing Wiener in front of complex mathematical formulas scribbled on a chalkboard.[57] Wiener had died the day before in Stockholm, on March 18, 1964.

Hilton clipped the article and highlighted a quote that stressed the need for more fundamental research. Three days later she sat down and wrote a long, grieving letter to the great man's widow, Margaret. Hilton told Wiener's wife about the void that her husband's death left in the world, and about her own loss of guidance. Hilton saw him not just as one of the greatest thinkers of the century, but as a prophet, she told Margaret: "He lived as prophets always do—ignored by most in his own country, venerated by his disciples, misunderstood by many, and only rarely believed." But Hilton, who had interacted with Wiener a few times, knew "his prophecies to be the truth."[58]

At the time, the automation debate was at a fever pitch. Since 1961, the country had experienced an unusually high rate of unemployment. Many assumed that the rise of the computer and of automated production was responsible for the layoffs. Four days after Wiener's death, on March 22, 1964, the Ad Hoc Committee on the Triple Revolution presented a report to the president of the United States, now Lyndon B. Johnson. The report was funded by the Center for the Study of Democratic Institutions. Launched with support from the Ford Foundation, the center was later supported mainly by Chester Carlson, a philanthropist who had cashed in after inventing the Xerox process. The thirty-two-member committee included technologists, economists, diplomats, historians, well-known private-sector executives, social critics, leading labor and civil-rights activists, a soybean innovator, a Nobel Peace Prize laureate (Linus Pauling), the longtime publisher of *Scientific American* (Gerard Piel)—and Hilton.

The three revolutions were the "cybernation revolution," the weaponry revolution, and a human rights revolution. The thirteen-page
report dedicated only two short paragraphs to nuclear weapons and
the civil-rights movement, respectively. Its main concern was the
revolution that cybernation was causing. The authors described it in
powerful language:

> A new era of production has begun. Its principles of organization
> are as different from those of the industrial era as those of the
> industrial era were different from the agricultural. The cyberna-
> tion revolution has been brought about by the combination of
> the computer and the automated self-regulating machine. This
> results in a system of almost unlimited productive capacity which
> requires progressively less human labor.[59]

The revolution was already under way, the commission argued in
1964: productivity "per man-hour" has increased "since 1960, a year
that marks the first visible upsurge of the cybernation revolution."[60]
After that, productivity jumped 3.5 percent every year. After 1958,
surplus capacity and "excessive" unemployment coexisted, which
meant 5.5 percent at that time. The result was a paradox: more production hand in hand with less work. Overcoming this cybernation-
caused paradox required fundamentally rethinking economic theory,
many of the report's authors agreed.

At present, the commission's report observed, jobs would provide
income to those who worked. That was about to change. The entire
industrial and capitalist system, the self-appointed group of activists
argued, assumed that more goods would be produced as efficiently
as possible. The system would then distribute the resources to buy
these goods "almost automatically." Employment granted the right to
consume. Soon this would no longer be the case. "Radically new circumstances demand radically new strategies," the commission told the

president. The link between work and income "now acts as the main brake on the almost unlimited capacity of a cybernated productive system."[61] Cybernation meant that society no longer had to impose "repetitive and meaningless toil" upon the individual. Technology would be ready to free citizens to make their own choices.

Three months later, on June 19–21, 1964, Alice Mary Hilton held a major cybernetic conference in New York, during three of the hottest days in the city that summer. The event took place in the Terrace Ballroom of the Roosevelt, an art deco hotel just off Park Avenue, a short mile from where Ashby and Bateson had debated the homeostat a dozen years earlier.

Hilton admitted that the newly discovered field of cybernetics shared its fate with its founder: it had become too popular too quickly, and it was often misinterpreted. Hilton dedicated her book *The Evolving Society* to Norbert Wiener, "whose wisdom and humanity is the foundation upon which the age of cyberculture shall be built."[62] The two (it is unclear if she ever met her idol) shared an extreme vision, but not the same interpretation of that vision.[63] For Hilton, it was all rosy.

Hilton tried to counter the dark fears so effectively peddled by Wiener in the popular press. By definition, the "terrifying monsters" that were sometimes even visually depicted, Hilton complained, were "actually quite *un*cybernetic." For fictional deadly machines don't respond to feedback from within the system they are part of: "The mechanistic monsters are not part of the system they devour and cannot, therefore, be considered cybernetic," Hilton argued in her book, not entirely convincingly. "Philosophically, there can be no doubt that there is, of course, only one closed system: the universe, since everything in the universe is inextricably interwoven."[64]

Of course, automation wasn't just a philosophical problem for ambitious mathematicians. Cybernation caused great concern on the left and in the civil-rights movement.[65] The predicted changes in the labor markets threatened to hit blue-collar workers and America's weakest populations first and hardest, especially African Americans. Indeed, Martin Luther King himself repeatedly highlighted employment in the machine age: "The full weight of the federal government must be employed to grapple with problems of joblessness in this age of automation and cybernation," the famous civil-rights leader demanded a few weeks after Hilton's conference in New York, and he repeated his dire warning in several speeches and interviews.[66]

Indeed, one of the conference participants highlighted the racial implications in similar ways, but drew the opposite conclusion. African Americans had been the scavengers of America's economy, pointed out James Boggs, an auto worker and member of the Ad Hoc Committee on the Triple Revolution: "the last hired and the first fired." Therefore, they would be in the best position to break with the dominant yet outmoded economic tradition. It would be absurd to think that African Americans, after being deprived for so long, would be able to catch up economically or achieve equality on a vocational basis. It would be equally "absurd," argued Boggs, to assume that "whites" would catch up with African Americans "in terms of political orientation or the concentration of human relations." African Americans, the labor activist was convinced, "by virtue of their past experience, are better prepared for life and leadership in the new cybercultural society than the whites."[67]

Perhaps the most prominent participant at Hilton's conference was the political philosopher Hannah Arendt. Arendt was fifty-eight when she attended the conference, and she agreed with the general view on automation. "Cybernation *is* a new phenomenon," she emphasized. The industrial revolution was about replacing the power of muscles. Now machines would be able to perform activities of the mind.

This momentous change called for a reevaluation of what thinking really means: "What, we must ask, is intellectual activity, as such?" The German-born philosopher then proceeded to discuss the sharp contrast between idleness and leisure. She concluded that even biblical commandments are being challenged: *he who does not work shall not eat*, for instance, would be a precept ideally suited to an agricultural society; *be fertile and multiply* was a rule that made sense in a scarcely populated country. Both were now obsolete. And "both are dangerous for a cybercultural society plagued by a population explosion and affluence," Arendt suggested.[68]

Hilton, wearing a dark-green dress, seemed to have lost touch with reality at her euphoric Manhattan conference. "The Sara Lee bakery in Chicago is cybernated; there are practically no workmen at all," she told the *New Yorker* over lunch between lectures in the ballroom. "Within not too long a time, the Carlsberg brewery, in Denmark, will be one of the most completely cybernetic factories in the world." She pointed out that this revolution went beyond mere automation. Automation still required people. So more production meant more jobs. The control systems she foresaw were different. Her machines didn't even have human-machine interfaces; "cybernated machines don't even have control panels," she said, munching on a handmade ham sandwich. "Cybernated machines run themselves, and people are superfluous."[69]

Even Marshall McLuhan was smitten by cybernation. The widely popular media theorist is best known for the idea of a "global village," of the world contracted into a small place by electrical information links. In November 1964, at the height of his fame, McLuhan presented a paper in Washington, DC, at a symposium on the social impact of cybernetics sponsored by three of the city's largest universities. In his presentation, titled "Cybernation and Culture," McLuhan spoke about the new technology with his trademark optimism: "The electronic age of cybernation is unifying and integrating," he argued, whereas the industrial age had fragmenting and disintegrating effects. McLuhan saw "all human cultures as responsive cybernetic systems."[70]

At the same symposium, another star author, John Diebold, spoke about the meteoric rise of cybernetics and the computer: in 1946, he recalled, the estimate was that twelve computers could do all the work needed to be done by computers in the United States. A year later the estimate was fifty. "Today we have 20,000 computers in operation," Diebold added, "and we shall double that figure very shortly."[71] Indeed, by 1968, approximately fifty thousand computers were installed in American businesses, government agencies, and universities.[72] The cheapest "minicomputers" were still about as expensive as a car. As recently as September 1969, the president of the steel workers union, I. W. Abel, proposed a four-day workweek for all labor and warned that machines would grab millions of American jobs.[73] But by the end of the decade, the fear and the glamour of automation had subsided.

The prophets of doom, led by Wiener, had predicted robot factories, mass unemployment, the disappearance of blue-collar jobs, the loss of dignity, and push-button nuclear war with machines making life-and-death decisions. The dreamers, led by Hilton, had foreseen unprecedented opportunity, the end of drudgery, more dignity, leisure for all, and the pursuit of the "good life" in peace.

The hype over automation was caused in part by a misinterpretation of employment figures, and how the new machine age would affect these figures. Unemployment in the United States steadily cycled upward throughout the 1950s, from 4.2 percent at the start of the decade to 5.8 percent.[74] By the end of the decade, economists had begun articulating concerns about structural unemployment. But from 1961 to 1969, employment in the goods-producing industries grew by 19 percent, and the service sector grew by nearly 30 percent. It also became clear that computers and control systems created new jobs. A leading trade magazine, *Automation*, commissioned a study of 3,440 industrial plants. Eleven percent were using advanced automation technology, such as computer control. Of those automated factories, only 10.4 percent reported a reduction of personnel; 41.5 percent reported no change; and nearly half of all automated companies told the magazine that *more* workers were needed to service the machines, not fewer.[75]

The distinctly 1960s hype about automation had two main drivers. The first driver was the Cold War. One of the lessons of World War II was that the nation with the superior production capacities is likely to win. Sperry's successes in wartime production were all but a glimpse of the coming Cold War at the home front. The United States had to outproduce the Soviet Union, and the industrial engine that powered America would be even more efficient with advanced automation. Diebold saw automation as a defensive measure against communist expansion.

Marxist-Leninist doctrine had recognized how technology could be leveraged for social change, and the Soviets were "positively *embracing* automation," Diebold feared by the mid-1960s. With panic seeping out from between the front lines, the free world's leading pioneer of automation recounted a statement by Nikita Khrushchev, the Soviet premier: "Automation is good. It is the means we will use to lick you capitalists."[76]

The role of machines in society was one of the very few questions where American conservatives and Soviet communists found common ground. Senator Barry Goldwater, a Republican from Arizona famously tough on "Reds," dismissed leftist ideas of regulating the age of cybernation. "To talk of controlling the advance of technology is as practical as trying to hold back the dawn," he told the *Washington Post* in response to the Ad Hoc Committee's report.[77] Diebold agreed: "It is only by increasing output per man-hour worked that we will be able to build effective defense against the aggressive powers of communism," he wrote.[78] Increasing productivity wasn't just in the interest of profit; it was in the interest of freedom: "Automation is the key to national survival," Diebold was convinced.[79]

The second driver for the '60s automation hype was the cybernetic myth of the inevitable rise of the machines. Wiener's Palomilla; the ubiquitous expectations of ever-more-intelligent machines; the stories of the sorcerer's apprentice and of the monkey's paw—all this carried a powerful subtext: the machines were autonomous, they were getting more like us, they were unpredictable, and they were increasingly competing with their creators.

One of the most remarkable articulations of this myth comes from Herman Kahn, one of America's most celebrated strategic minds. In a notable book published in 1967, *The Year 2000*, Kahn made a number of bold, yet stunningly accurate, predictions about the end of the millennium. He foresaw the widespread use of nuclear reactors for power, new techniques for birth control, commercial oil extraction from shale, "pocket phones," and the use of home computers. Nineteen sixty-seven was one year before Intel was founded and two years before ARPANET was launched, the famous predecessor to the internet initially funded by the Pentagon's Advanced Research Projects Agency Network, later called DARPA (the Defense Advanced Research Projects Agency).

Yet Kahn saw the writing on the wall: the computer revolution— as this man, the Rand Corporation's foremost nuclear thinker saw it—was the most important, most salient, and most exciting aspect of modern technology.[80] Remarkably, Kahn even suspected that "each user" might have a private file space in a central computer, for such uses as consulting the Library of Congress. Computer access would be used to reduce crime, as police can check immediately the record "of any person stopped for questioning."

Still, Kahn and his coauthor veered off course when predicting the path of "cybernation." By 2000, they suspected, artificial "laboratory men" who were "indistinguishable from ordinary men" might be treated differently on moral grounds, just as slaves had been treated differently on moral grounds on the basis of race a century earlier. Kahn expressed the hope that these "manufactured" and "specialized" workers would be granted the same human rights as everybody else.

By the year 2000, such discrimination of non-natural men, the strategic thinker surmised, would be likely if these manufactured laboratory creatures would differ in outlook from humans, even if artificial beings would possess an ability to reason.

This problem would be unavoidable "if bionic computers are made that perform many of the tasks of men and develop creative capabilities," the futurists wrote. The authors treated it as a given that such humanoid machines would rise sooner or later. "As the distinction between man and lesser creatures and machines begins to shade off, the uniqueness of man and the rights that are attributed to this uniqueness may begin to attenuate."[81]

Kahn was touching on the myth of machines coming alive, acquiring organic characteristics.

4. ORGANISMS

THE VISION OF MACHINES AND ORGANISMS INTERACTING in novel ways had emerged already in the late 1940s but gained strong momentum in the early 1960s. The Cold War and technological competition between the United States and the Soviet Union played a major role in driving American innovators.

Organic machines could be realized in one of two ways. One was bolting machine parts onto existing biological organisms. From the beginning, the goal of merging the artificial and the natural was to enhance the performance of the organism. Machine modification could help make an animal—and ultimately humans—fit to survive and operate in previously hostile environments, such as outer space or the deep seas. Life was no longer bound by evolution. The resulting modification was a *cyborg*, shorthand for "cybernetic organism."

The second possibility was even more ambitious: creating living machines without an organic base. Stand-alone machines, entirely without tissue, could be endowed with features of living organisms, such as the ability to reproduce, to mutate, to evolve, and to think—or to fight and kill autonomously. Endowing machines with lifelike attributes raised two hairy questions: when and if machines could come alive, and when and if machines could outperform human beings. Only the distant future would hold answers to these questions, if there were, in fact, answers. These were more philosophical than technical themes. But their appeal was bound to grow as technology advanced and as cybernetics offered an inspiring vocabulary for coming to terms with machines as organisms.

Already in 1943, Norbert Wiener discussed some of these questions with John von Neumann. The two debated similarities between the brain and computers in an interdisciplinary meeting with neuroscientists and engineers at Princeton.[1] That year, Wiener and von Neumann jointly founded the "cybernetic circle," which led to the influential series of Manhattan meetings supported by the Macy Foundation. The two men, perhaps America's two most resourceful mathematical minds at the time, shared a similar scholarly trajectory. But they were quite different in temperament: Wiener could be diffuse and incomprehensible; von Neumann was immaculate and paid thorough attention to minute details.

At that time, just as the war was ending, von Neumann was getting involved in the work on the ENIAC (Electronic Numerical Integrator and Calculator), a 30-ton, 80-foot-long giant machine powered by vacuum tubes to calculate artillery firing tables for the army. The era-defining device was built in 1944–45 at the University of Pennsylvania's Moore School of Electrical Engineering.

The ENIAC presented a new problem to engineers: it could calculate faster than instructions could be read into the machine. Frustrated and inspired by this memory problem, von Neumann wrote a conceptual paper that is widely seen as the founding document of modern computing, "First Draft of a Report on the EDVAC," dated June 30, 1945. The EDVAC (Electronic Discrete Variable Automatic Computer) was the successor to the ENIAC.[2] In 1946, Julian Bigelow—who, with Wiener, had tried in vain to predict flight patterns of pilots under stress—became chief engineer of von Neumann's computer project at Princeton.[3]

By the end of 1946, von Neumann was irritated by the cybernetic research he had been discussing with Wiener for three years already. He felt that the human brain was simply too complex to study as a template for computers. So, in a remarkable letter to Wiener, von Neumann suggested narrowing the focus of their research.

"Dear Norbert," he started, and he suggested a personal meeting a few days later. He then tried to nudge the father of cybernetics away from the human nervous system: "In trying to understand the function of automata and the general principles governing them, we selected for prompt action the most complicated object under the sun—literally." He was referring to the human brain.[4]

Their work had made good progress, von Neumann admitted, despite its excessive ambition. But a breakthrough was unlikely. Instead, von Neumann suggested studying simpler organisms, organisms even simpler than single cells: "Viruses," he proposed to Wiener, "possess the decisive traits of any living organism: they are self reproductive."

"I did think a good deal about self-reproductive mechanisms," von Neumann told Wiener. He was convinced he understood some of the main principles involved. The virus was an entity in the gray zone between the living and the nonliving. It seemed to von Neumann that this trait made it an ideal subject to study. "I want to fill in the details and to write up these considerations in the course of the next two months." That time frame turned out to be too optimistic.

Coming up with a "Theory of Self-Reproducing Automata," as he called a series of lectures on the subject, took von Neumann about two years. He outlined how machines could build other, similar machines from elementary parts. In the lectures, von Neumann liberally jumps from describing machines in organic terms to describing living beings in mechanical terms. This switch in perspectives wasn't sloppiness; it was creativity at work.

"Anybody who looks at living organisms knows perfectly well that they can produce other organisms like themselves," he told a small group of colleagues and friends in June 1948 at the Institute for Advanced Study.[5] He couched his theory in cybernetic terms, blurring the boundary between machine and organism. Plants and animals produce offspring. But the reproduction of life over generations was doing more than simply reproducing the same life. Natural reproduction introduced a steady stream of errors and modifications. The result was improvement: "It's equally evident that what goes on is actually one degree better than self-reproduction," von Neumann said. Nature wasn't just reproducing the design of life. It was evolving and improving existing designs. "Evidently, those organisms have the ability to produce something more complicated than themselves."

Von Neumann then switched the perspective, thinking not machine from organism but vice versa. But reproduction from a mechanical point of view led to the opposite conclusion: organic self-reproduction was evolutionary; mechanical self-reproduction was degenerative. The reason for the two opposing logics was simple: "Everyone knows that a machine too is more complicated than the elements which can be made with it," he told his students. The machine designed to make other machines like it must contain its own component parts, the design description of the new machine, and the parts and tools for assembling the new machine. The parent machine, in short, was bound to be more complex than the child machine, von Neumann reasoned: "An organization which synthesizes something is necessarily more complicated, of a higher order, than the organization it synthesizes."[6]

How, then, could a machine build another machine that was at least as complex as itself? It was a theoretical question, of course. But building machines that could output fertile offspring was a tough nut to crack, even theoretically.

In theory, the machine needed eight parts. The biblical number in von Neumann's plan for creation was probably coincidence. The eight parts included a "stimulus organ," a "fusing organ" to weld or solder separate parts together, a "cutting organ" to unsolder a connection, and "a muscle" to produce motion. The professor then outlined the assembly in abstract mathematical terms. Significantly, he considered mutation, that crucial feature of evolution. "By a mutation I will simply mean a random change of one element anywhere," von Neumann told his small audience at Princeton. After such random change, he said, "the system will produce, not itself, but a modification of itself."[7]

As in nature, the outcome would, in most cases, be negative, not positive—a degeneration, not progress. But the possibility of random change emerged.

> So, while this system is exceedingly primitive, it has the trait of an inheritable mutation, even to the point that a mutation made at random is most probably lethal, but may be non-lethal and inheritable.[8]

John von Neumann did not discuss the question of whether his machines would be alive, like the virus that he suggested as a case study of self-reproductive design in his letter to Wiener—or indeed whether machines were alive like humans, the master inspiration of cybernetic pioneers. But he was certainly prepared to think that his primitive automata could die in some sense from "lethal" mutations.

The 1950s were an optimistic time. Utopia was more attractive than dystopia, perhaps because the dark memories of World War II were still too fresh and too close for comfort. Blue-sky thinking was a form of escape. Self-reproducing automata didn't have to be killer robots; nor did they have to resemble predators, or even animals such as worms. Self-reproducing machines might as well be more benevolent plants. And not just weeds, but useful plants that, in time, could be programmed to generate the desired harvest. Edward Moore suggested this design in 1956, in a much-noted *Scientific American* article.[9] Moore was a lecturer at MIT and Harvard simultaneously; then he moved to Bell Labs, which a dozen years earlier had revolutionized the business of ballistic prediction and gun laying.

Moore acknowledged that von Neumann had demonstrated the feasibility of self-reproducing machines. A thought experiment like that suggested by von Neumann was fine. But such automata, Moore thought, could serve an actual purpose. The machine could be useful: "It would make copies of itself not from artificial parts in a stock room but from materials in nature."[10]

Like a shrub in an English garden, Moore's mechanical organism would grow and reproduce best if placed in just the right spot: direct sunlight, no frost. For that purpose, Moore suggested a beach: "For the first model of such a machine, a good location would be the seashore, where it could draw on a large variety of available materials." The breezy air would provide nitrogen, oxygen, and argon; the seawater offered hydrogen, chlorine, sodium, magnesium, sulfur, calcium, carbon, and other elements; and the sand and soil had silicon, iron, and aluminum. "From these elements the machine would make wires, solenoids, gears, screws, relays, pipes, tanks and other parts," Moore wrote, "and then assemble them into a machine like itself, which in turn could make more copies."[11]

Such a machine, Moore suggested, could be harvested for materials that it had extracted or synthesized from the soil or water or air, in the same way that cotton, mahogany, or sugarcane was harvested from natural plants. Except the artificial plant could be designed to produce "any desired crop," not just the limited crops that nature happened to be providing. Moore predicted the cultivation of freshwater, or grow-ing gold from seawater. Even Antarctica, the unused continent, could be brought into production.

One important aspect in designing the plants was time: How long would it take for the population of artificial living plants to double itself? Algae in a pond, for instance, can double in size in a week; a population of sequoias, by contrast, can take centuries. If the machine's reproduction rate was fast enough, the investment in designing and building it would pay off handsomely. Any such calculation, Moore explained, had to take into account mortality among the machines. "A certain fraction of each generation would 'die' because of internal failures, degeneration or natural catastrophes," he wrote.[12]

Moore was conscious of the limitations, of course. The artificial machines need not be made from ferromagnetic materials and electri-cal motors with gears and screws, wires, and valves. They could as well be made from organic materials, Moore knew. The only problem was that 1956 organic chemistry wasn't yet advanced enough. The same applied to theoretical genetics: the human understanding of evolution wasn't yet advanced enough "to enable us to endow a machine with evolutionary abilities."[13] Thus, the machines, for the time being, had to be improved by their makers.

The most important limitation was the price. Designing such machines would be hard and expensive. But it was not as hard and expensive an achievement as other ambitious projects of the day, such as transporting humans to the moon and other planets. "The whole design problem could probably be solved in five to 10 years, for as little as $50 million to $75 million."[14]

By 1961, Wiener was entertaining ever wilder ideas about living machines. "Can machines give birth to machines?" he asked. The second edition of *Cybernetics* came out that year. It included a sum-mary of what had happened during the 1950s. Wiener's conclusion: machines were on the cusp of acquiring two features of living systems—the power to learn and the power to reproduce—he told the *Christian Science Monitor* when the second edition of his best seller hit the book shops.[15]

John McCarthy, one of Wiener's colleagues at MIT, agreed that "self-producing" machines would be entirely possible. A machine could be installed on a mountain of granite, he speculated. The machine could then melt the granite, refine any needed materials from the rock, and build other machines in its likeness. "Each machine would carry a sort of tail bearing a code describing how to make the body," the *Christian Science Monitor* quoted McCarthy, somewhat incredulous in tone.[16]

Perhaps the best-known and one of the most influential articulations of machines coming alive is not science, but science fiction: Arthur Clarke's monumental story *2001: A Space Odyssey*, about technological progress and machines acquiring human characteristics. Clarke was mesmerized by Norbert Wiener's work. The British science fiction writer had read an essay published in *Science* by the father of cybernetics during the summer of 1960: "Some Moral and Technical Consequences of Automation." In that essay Wiener wanted to take stock of the debate a dozen years after he started it. His tone was professorial, if not arrogant. Wiener sharply attacked "the man in the street" for assuming that machines cannot possess originality. Such a view was clearly naïve. For ordinary people did not understand modern machines. Wiener:

> It is my thesis that machines can and do transcend some of the limitations of their designers, and that in doing so they may be both effective and dangerous.... As is now generally admitted, over a limited range of operation, machines act far more rapidly than human beings and are far more precise in performing the details of their operations. This being the case, even when machines do not in any way transcend man's intelligence, they

very well may, and often do, transcend man in the performance of tasks.[17]

This was the kind of bold prediction of the future that would make Wiener's books and articles such popular reading material among artists and science fiction authors, and Clarke was duly impressed when he read these lines. "The tool we have invented is our successor," Clarke wrote in *Playboy*, in July 1961, just overleaf of an article that featured iconic midcentury furniture and interviews with designers Charles Eames and Eero Saarinen. Biological evolution, Clarke believed, had given way to a far more rapid process, technological evolution. "To put it bluntly and brutally, the machine is going to take over."[18] *Playboy* illustrated the story with an evolutionary tree that showed the progress of evolution: from microbe to fish, dinosaur, monkey, Neanderthal, human, and eventually machine.

Clarke wrote these lines for the first time about seven years before he published *2001*. He was writing twenty years after the first electronic computers appeared, but two years before the network that preceded the internet was first suggested. Clarke was deeply impressed by the fast advance in sheer computing power, and he foresaw the possibility of building a machine that could pass as human itself: "We are still decades—but not centuries—from building such a machine," he suspected.

Yet he felt enough confidence to ridicule the naysayers and skeptics who made the argument, common at the time, that no machine could possibly be more intelligent than its makers, that nothing could come out of the machine that had not first been put into it. "The argument is wholly fallacious," Clarke was sure. Those who would still argue along such lines would be stuck in the past, "like buggy-whip makers who used to poke fun at stranded Model Ts."[19]

To bolster his confident case, the science fiction author quoted the cybernetic master himself: "as a careful reading of these remarks by Dr.

Norbert Wiener will show." Clarke understood that machines could escape human control even if they were less intelligent than humans, simply by virtue of the sheer speed of their operation. And Clarke saw many reasons why machines would become not just faster, but also "much more intelligent" than their creators, and already "in the very near future." Machines that learn by experience already existed, and unlike human beings, they learned properly, never repeating their mistakes. All intelligent machines, Clarke argued, were inspired by what we know about the human brain, "the only thinking device currently on the market."[20] Clarke was practically parroting Wiener.

But of course, Clarke had the mind and the style of a science fiction author, and he was writing for *Playboy*, not *Science*. So he could reveal himself a little more freely. "It will take a little while for men to realize that machines can not only think, but may one day think them off the face of the Earth," he wrote.[21] Clarke foresaw entirely new and yet unimagined forms of human-machine interaction. "I suppose one could call a man in an iron lung a Cyborg." But that alone wasn't too remarkable. Man-machine interaction had far wider implications: "One day we may be able to enter into temporary unions with many sufficiently sophisticated machines," the science fiction author wrote in *Playboy*, predicting that future generations would be "able not merely to control but to become a spaceship or submarine or TV network." The idea of becoming a spaceship formed the basis of the story that would define Clarke's career. It would take decades for this thought to be articulated in more detail. But by 1961, Clarke was suggesting that networked machines could change not what humans do, but what they are.

The cyborg had been born in Texas barely a year earlier, at the end of May 1960, at Randolph Air Force Base. The challenge of flying at new altitudes is what led to the rise of the man-machine.

World War II had elevated the special field of aviation medicine. Flying at ever-higher altitudes and faster maneuvers during flight presented aircrews with unknown physiological and psychological problems. How much gravitational force could the human body take? What were the effects of low cabin pressure on cerebral activity? How would low or no gravity affect astronauts?

Cybernetics was all the rage among engineers at the time. In the 1950s, the School of Aviation Medicine at Randolph Field, in San Antonio, Texas, was one of the air force's foremost research centers to explore these novel questions. As early as 1948, scientists at Randolph Field—a ten-hour drive from where the army was hosting the Third Reich's leading missile engineer at White Sands Proving Ground—held meetings on topics as visionary as "aeromedical problems of space travel." The Cold War nuclear arms race was on, and the accelerating space race added urgency to bleeding-edge aviation research.

On October 4, 1957, the Soviet Union successfully launched Sputnik 1, the world's first artificial satellite, causing shock and consternation in the United States. NASA was founded the following year. Space travel came with a host of challenges, one of the most intricate and important of which was adapting the human body to extraterrestrial conditions. On May 26 and 27, 1960, the School of Aviation Medicine hosted its fourth space flight symposium to explore the physics and medicine of the upper atmosphere and space flight.

That May, two improbable researchers from the Rockland State Hospital presented a bold idea. The institute in a rural suburb of New York City seemed an odd place for a mesmerizing invention. New York State's government had established a research facility at the previously neglected hospital in Orangeburg, just north of Manhattan, to boost morale. Nathan Kline, a doctor, was the hospital's dynamic, young, and well-connected director of research. He was a dominant figure in psychopharmacology, a new discipline with a reputation at the time of rough treatment methods for the mentally ill. In 1955, Kline hired a highly gifted Austrian émigré, Manfred Clynes. Clynes had studied engineering and music at the University of Melbourne in Australia.

Two years earlier he had performed Bach's *Goldberg Variations* across Europe to critical acclaim, even performing solo in London's newly built Royal Festival Hall.

The Austrian-born inventor and artist took over Rockland's Dynamic Simulation Lab. Kline bought a computer for Clynes in 1955, when such a machine was far more expensive than a decent family home. The ambitious engineer soon put the new device to work on calculations pertaining to the body's nervous system and cybernetic control. In the coming years, Clynes would file eight patents in the fields of ultrasound, frequency modulation, and telemetering. Clynes was extraordinarily energetic and productive, both as a scientist and as a pianist. In 1960 he published an article in *Science* on the control of heart rate through respiration: "Computer Analysis of Reflex Control and Organization." In the article, Clynes applied automatic control system theory to the body. Clynes had been fascinated by Norbert Wiener's ideas on cybernetics and even discussed cybernetics with the famous MIT professor in Ukraine.[22]

Clynes and his boss considered presenting their findings at the symposium in San Antonio, Texas. After working on the paper, Clynes suggested using the word "cyborg" in the title. He asked Kline for his opinion. "Oh, this sounds kind of interesting," Kline responded, "But it also sounds like a town in Denmark."[23]

The basic idea of the cyborg was intuitive. On Earth, most of the body's regulatory functions just work. We don't have to remember to adjust our blood pressure. We don't have to remind ourselves to breathe. The goal was to enable the same unconscious, automatic regulatory behavior in outer space. The goal was to liberate the astronaut from the limitations of the human body. In their presentation at Randolph—"Drugs, Space and Cybernetics: Evolution to Cyborgs"[24]—Clynes and Kline outlined a bold idea that would solve that problem by automating those newly required body functions.

To illustrate their point, the two scientists invoked a fish. Not just any fish, but a particularly intelligent and resourceful fish. If this resourceful fish wished to live on land, it could do so. It would need some background in biochemistry and physiology, and it would have to be a "master engineer and cyberneticist," with excellent lab facilities available to it. If all those requirements were met, "this fish could conceivably have the ability to design an instrument which would allow him to live on land and breathe air quite readily," they suspected. Humans in space were like that fish on land.

Their entire presentation was laced with ideas borrowed from cybernetics: the man-machine entity would improve "man's homeostatic mechanism." Implants into the lungs, heart, the nervous system, and various other organs would extend the self-regulatory control of an organism into a new environment, outer space. Drugs would be injected into the bloodstream from within the body. The implanted machines would even regulate the astronauts' sleep and sensory input. Problems would be solved automatically, "leaving man free to explore, to create, to think, and to feel."[25]

Clynes and Kline invoked America's frontier spirit, a popular comparison to outer space at the time. Space was the "new frontier," and cybernetics would help the pioneers colonize this mythical space that only very recently had seemed entirely out of human reach.[26]

A few months later they published the paper as "Cyborgs and Space" in *Astronautics*, a leading journal on America's space program.[27] The article included a picture of the first cyborg, a white, 220-gram lab rat with an osmotic pump implanted under the skin of its tail. The implant made the tail look like a white ball that was tied to the rodent's backside. The pump allowed continuous injection of chemicals into the rat's bloodstream—controlled by the machine, not the animal.

A range of problems could be solved during flight through machine-controlled counteraction: sensors could detect radiation and automatically inject pharmaceuticals into the pilot's body to counter the effects of radiation, for instance. Sleep could be automated, as well as fluid intake and output, cardiovascular activity, and body temperature. Clynes and Kline were conscious of the limitations of their suggestions. They pointed out that they had not discussed motion sickness. They also didn't discuss "erotic requirements" during space flight in their ambitious five-pager, apparently deciding that suggesting some sort of sex machine would be an idea ahead of its time. Some of their solutions, they understood, would have appeared "fanciful" in 1960.

The presentation must have appeared fanciful indeed: so far, no human had even been in outer space. Yuri Gagarin would complete an orbit of Earth only eleven months later, on April 12, 1961. Yet the cyborg pioneers were acutely aware of the high stakes, precisely because the enemy of the free world seemed so far ahead in science and technology: "There are references in the Soviet technical literature to research in many of these same areas," they pointed out, still Sputnik-shocked. The researchers were confident of one necessity: "adapting man to his environment, rather than vice versa." This advance would mark not just a significant step in human scientific progress; the cyborg, they hoped, "may well provide a new and larger dimension for man's spirit as well." If the human body could be machine-enhanced, enhancing the mind was only a question of time.

The two researchers from Rockland had captured the Zeitgeist. The idea would inspire the design of an entire range of machines and even philosophical inquiries in ways that even the two inventors from Orangeburg could not have imagined. *Life* magazine visited the Rockland laboratory, interviewed the two researchers, and covered their work in a story.

Cyborgs will wear sealed skintight suits but will travel in unsealed cabins exposed to the near vacuum of space. Ordinarily, at these low pressures, the blood would boil and the lungs explode. But cyborgs' lungs will be partly deflated and their blood will be cooled. To keep from getting numbed their brains will be warmed or fed energizers. Their messages to one another will be picked up electrically from their vocal nerves and transmitted by radio. Their mouths will be sealed and unused. Concentrated food will be piped direct into their stomachs or blood streams. Wastes will be chemically reprocessed to make new food. Totally worthless end-products will be kept in small canisters on their backs.[28]

The magazine illustrated the story with a large picture of two cyborgs working on the moon, complete with sealed mouths and waste canisters. Clynes mounted a big photograph of the story on his wall and had it up for years. He highlighted his artistic background in the interview. "Imagine," he told *Life*, "what leaps a ballet dancer could take on the moon."[29]

These ideas were controversial among serious scientists, yet they didn't go far enough for some engineers who were in the business of beating back the Soviets technologically. Martin Caidin, a prominent space-aviation author, captured the mood in the early 1960s when he wrote of a new undercurrent: "That undercurrent is one of urgency."[30] The fear was that the Soviets might get to the moon first. Michael Del Duca, the chief of biotechnology at NASA headquarters in the early 1960s, had a reputation for far-fetched ideas about life-support systems. Dr. Del Duca felt that the cybernetic possibilities for space exploration were quite literally boundless. Space would no longer be hostile and inaccessible.[31]

By May 1963, NASA came to somewhat more nuanced conclusions in the final report of contract NASw-512, an experimental project that explored how humans could be reengineered for extraterrestrial environments, titled *Engineering Man for Space: The Cyborg Study*. The project team concluded that the potential of fully artificial lungs, kidneys, and "extracorporeal pumps" was limited. But NASA remained more optimistic that astronauts could be "modified" through biocybernetics—for instance, by artificially cooling their body temperature and by managing sensory deprivation—to ensure "the success of prolonged space flights or interplanetary exploration."[32]

II

The cyborg had obvious uses not just in space, but also on Earth. The military, naturally, was keen on the idea. One man in particular understood the potential, certainly the fundraising potential, of military cybernetics: Ralph Mosher, a General Electric engineer. He would rake in millions of dollars of funding for merging man and machine, over more than a decade, from all services—first the air force, then the army, and finally the navy.

Mosher's streak had begun back in 1955. That year, General Electric started developing experimental atomic aircraft engines for the Aircraft Nuclear Propulsion program, run jointly by the Atomic Energy Commission and the air force. The Soviets, it was known, also tried to equip their bomber fleet with atomic reactors. GE had two experimental nuclear-powered gas turbine projects—the X-39 and the larger X-211 engine—that were powered by gigantic experimental reactors mounted on railcars to move them to remote test locations.[33]

The US Air Force had already designated a new super-long-range bomber, the B-72, as a nuclear-powered aircraft, able to fly for weeks at a time and at higher altitudes than most other aircraft could.[34] The biggest design problem was protecting the aircraft's crew and engineers from the onboard reactor's radiation. The experimental test plane had a 12-ton lead-shielded crew compartment with 10-inch-thick leaded-glass windows. The reactor would have been cooled by the airflow through the engine in flight, which meant that aircraft maintenance on the ground would be a vast engineering challenge.

General Electric's research, as well as the predicted ongoing maintenance of the aircraft, required so-called manipulators, simple remote-controlled claws—with enough dexterity and sense of touch to turn screws, fit parts, and assemble components in high-radiation environments. In 1958, GE turned to one of its best engineers for help. Mosher, then thirty-eight years old, was a tall and husky man, sporting a crew cut and a smart outfit, usually a white shirt with dark tie. He worked at GE's Schenectady factory in eastern New York. It was the same factory, close to where the Mohawk and Hudson Rivers meet, that had inspired Kurt Vonnegut to write *Player Piano* ten years earlier.

Mosher, like Vonnegut, was familiar with Wiener's ideas on cybernetics. But the ambitious engineer got a very different kind of inspiration: "I realized that after a certain point improvements in mechanical dexterity added little to a manipulator's performance," Mosher remembered. He wondered why people were so efficient with their hands and why robots were so awkward. "Soon it was obvious," Mosher recalled. "The manipulator's operator was missing what he ordinarily experiences: a sense of feel."[35]

Feedback was missing. Mosher understood that a kinesthetic sense mattered, the sensing of forces in the body's bones and muscles. A man could open a door in the dark because he sensed the doorknob, sensed how it turned, and then sensed the door's circular opening motion. A robot would risk ripping the door out of its frame because it didn't sense the circular motion. So Mosher came up with the idea of force feedback for high-performance robots. "Such a device, possessing the properties of feedback and kinesthesis, can be described as a cybernetic anthropomorphous machine," he wrote in *Scientific American*.[36] That was a mouthful. Mosher suggested a shorthand, CAM. The results were dramatic. Touch and feel worked wonders. "We didn't just make a better manipulator," Mosher said about the new CAMs. "Adding touch created an entirely new *kind* of robot."[37]

The result was Handyman. Handyman was a pair of powerful mechanical arms, slightly longer than human arms but with a similar structure: shoulder joints, elbow joints, and two-fingered claws that could be twisted at the wrist. Each arm was capable of ten motions in three-dimensional space. The two tools were protruding from a black box with "General Electric" proudly printed on the front. The box was fed by bundles of hydraulic cables. The arms were made of black steel from the elbow downward, the biceps coated in thick black rubber. The claws were highly dexterous: one claw could pick up a thin wooden hammer while the other one held on to a block of wood with a nail sticking out, and then hit the nail with the hammer.

A man strapped into a harness controlled the mechanical claws. The harness looked like an exoskeleton for Mosher's arms. It was the equivalent of a mouse and keyboard—the human-machine interface, what the GE engineers then called an exoskeletal master station. "The cybernetic control method requires an exoskeletal master station that has precise spatial correspondence with the operator," the final report to the army explained in technical jargon.[38]

The hydraulic claws precisely mimicked the actions of the man's arms and hands. The man, in turn, received tactile feedback from the steel claws, in effect coupling the machine with the man's sensory and motor systems. GE called the harness the "follower rack" because the machine simply followed the man's motions. The hope was that the movement could become natural, so that the operator would not have to think about using the machine—in effect, as the engineers wrote, "merging man and machine, using the best capabilities of both."[39]

Handyman was designed for the air force, to handle "hot" radiation material inside a propulsion laboratory to build an experimental long-range bomber aircraft that was nuclear fueled and nuclear armed. But

General Electric, keen to garner upbeat media coverage for its fancy devices, preferred showcasing more benevolent uses. At the debut press conference, Mosher was smartly dressed, strapped into the follower rack, with his remote iron claws twirling Hula-Hoops, two little girls in pretty dresses looking on in awe. In one story, *Life* showed an attractive young brunette with two GE steel claws helping her into her coat—Mosher in the background, several yards removed, strapped into the follower rack, smiling.[40]

In 1961, John F. Kennedy ended the air force's program for a nuclear-powered bomber. But he also escalated the Vietnam War. This meant that Mosher would continue his work for a new client; as the air force walked out, the army walked in. By the early 1960s, the army was experiencing new and unexpected tactical problems in Vietnam. Tanks, trucks, and artillery guns were too clumsy for jungle warfare. Only infantrymen on foot—along with mules—were able to negotiate narrow trails, steep potholed roads, dense forests, swamps, and rice paddies. Worse, the Viet Cong preferred hit-and-run ambushes on US troops precisely in those remote locations. Then, in the early 1960s, the army's top brass heard about GE's Handyman.

The army faced a special problem. For the air force and the navy, mobility was easier. The nation's ground force wasn't particularly innovative in how it traveled in combat: airmen were looking at nuclear propulsion and space travel, and sailors had submarines and carriers; meanwhile, infantrymen were still wading through the mud. The reason was straightforward: air and sea were predictable and consistent travel mediums; terrain offered endless variety. It was therefore much easier to navigate ships and aircraft autonomously. GE promised a way out of this conundrum: a giant walking machine for jungle warfare, a "profitable symbiosis of man and machine."[41]

The goal was to equip infantry units in the deep mountainous jungles of Southeast Asia with armor and heavy equipment while remaining highly mobile and versatile. Officials at the US Army's Tank-automotive and Armaments Command (TACOM) in Warren, Michigan, were intrigued.

Balancing a large walking machine, however, wasn't trivial. So TACOM's Mobility Systems Laboratory funded an experiment to test whether a giant, two-legged walking tank could be kept in balance.[42] The outcome was the Pedipulator.[43]

The Pedipulator, built in 1964, was an 18-foot-tall experimental biped. The machine looked like a prototype of a *Star Wars* biped, thirteen years before the science fiction film came out: a cabin with a large front window, a little larger than a telephone box, was balanced on two thin legs. Eighteen feet was high. Some people refused to try the machine, because of its height.[44]

One reporter from *Popular Mechanics*, in a dark suit and skinny tie, came to test-drive the biped. The GE engineers helped him climb into the driver's cabin, showed him the skateboard-like control panel to stand on, fixed his torso between two bars, and then fired up the biped: "With a loud sigh of hydraulic valves the automaton I was commanding sprang to life," the reporter recalled. The machine started mimicking his moves. With no practice in balancing a giant biped, the reporter pitched too far forward on the swivel board:

> Using my toes for leverage, I frantically tried to stop our headlong plunge by throwing myself backwards. The robot's reaction was as quick as it was violent. Accompanied by a piercing shriek of valves, the automaton shuddered to a halt, then swiftly heaved back. Before I could react, it had crashed down on its heels with a jolt that rattled every bolt in its body.[45]

The prototype was tied to rails on the ground and could not actually walk, let alone topple over. Its purpose was to test the balance in a giant human amplifier. The *Popular Mechanics* reporter-turned-balance-tester quickly learned his moves and soon was able to go through a series of motions "as outlandish as the latest discotheque dance." Balancing the biped wasn't so difficult after all.

Mosher had become acutely aware of the limitations of robotics while working on the air force contract. "Compared with the versatility of man," he understood, "the things a machine can be programmed to do are extremely limited." It was impractical, he acknowledged, to build a machine that could walk on sand, mud, and rocks, as well as through a forest on its own. Machines were good at repetitive tasks that didn't change at all. But walking in rough terrain meant that every step was different. Moving in such environments was simply too complex and dynamic for microprocessors at the time. "But a man can do this," Mosher suggested, "and if you join man and machine, using the best capabilities of each—man's brains and the machine's great strength—then a machine can do it, too."[46]

"Cybernetic mechanisms," the GE engineers understood, had a range of advantages over conventional vehicles: effective man-machine integration eliminated levers, brake pedals, and clutches; it made programming obsolete; it required very little training; its force feedback reduced risk; and cybernetic machines would free operators to focus on the actual problem at hand. "The operator is able to react in such a natural manner that he subconsciously considers the machine as part of himself," Mosher told the army's transportation experts in Michigan.[47]

The army's vision was to create some sort of intelligent full-body armor, turning the soldier, in effect, into a walking tank. When the experiments with the limited-motion Pedipulator were finished, GE mailed two enthusiastic reports back to the army: humans could indeed balance the machine, position it quickly and accurately, and retain "nearly perfectly" how they learned to operate the new gear.[48]

TACOM was impressed by the Pedipulator. But the Department of Defense was concerned that a biped could be knocked over with simple means during battle in forest environments and would not be able to get up again on its own. So the army decided to fund a four-legged walking machine. A quadruped was more stable; it was also lower, which made walking in jungle underbrush easier; and four legs could simply carry more load than two legs could.[49] In addition, quadrupeds just made more intuitive sense to cavalry officers.

Just before Christmas 1969, in a cluttered machine workshop, GE engineers had erected an 11-foot-tall beast of burden weighing 3,000 pounds. The Schenectady engineers called it the "walking truck." The machine's sturdy skeleton was made of aluminum beams. Bulky hydraulic muscles powered its four legs. Each leg had a hip joint, a thigh, a robust knee joint, calves, and small feet, but no ankle joints. The hips could move the thighs in all directions; the knees were limited to fore and aft motions.

One of GE's innovations was the walking truck's human brain: the operator had to climb into the machine's belly on a small metal ladder that could be flipped down, suspend himself inside the skeleton, slip his feet into a pair of metallic holsters, and hold on to two joystick arms with handles and a number of triggers. The rider then revved up the 90-horsepower gasoline engine, pumping hydraulic fluid into the cyborg's body and legs, through a tangle of tubes and gauges and valves, bringing it to high-pressured life. When the rider raised his right leg, the machine raised its right hind leg. When he turned his left forearm, the machine turned its left foreleg.

Controlling the quadruped was less intuitive than controlling the experimental biped. It took about ten hours to learn how to operate the machine. Mosher got rather good at it. With a bit of practice, the walking truck could go where no wheeled vehicle could go, across fallen trees and rocks lying in its way, at about 5 miles per hour. A person, amplified by machine, could reach with one arm and kick a 1,500-pound rock out of the way, toss a jeep out of the mud, or even push a small military vehicle over an obstacle. The metallic beast could walk forward as well as backward, and even balance on two legs. "What's 11 feet tall, walks on four legs and drinks gasoline?" asked GE in an ad.

Despite its size and power, the quadruped wasn't a brutal monster. The GE cyborg had tactile force feedback built in. Mosher, in his white lab coat and white helmet, could feel what the purring vehicle "felt." When an aluminum foot touched the ground, the operator would feel, via feedback from sensors, the heavy leg touching the ground in his holster. The machine was capable of "great gentleness," as one TV documentary put it at the time.[50] In the lab, GE demonstrated the cyborg's haptic skill by having it step on a glowing lightbulb resting on a red pillow, gently touching it without breaking the glass. Yet by simply turning his wrist, the operator could also shove 175-pound railroad ties out of the machine's way as if they were toothpicks.

Once an operator has enough practice, Mosher said about his four-legged vehicle, he "begins to feel as if those mechanical legs are his own." The engineer had been the machine's primary driver for a while at that point. It indeed started feeling natural to him: "You imagine you are actually crawling along the ground on all fours—but with incredible strength."[51]

But operating the walking truck was harder than the engineers had hoped, because the rear legs were out of the driver's line of sight. Walking by machine also was extremely tiring. It became difficult to concentrate after fifteen minutes. Another problem was that the machine's high volumes of hydraulic fluid required external hookup even in advanced versions, when testing had already moved outdoors. TACOM was disappointed. GE's promotional material had looked so promising, with a column of quadrupeds trotting along, crossing a jungle creek under giant tropical trees. But there were no hydraulic hookups in Vietnam's underbrush. Only one cybernetic walking machine was ever built for the army.[52]

GE built the most visually stunning remote manipulator for the air force, to service the flying nuclear reactors of the planned long-range bombers: the "Beetle," an 80-ton machine that looked like a giant tank on extra-large caterpillar tracks with two humongous mechanical arms and claws. The driver was shielded from hot radiation in a cabin behind 2-foot-thick lead glass.[53] Even Mosher thought the machine was "monstrous."[54]

GE's work pushed the engineers into philosophical terrain. There was a subtle difference between human control and automatic control.[55] For Mosher, a simple inert hand shovel used in the garden was a cybernetic anthropomorphic machine; it extended the human body and senses and could be used without training. "This simple device qualifies as a CAM!" he wrote about the shovel, excited by this fundamental insight.[56] The shovel, like the chisel of Ross Ashby's sculptor, perfectly extended the user's arm, functionally becoming a part of the operator's body. But more complex machinery—for instance, a crane—broke up this union, cutting the operator off from "continuous sensory appraisal."

For optimal control, the user needed to *sense* force, surfaces, position, speed, and direction—not simply see the end of the crane's arm from a remote cabin. Operating a crane was a bit like trying to catch a ball while looking at yourself and the ball in a mirror; it was difficult and clumsy. Operating a cybernetic machine was like being a more powerful version of yourself, simply catching the ball—almost like being a spaceship or a TV network.

In November 1965, GE launched its boldest cybernetic project to date, combining all previous ideas into one machine: a fully functional exoskeleton for heavy loading. The device, again, looked like something that would be brought to the screen twenty years later by a cult film: James Cameron's 1986 science fiction horror movie *Aliens*, which features Sigourney Weaver battling an alien in a power-loader exoskeleton, the fictional Caterpillar P-5000. Again, the idea wasn't new.

GE's power loader was called Hardiman ("man" was GE shorthand for "manipulator"). The US Navy's Office of Naval Research and the

US Army's Natick Laboratories in Massachusetts jointly funded the development of this extravagant machine. Like Cameron's rip-off in *Aliens*, it was designed for handling heavy material in extreme situations: bomb loading under the wings of fighter aircraft, underwater construction, and manual work during space travel. The company envisioned the exosuit in different sizes, from life-sized to a 50-foot giant, as tall as a five-story building. Force and position information could easily be scaled up or down, the engineers believed, with oil-powered hydraulic servos pressurized at 3,000 psi.

The exosuit's arms would be mounted off the waist, for handling heavy loads and for ruggedness. "Load handling tasks such as walking, lifting, climbing, pushing and pulling can be performed with a lift capacity of 1,500 pounds," GE wrote, matter-of-factly.[57] The force ratio for the prototype was 25:1, so a man picking up a load of 1,500 pounds would feel only 60 pounds. The operator's hands were protected inside what the engineers called the slave housing.

By early 1967, the company expected the exosuit to be ready for testing and evaluation one year later, in the spring of 1968.[58] But the military funders weren't fully convinced of the near-term feasibility of the cumbersome exoskeleton. GE's work was never completed. Only an arm was built to specification, with nine joints. The navy contract expired after nearly six years, on August 31, 1971.[59] The Hardiman became yet another cybernetics-inspired project that failed in the initial development phase.

Mosher was undaunted. The engineer was already thinking about the next steps. "There's no reason why the operator has to be inside his CAM," he suspected, referring to his cybernetic machine. "You could link the two by radio."[60]

The Philco Corporation, headquartered in Philadelphia, was a major electronics contractor for the NSA, the Department of Defense, and NASA in the late 1950s and early 1960s. The company understood that putting human operators into space or the deep seas was too complicated and too expensive, machine-modified or not.[61] Instead, William Bradley and some of his colleagues at Philco suggested building a long-distance cyborg: creating a mock-up of the interior of a space capsule, or of an undersea environment, would be more elegant and efficient than sending a person into such hostile environments. The Philco engineers envisioned the remote machinery with elaborate sensors, recording sound and tactile sensations in real time. An operator could then stand at the remote-control interface and "see" and "hear" and "feel" the movements of a remote arm or hand.[62] The idea of using remote-controlled robots for hazardous tasks wasn't new. But Philco's implementation of the remote presence was new.

In 1961, two engineers at Philco—Charles Comeau and James Bryan—published some early results. They had built the first binocular head-mounted display, calling it "Headsight." The basic idea was simple: link a CCTV surveillance camera to a forehead-mounted monitor. Getting it to work, however, wasn't so simple.

The helmet was almost stylish, in keeping with the design of the time: a slick, black leather shell, a few black cables snaking down the neck, with a small antenna at the forehead for orientation, and a relatively small screen in front of the eyes. Comeau and Bryan's system used a spherical mirror close to the user's face to project a virtual 10-inch-high image that seemed to appear one and a half feet in front of the user.

A TV camera was slaved to the device. Three servos controlled the camera's movement in three dimensions: rotation, nod, and tilt. When the operator pointed his head up or down, or left or right, the camera followed at exactly the same angle. When the observer tilted his head, the camera also tilted, maintaining a constant horizon level. Through all of this manipulation, the operator's hands remained free.

There were, however, two rather difficult problems. One was that the camera and display needed to be spatially in sync. If the viewer looked to the top right, the camera also needed to look to the top right—in exactly the same position. To couple display and camera, the Philco team set up rotating magnetic fields around both the helmet and the lens. Position-detecting coils could then sense the position of both camera and head, and provide precise coordinates. The viewer's head and the camera's spatial position were then compared. If an error was detected between the two, the camera's motors would whirr into action and reduce the error to zero, bringing eye and camera in sync again. This was negative feedback at work.

The second problem was that all this took a bit of time. When the operator moved his head, his field of vision changed and his eyes swiftly refocused. The CCTV camera had to orient, make the same move, and focus. Doing all this created a significant lag, which was cumbersome, dizzying, and tiring. NASA research into displays would later reveal that lags of more than fifteen milliseconds caused dizziness and nausea.

Nevertheless, head-mounted camera control, the engineers at Philco Corporation found, was more precise than navigating a camera with a joystick. Another Philco engineer, Stephen Moulton, took this vision a proverbial step further. He installed the camera on the roof of a company building in Philadelphia. When he moved his head, the camera would move with it, relaying the city vista back to the head-mounted screen downstairs. Wearing this helmet, the viewer had the impression of being on top of the building and looking around the city.[63]

When Moulton, in the safety of a Philco lab with his helmet on, leaned over, looking down, it was "kind of creepy." Moulton then started playing with his new gadget. One of the best effects he achieved was amplifying the control movement: he put a two-to-one distortion on the neck twist. So when the viewer wearing the helmet would turn his head by 30 degrees, the normal range of a head turn, the roof-mounted eye would turn twice as far, by 60 degrees, giving the viewer the impression that he had a rubber neck, like a woodpecker's.[64]

Yet head-mounted displays were serious business, deadly serious. The most powerful application, as the engineers illustrated with a large graph in one publication to showcase Philco's achievements, was "to mount camera in drone or rocket." The operator could then sit on a chair "300 miles" away from a camera flying at double the speed of sound in the cone of a missile homing in on its target. Alternatively, a drone would make the technology reusable, and perhaps enable the surveillance of military combat. "The viewer, at home base, has the sense of being in the drone and can survey remote areas in complete safety," the defense contractors explained.[65] Other applications were to explore space or ocean depths, or to work in radioactive areas. By 1963, Philco was providing the display and control systems at NASA's Mission Control Center in Houston, Texas. The firm's visual control interface linked humans in Houston to the computers aboard spacecraft.

By 1965, the cyborg had begun to capture the popular imagination. "What is man?" was the opening line of the first full-length book on the subject, *Cyborg*. Its author, D. S. Halacy, took a grand and ambitious view, portraying the evolution from ordinary man to "superman," as the book's subtitle promised, in direct reference to Friedrich Nietzsche's Übermensch, as the author made clear in the text.[66] For millions of years, the evolution of humans had been left to nature. Now, by the early 1960s, humans had taken evolution into their own hands. Human progress wasn't any longer driven passively by evolution.

"Participant evolution" meant that man himself was now an active factor in his own development—in the masculine language of the 1960s. Radical changes would become possible to adapt the body to extreme environments: nose and mouth could be permanently sealed to enable life in the vacuum of the space beyond Earth's atmosphere, while a purpose-built implant would oxygenate the astronaut's blood. But there wasn't just outer space; the planet's "inner space" in the deep seas was equally promising. Already, ocean divers could breathe gases other than air; "a more drastic approach is that of learning to breathe *water*," Halacy wrote.[67] In fact, these changes were so drastic that the idea of human evolution itself was probably obsolete. Yes, there was an "evolution to the cyborg"—but then came the *cyborg revolution*.

The human urge to fight had already created primitive cyborgs: cavemen with clubs, lancers, swordsmen, and frogmen. Halacy was especially taken by the medieval armed warriors of King Arthur's day: lance at the ready, sitting astride a horse, protected by a cover of chain mail, the knight represented "a complex development of the military cyborg." For man had started to modify his body by adding protective coating, by converting the arm into a lethal weapon, and by "supplanting" his own legs with far sturdier ones.

Halacy's vision drips with the deeply modernist belief in progress: artificially produced human beings would have a body "superior to natural man," with none of the natural weaknesses and susceptibilities to disease and decay. Machines could even stop humans from ageing: "The cyborg will live not just a better, healthier life, but a much longer one as well." Modern life was prolonged and improved by ceramic hip joints, titanium bones, silicon breasts, electronic bladders, pacemakers, plastic corneas, and lifelike mechanical hands.

By 1970, David Rorvik, a science writer for *Time* and the *New York Times*, foresaw markets to "trade in" body parts for more durable, "if not immortal," mechanical spare parts. An individual with a family predisposition for heart disease could decide not to wait until fate strikes, choosing to preemptively purchase a perfected plastic heart "rather than risk middle age with a vulnerable flesh-and-blood pump."[68]

Many incurable defects could be fixed. Those with failing sexual organs, Rorvik dreamed not long after San Francisco was celebrating the summer of love, would be able to buy "youthful potency" over the counter at a medical spare-parts consortium. Naturally, amputees would benefit as well, and the United States had too many, since badly injured Vietnam veterans were returning home by 1970. Man would abandon part of his old identity, "melting so that he can be forged anew," Rorvik foresaw. A new man would be created, "welded to machines that amplify his senses, extend his grasp, deepen his under-standing of himself and his world."[69]

Meanwhile, the cyborg as a scientific idea had died of dry rot. That same year, 1970, *Astronautics* invited cyborg pioneer Manfred Clynes to write another article about his original idea. How far has technol-ogy come in the past dozen years in simplifying man's approach to space travel? "Not very far yet," Clynes thought. The human organism hasn't "yet" been engineered to use sunlight "even like a plant" as a source for organic chemical energy. Automatic recycling of oxygen in the bloodstream remained impossible, despite the aim of "cyborg tech-nology" to bypass the lungs' in-out tidal breathing flow and oxygenate the blood directly through an implanted fuel cell. And even the body's own regulatory system was still "floating" along, unstable, without improved and superior machine controls.[70]

Yet space exploration had made vast progress since 1960: In 1961 the first ape, a chimpanzee named Ham, and then the first human, entered orbital flight. By the mid-1960s, several planetary flybys had been accomplished, and one mission had reached Venus. The Soviet Union had sent a satellite around the moon. At the end of the decade, on July 21, 1969, the first humans landed on the moon, with the first manned orbital observatory to follow two years later.

Modifying humans for life in space, though, remained a distant dream. "There is a strange technological imbalance between man's development of his tools and machines for the penetration of the nature of space, and his lack of progress in cyborg technology," Clynes noted with frustration in 1970.[71] The first thing the astronauts had to do before landing on the moon was something as mundane as sleep-ing eight hours, the would-be body engineer observed: "We do not know why man needs to sleep." Machines don't sleep. And cybernetically engineered humans wouldn't have to sleep either. In that respect, naked man was inferior to the very machine he had built to reach the distant destination in the sky. "If the spaceship had such pervasive unknown needs, it surely never would have made it to the moon!" To add humiliation to defeat, *Astronautics* rejected Clynes's article and refused to publish it, without giving an explanation.

III

Man-machine interaction, of course, wasn't limited to human physical capacities but could very well apply to human intellectual capacities. Consequently, the computer itself became the subject of man-machine interaction. Perhaps the most influential thinker and technologist to tackle this specific question was J. C. R. Licklider, one of the path-breaking early pioneers of the internet and a participant in the Wiener circle. Licklider, remarkably, used the world's most ambitious automation project to show the limits of automation.

Licklider was deeply familiar with cybernetics, as well as with the ever-more-urgent air defense problem. "There was tremendous intellectual ferment in Cambridge after World War II," he recalled after participating in Norbert Wiener's weekly cybernetic circle. "I was a faithful adherent to that."[72] Licklider was a researcher and faculty member at Harvard University at the time. Nevertheless, he audited one of Wiener's seminars at MIT: "There was a faculty group at MIT that got together and talked about cybernetics and stuff like that. I was always hanging onto that." The discussion circle was so significant that Licklider would later try to emulate a "miniature Wiener circle" to discuss projects at the Air Force Office of Scientific Research. Licklider even presented a paper at the last Macy conference that Wiener attended.

Licklider was also intimately familiar with air force research on man-machine interaction for improved command and control. In 1951 he had consulted as a psychologist on the project that later became SAGE at MIT. The air defense network shaped Licklider's thoughts about information processing, and he had the idea of a "network of thinking centers" through the network.[73] Licklider's own work on the subject wasn't just funded by the air force, but was inspired by the problems of air defense. Already in 1957 he had written an unpublished essay: "The Truly SAGE System, or, Toward a Man-Machine System for Thinking."[74] Licklider served on the Air Force Scientific Advisory Board for about six years, leading up to 1962.[75]

Licklider saw that "the problems of command and control were essentially problems of man computer interaction." But viewing the computer as a more powerful abacus did not make sense, he was sure. The stress of battle didn't allow for preprogrammed scripts. Batch processing was a misguided approach; "I thought it was just ridiculous to be having command control systems based on batch processing," said Licklider.[76] In practice, chance and friction dominated the battlefield and commanders would have to react to the unexpected at every turn. It was simply impossible to preprogram the chaos of fighting, Licklider knew: "Who can direct a battle when he's got to write the program in the middle of the battle?"[77]

Yet this was exactly what SAGE had tried to do. "The main experience we have had with a large-scale man-machine system for situation analysis and control has been provided by the SAGE system," Licklider told an air force committee in late November 1958, just a few months after the system had become operational.[78] The air defense network was originally designed to be "very largely automatic," he explained. The many human operators were brought into the air force system to handle tasks that could not be automated at the time, so the air force treated human operators as a second-best machine part. Licklider implied to the air force that this had been a design flaw: SAGE was "too much a matter of men aiding the machine, and not enough a matter of true man-computer symbiosis."[79] Therefore, the network was not a very good preview of the air force information-processing and control systems that he hoped would be built in the future. Man-machine symbiosis, in short, was superior to automation.

Licklider didn't want to advocate more automation, or to delegate ever more decision authority to machines. He was sharply critical of the automation enthusiasts of his day. The very concept of mechanical extension, as he saw it, led to the idea that humans could and should be replaced by machines, that "the men who remain are there more to help than to be helped." Licklider wasn't opposed to this vision in principle. But it was impracticable, "fantastic," he thought.[80] Like Mosher at General Electric, who worked on a very different problem for the army, Licklider realized that the best systems were a blend of the best of both humans and machines. His suggestion for a "truly SAGE" system clearly spelled out this vision in 1957:

> The scientists and engineers combine their cerebral data-processing with the facilities of the machine to constitute a more effective system than either the human or the mechanical parts alone could make.[81]

Human and machine were not in competition; they complemented each other. Their partnership had a template in nature: a symbiosis. Licklider's highly influential notion of "man-computer symbiosis" emerged in response to SAGE, during the summer of 1958 at committee discussions with Licklider's air force funders on the future of command and control. Licklider articulated his vision in a famous paper, "Man-Computer Symbiosis," in 1960. "The hope is that, in not too many years, human brains and computing machines will be coupled together very tightly," he wrote, using a phrase that would become common in human-machine engineering: "coupling tightly."[82]

Unsurprisingly, Licklider's opening example was the fig wasp, a small insect that has evolved along with the fig tree over millions of years. The wasp's larvae live in the plant's ovary. The wasp needs the tree for survival, and the tree in turn needs the wasp for pollination and reproduction. Such a mutual and existential dependence between humans and machines would not yet exist, Licklider observed. But he hoped it would soon come about. Humans, he believed, were better at formulating questions and answers, at detecting relevance, and at reacting to unforeseen exigencies; machines, by contrast, were better at storing and retrieving large quantities of information precisely, at calculating rapidly, and at building and remembering a repertoire of routines.

"The intellectual power of an effective man-computer symbiosis will far exceed that of either component alone," Licklider wrote in 1962.[83] By then, even military commanders who had simulated semi-automated maneuvers were eager to regain the initiative and flexibility they felt they had lost to machines in computer-centered command-and-control arrangements, and Licklider knew this. But air force officers would also want to retain the storage and processing capabilities of their computers. Symbiosis was the way forward.

Several problems, however, needed to be solved first. One was what was then known as "time-sharing," dividing the processing resources of extremely expensive supercomputers among a number of human users. A second problem was improving the severely limiting input-output interfaces of these computers; electronic typewriters and SAGE-style light guns weren't good enough. A third problem identified by Licklider was the speedy storage and retrieval of vast quantities of information and data.

Licklider suspected that graphical interfaces and speech recognition would be highly desirable. A military commander, for instance, would need fast decisions. The notion of a ten-minute war would be overstated, yes, but it would be dangerous to assume that leaders would have more than ten minutes for critical decisions in wartime. Only speech recognition was fast enough as a human-machine interface; an officer in battle or a senior executive in a company could hardly be taken "away from his work to teach him to type," Licklider quipped. It would probably take five years, he concluded in 1960, to achieve practically significant speech recognition on a "truly symbiotic level" of real-time man-machine interaction.[84]

In 1962, Licklider moved on to the Pentagon's Advanced Research Projects Agency, ARPA. He became the first director of ARPA's newly founded Information Processing Techniques Office, a research and funding organization tasked to improve military command-and-control systems. At ARPA, Licklider continued to work toward improved man-machine communication. He especially supported university-based research projects working on time-sharing over long distances, just as the air force had done. Soon the vision of a global computer network began to take shape.

On April 25, 1963, Licklider wrote a famous memorandum, addressed to "members and affiliates of the Intergalactic Computer Network."[85] This was meant in irony, "as you may have detected in the above-Subject," he wrote to his colleagues and collaborators in the memo. "I am at a loss for a name." Then Licklider articulated in more detail what this computer network was supposed to be all about: the advancement of the art of information processing, and "the advancement of intellectual capability (man, man-machine, or machine)," Licklider wrote. The memo went out to ARPA-contracted researchers at Stanford University, UC Berkeley, UCLA, MIT, the Rand Corporation, and several contractors in industry. To make progress in these endeavors, he reckoned, each researcher needed hardware facilities, as well as a software base more complex and more extensive than one person alone could build.

The only solution was a network of computers, a network of individual "thinking centers," as he called it. The researchers played a key role in conceiving and funding ARPANET. But the kind of network that Licklider suggested in 1963 would take almost exactly twenty years to mature into what would later be called the internet. By the end of the '60s, the myth of cybernetic organisms and living machines had begun to retreat into science fiction—and critical theory.

The notion that machines could outthink humans was still hot among scientists in the 1960s. Irving "Jack" Good was a leading UK mathematician, then based at Trinity College, Oxford, and the Atlas Computer Lab in Chilton. He had worked as a cryptologist at Bletchley Park with Alan Turing during the war, and later at GCHQ until 1959.[86] Good had become convinced that "ultraintelligent machines" would soon be built. "The survival of man depends on the early construction of an ultraintelligent machine," he enigmatically opened his most-read paper, in 1965. In Good's view, a machine was ultraintelligent if it could "far surpass" all the intellectual activities of any human being, however clever. Once this was achieved, Good reasoned, then a singular moment in human history would have arrived. Humans would not be at the top of creation any longer.

> Since the design of machines is one of these intellectual activities, an ultraintelligent machine could design even better machines; there would then unquestionably be an "intelligence explosion," and the intelligence of man would be left far behind. Thus the first ultraintelligent machine is the last invention that man need ever make, provided that the machine is docile enough to tell us how to keep it under control.[87]

The notion that man might be able to create other new intelligent beings is as bold as it is old. Much bolder and grander was the idea that man could create an even better creator than himself or, indeed, than whoever had created him. Good took off where Wiener, who had died the year before, had left it. He didn't want just to play God, but to create an even better God.

By the 1970s, such ideas had found a home in science fiction—not least, thanks to Good himself, who had served as a consultant for Stanley Kubrick's *2001: A Space Odyssey*. One of the prime meeting places for science and science fiction at the time was *Omni* magazine. Founded by Bob Guccione, who also started *Penthouse* magazine, it was published in print between 1978 and 1995. Many futuristic ideas were either born or buried in *Omni*'s brightly illustrated pages.

One example was science fiction writer Vernor Vinge's word "singularity." Having read Good, Vinge chose to describe the British scientist's intelligence explosion as a "singularity," the expected future moment when machines would overtake humans in their intellectual capacities.[88] Vinge, himself an unsuccessful scientist at San Diego State University but a quite successful science fiction writer, compared that moment to a black hole: "When this happens," he wrote in *Omni* in early 1983, "human history will have reached a transition as impenetrable as the knotted space-time at the center of a black hole, and the world will pass far beyond our understanding."[89]

Meanwhile, a student in Germany was occupied with similar thoughts. Jürgen Kraus mentioned and analyzed computer viruses for the first time in his master's thesis, written in the late 1970s: "Selbstreproduktion bei Programmen."[90] Biology saw reproduction and mutation as crucial features of all life. Computer programs had some of the same features. "Is it perhaps even possible to speak of living programs, as is done in biology?" Kraus asked.

Large mainframe computers, Kraus wrote in 1979, would already form a "universe" of circuits and bits.[91] The complexity of these systems was comparable to the complexity of a young planet Earth. Software, Kraus observed, never operated at 100 percent accuracy; it was never perfect. Therefore, the possibility of mutation among programs was real. The search for life among computer programs, Kraus was sure, was a matter of philosophy and theoretical biology.[92]

Kraus traced the parallels of biological systems and computers over 228 pages. He was obsessed by the question of whether programs, like biological organisms, were alive and whether they could *evolve* as all living entities do. A self-reproducing program, he reasoned, resided in the "environment" that is the computer, its hardware and software, and its memory. The environment, Kraus argued, was alive, "*belebt*": the self-reproducing programs would put competitive pressure on each other, including "conflict behavior" and selection pressure. In effect, then, the "evolution" of self-reproducing programs had become a possibility.[93]

Kraus highlighted the technical differences between the biological virus and self-reproducing software: A biological virus "actively initiates its reproduction by intruding into the energy-providing system 'cell,'" Kraus wrote. "A self-reproducing program can't do that." A computer virus, even if it was already inside the "memory-space and energy-providing system 'computer,'" required activation on the part of the machine's operating system.[94]

Kraus helped coin an expression that has since become ubiquitous: computer virus. But his entire analysis was off base, as he admitted later. His discussion was also somewhat quaint by the time he published it, inspired by ideas that had already fallen out of fashion among his English-speaking colleagues in computer science and engineering. The old cybernetic idea that machines—hardware or software, alone or networked—could come alive, self-reproduce, mutate, and turn on their creators would continue to appeal to science fiction authors and screenwriters. But it had only a short moment of scientific attraction. In science and engineering, the "cybernetics moment" had by the early

1970s passed.[95] Scientists and engineers moved on, no longer captivated by the idea of creating cybernetic organisms.

The cyborg was reborn as a powerful myth and metaphor a dozen
years later. This new discourse on cyborgs quickly became pervasive
and dominant. So dominant that by far the most widely read and most
widely cited nonfiction articles and books on cyborgs are on feminism
and postmodern thought, not engineering.[96]

Donna Haraway is the best known of these thinkers. Like Alice
Mary Hilton a generation earlier, Haraway drew her inspiration
from cybernetics and launched into an ambitious project of cultural
engineering with socialist leanings. In 1985 she was a newly minted
professor in the history of consciousness at UC Santa Cruz. That
year, Haraway published an essay in the *Socialist Review* that gave
her a name: "A Cyborg Manifesto," its title and its theme of a call to
action drawn from *A Communist Manifesto*. The text's success fell only
slightly short of its title's youthful ambition.

Postmodernism was about blurring science and fiction. Haraway,
from the start, said she believed that the boundary between science
fiction and social reality was "an optical illusion."[97] The goal of the
radical philosopher was to advance what she dubbed "socialist-fem-
inism." Reality was full of contradictions. And one of the best ways
to deal with contradiction was irony. With this warning, Haraway
opened her manifesto: "At the center of my ironic faith," she wrote, "is
the image of the cyborg." The image mattered to her, the comparison,
the metaphor—not actually engineered humanoid robots.

"A cyborg is a cybernetic organism, a hybrid of machine and
organism, a creature of social reality as well as a creature of fiction,"
she wrote. To Haraway, the cyborg was everywhere. Science fiction,
of course, was full of cyborgs. So was modern medicine, where it was
normal to hook up organisms to machines. Industrial production
was full of humans merged with machines. So was sex. "And modern
war is a cyborg orgy, coded C3I, command-control-communications-

intelligence," she wrote, referring to a military abbreviation that was then in fashion. We are all cyborgs, she insisted.[98]

The cyborg was highly attractive to postmodern philosophers; it embodied many of their ideals. Cyborgs, as the feminist philosopher saw it, were about "breached boundaries." Such boundary-breaching beings would "confuse" historical stories, question identities that are taken as given, and blur what counted as distinct categories. The cyborg breached borders left and right: between body and machine, human and nonhuman, mind and body; between nature and culture, man and woman, maker and made; between active and passive, total and partial, agent and resource. It even broke up distinctions of consciousness and simulation, of the natural and the artificial, of right and wrong, of truth and illusion, and—perhaps most important—of God and man. In the postmodern view, these dualisms of the dreaded establishment underpinned much of what was wrong: patriarchy, imperialism, capitalism, even militarism.

As Haraway put it, "High-tech culture challenges these dualisms in intriguing ways." With machines becoming ever more creative, it is no longer clear who makes whom. It is also no longer clear what is mind and what is body "in machines that resolve into coding practices," she wrote, cryptically.[99]

The cyborg wasn't susceptible to bourgeois values. It wasn't made out of mud and therefore could not dream of returning to dust. Unlike Frankenstein's monster, it didn't expect its father to "restore the garden" by creating a female mate for it. It didn't dream of the organic family as its preferred form of community. The cyborg was deeply subversive.

"The main trouble with cyborgs, of course, is that they are the illegitimate offspring of militarism and patriarchal capitalism," Haraway lamented. She knew of Clynes's Cold War vision and the military funding that had gone into exoskeletons at General Electric. But she could live with that. "Illegitimate offspring are often exceedingly unfaithful to their origins," she pointed out. "Their fathers, after all, are inessential."[100]

"Our machines are disturbingly lively, and we ourselves are frighteningly inert," Haraway wrote, mystifyingly.[101]

Haraway was a self-described vegetarian, feminist mystic, deeply suspicious of what she saw as the Cold War's military-industrial complex and its "patriarchal" technologies, especially that "heavy brew of cybernetics in the 1950s and 1960s."[102] Viewed from Haraway's elevated academic perch, the jump from Manfred Clynes's enhanced rat to Arnold Schwarzenegger's Terminator was preordained:

> Cyborgs do not stay still. Already in the few decades that they existed, they have mutated, in fact and fiction, into second-order entities like genomic and electronic databases and other denizens of the zone called cyberspace.[103]

The feminist philosopher employed a scholarly trick: she clandestinely flipped the causality on its head. Cyborgs were blurring boundaries, but blurred boundaries were also creating cyborgs. Hence, she argued, we are all cyborgs.

At first glance, such a bold statement seemed a stretch, to say the least. The editors of the *Cyborg Handbook*, a jargon-heavy tome of postmodern free-association writing, saw their favorite creature not as a subject of geeky science fiction, but as a common fact: "It's not just *Robocop*, it is our grandmother with a pacemaker."[104] By 1995, the *Handbook* would point out, about 10 percent of the US population were estimated to be "cyborgs in a technical sense."[105] This estimate included people with an array of implants, such as artificial joints, electronic pacemakers, insulin pumps, new corneas in the eye, or artificial skin or limbs.

But 10 percent is still a small number. A much higher percentage were involved in some sort of labor that made them "metaphorical cyborgs," the *Handbook* continued. That could be a worker at a computer keyboard joined "in a cybernetic circuit" with the machine, or a neurosurgeon guided by fiber-optic microscopes on the job, or teenagers merged with the video console in a local arcade. Even the man who had invented cyborgs agreed: "*Homo sapiens*, when he puts on a pair of glasses, *has* already changed," Clynes told the handbook editor in 1995. "When he rides a bicycle he virtually has become a cyborg."[106]

Everybody was a cyborg. Therefore, Haraway's manifesto wasn't written as a manual for androids; it was a manual for everybody. The boundary-smashing *Terminator* had wider symbolic significance. It didn't just represent the rise of the machines of Skynet; it represented the fall of dearly held dividing lines for everybody. Perhaps it is not surprising, then, that several postmodern scholars actually focused their attention on the T-1000, the evil robot made of liquid metal in James Cameron's *Terminator 2*, played with steel-blue eyes by the handsome American actor Robert Patrick.

In one scene the fictional T-1000 changes into three shapes in a matter of a few seconds: from a female character (heroine Sarah Connor's mother), to a metallic proto-android in neutral shape, and then to a uniformed male police officer. To some postmodern scholars, the scene had a deeper philosophical meaning. It showed "melting boundaries," how "sexual identities can be reformulated." The scholars note the attraction of seeing the fictional liquid-metal robot from the future change shape: "There is a highly erotic aspect to watching a virtuoso display of changes in human body-image."[107]

Another feminist thinker, also writing about Cameron's second *Terminator* movie, pointed out that the fashion preferences of the killer machines, both Terminators—the T-800 played by Arnold Schwarzenegger in black leather and the T-1000 played by Robert Patrick in cop regalia—"exaggerate homosexual types" and would therefore subvert the normative heterosexual authority structures.[108]

The phenomenal success of postmodern subversive cyborgs was lost

on Clynes, who had presented the original idea at the air force conference at Randolph Field in Texas in 1960. Clynes suggested reengineering human bodies, but reengineering human identities didn't jibe with
him. When the original *Terminator* was released, in late 1984, Clynes
was experimenting with computers interpreting classical music, and
he tried to link one of Haydn's sonatas to human expressions of emotions through touch. The maker of the first cyborg watched *Terminator* and was horrified. "Schwarzenegger playing this thing," Clynes
recalled, "dehumanized the concept completely." It made a monster
out of something that wasn't a monster, as the former researcher from
Rockland State Hospital saw it. "This is a travesty of the real scientific
concept we had."[109] That, of course, was the point.

5. CULTURE

THE COMPUTER, THE "THINKING MACHINE," WAS SO NEW
and unknown in the years following World War II that progress
seemed unlimited. The new thinking machines could calculate how
to build skyscrapers, how to run stock exchanges, and how to fly to
the moon. The only limit was the imagination. The "big brains" were
a miracle in the waiting that would change everything: war and work
would become automated; organisms and machines would merge,
creating new forms of life. But many of these midcentury visions
of the modern future were decades ahead of the actual technology
of the time. The vast, room-filling IBM machines had very limited
computing power. The giant thinking machines of the 1950s were far
dumber than tiny smartphones half a century later would be.

But the machines, it turned out, had nearly unlimited metaphorical power. The most obvious comparison was the human brain. If the
thinking machine was a simplified brain, then the reverse question
was practically asking itself: Wasn't the actual brain just a complex
machine? The mind suddenly became something that could be understood and described and analyzed with language borrowed from engi-

neering. And cybernetics provided that language: input and output, negative feedback, self-regulation, equilibrium, goal, and purpose. All this had a literally spiritual, drug-like appeal.

Seeing the mind as a machine was liberating. The reason was simple. Man could understand machines, make them, control them, tweak them, fix them, and improve them. If the mind was simply a kind of machine, then humans could understand it, control it, tweak it, fix it, and improve it. Doing so was only a matter of finding the right levers to pull and cogs to turn. No longer was human psychology something mysterious, something unknown, something beyond the comprehension and imagination of ordinary people.

The next logical step was to extend that comparison. If the individual mind was a self-regulating system that could be tweaked by oiling the feedback loops, what else was?

Norbert Wiener's and Ross Ashby's ideas immediately had a spiritual and quasi-religious appeal that went far beyond the fear of automation, or the fear of machines organically merging with humans. Soon, creative minds beyond the confines of the hard sciences discovered the power of cybernetics—especially in counterculture. By the late 1970s, cybernetics had gone viral. Sometimes in disguise. Entire communities functioned like whole systems, many in the countercultural avant-garde came to understand: there was a different way of seeing things, a circular way, where everything was connected, connected by feedback, kept in balance, in touch with the environment, even with animals and plants and rocks, in unity, as one single whole, one planet, shrunk into a village by communication technology. A veritable cult emerged. Seeing communities as self-regulating feedback systems was liberating, driven by a theory of machines that was quite literally "out of control," in the memorable phrase of the founding editor of *Wired* magazine, Kevin Kelly.[1]

The cybernetic myth had a major cultural impact. Wiener's work, in its countercultural and highly symbolic reading, forms one of the oldest and deepest roots of that firm belief in technical solutions that would later come to characterize the culture of Silicon Valley.

One of the earliest and most eminent writers roused by cybernetics was L. Ron Hubbard, then an immensely prolific science fiction author. Hubbard was fascinated by the new science, and especially by the idea that the mind was a thinking machine.

Hubbard was also a member of the Explorers Club, a New York–based society dedicated to expeditions to the world's unknown frontiers. The club had its own journal, the *Explorers Journal*. In the winter issue that came out in late 1949, Hubbard set out to explore the human mind, that vast and unknown frontier "half an inch back of our foreheads." Hubbard read Wiener and was hooked. Inspired by the MIT professor, the would-be prophet considered the brain "an electronic computing machine" and told fellow explorers that his own approach was a "bridge" to "Cybernetics," as he explicitly pointed out. Engineering equipped him with the tools that he needed to penetrate that half inch of skull bone.

The mind could do all the tricks of a computer, Hubbard wrote—the tricks of a good computer. But going even further, he said, "The analytical mind is not just a *good* computer, it is a *perfect* computer."[2] Therefore, the mind would be incapable of error, Hubbard reasoned. Error and imperfect human behavior would be introduced through "wrong data," just as in the case of an actual machine. The human mind, he implied, wasn't flawed in itself. It was indeed perfect. But like any other tool, it was subject to user error.

Hubbard later expanded the argument first outlined in that article into a book, *Dianetics*. Published in 1950, *Dianetics* would become one of the most translated books of all time, with editions in sixty-five languages, and allegedly more than twenty million copies sold. The 680-page tome is one of the best-selling self-help paperbacks of the twentieth century. It is also one of the founding texts of the Church of Scientology.[3]

Hubbard's example, right at the heart of his argument, was simple addition with a calculating machine. Input 6 times 1 and you get 6. This is the situation when the machine is receiving correct data. But what if the "7" key is stuck? If you input the same command, the calculator will receive the wrong data (6 times 7) and display the answer 42 instead. Any calculation entered on that keyboard, with the "7" stuck, would produce the wrong result, the wrong output.

In the same way, Hubbard was convinced, faulty data would trick the human mind. "Incorrect data gets into the machine. The machine gives wrong answers. Incorrect data enters the human memory banks, the person reacts in an 'abnormal manner,'" Hubbard explained. The situation was plain and simple. And it was liberating. Abnormal human behavior wasn't the result of a construction error in the unalterable mechanics of the mind, the central text of the Church of Scientology implied. Aberration, Hubbard preached, was a result of bad input data. Bad input equaled bad output. "Essentially, then, the problem of resolving aberration is the problem of finding a 'held-down 7,'" he wrote.[4] And so on.

The cyberneticists of the first hour were appalled. Hubbard and some of his employees had implied that Wiener enthusiastically endorsed the new science.[5] When Wiener found out, he was incensed. He had not the slightest confidence in Hubbard, he wrote in a letter, and he had doubted Hubbard's good faith from the beginning.[6] In other letters he compared Dianetics to voodoo and mesmerism, an eighteenth-century belief in the healing forces of animals.[7] Wiener had no patience for the man he considered a charlatan. On July 8, 1950, he wrote a short and terse letter to the later founder of the Church of Scientology, addressing him simply as "Sir."

"Be advised that I most definitely do *not* endorse your pretended science, your book, your system of therapy, your foundation, yourself, nor Dr. Powell," Wiener wrote in an enraged tone, referring to one of the foundation's representatives who had falsely implied he, Wiener, supported their cause. Wiener was so upset he couldn't type straight: "I hereby forbid either of you to make any explicit or implicit use of my name as endorser of any or all of these."[8]

Hubbard responded two weeks later. "My dear Dr. Wiener," he started out, "I am sorry if any action on the part of dianetics has disturbed you."[9] Hubbard promised not to use Wiener's name in the future. But he was a fan and could not resist pointing out how important cybernetics had been for his own work. "Some of your conclusions have been useful to dianetic development," Hubbard told Wiener, and he assured the professor that he would "set policy" not to mention his name. Nevertheless, Hubbard saw his own new field—what a few years later would form the intellectual foundation of the Church of Scientology—as "an engineering type science" that "dovetails nicely with Cybernetics" as he told another leading MIT mathematician in a letter in December 1949.[10]

This link wasn't news to Wiener. He had skimmed Hubbard's book, and the superficial similarities to his own work didn't escape him. "DIANETICS sounds like the attempt of an illiterate to capture the swing of CYBERNETICS," Wiener wrote to a colleague in July of 1950.[11]

Two months after the terse exchange with Hubbard, Wiener published a thoughtful and polemic warning against the temptations of pseudoscience. He lashed out against "the confused work of amateurs" who were using precise mathematical language without understanding what they were talking about. Wiener reserved his full ire for a set of ideas that he called "wholism," the notion that a system could be understood only as a whole system, as an entity that is more than the sum of its parts. That, to him, was the worst kind of false science.

"Let me lay the ghost of another pseudo-scientific bogy: the bogy of 'wholism,'" he wrote in a prominent philosophy journal in September.[12] "If a phenomena [sic] can only be grasped as a whole," he argued, "it is completely unresponsive to analysis." And if whole systems were unresponsive to scientific analysis, then whole systems were simply not available for serious inquiry. "The whole is never at our disposal."[13] The MIT professor, perhaps because he himself felt guilty of the crime he was accusing others of, closed his polemic against the faux use of mathematical language with what seemed like faux outrage: "Let us have done with this sorry profanation."[14]

W. Grey Walter, one of Britain's leading figures in the new field, had a hard-nosed scientific view of the world, and equally little patience for pseudoscience and charlatans. He knew about the frequent attempts to integrate the sundry schools of scientific disciplines into a larger whole. And he had seen the seductive power of cybernetics. Walter, who three years later cofounded the International Association for Cybernetics, was vigilant and circumspect. The "loop of cybernetics," he warned in 1953, had cultish potential. The science of feedback loops had such seductive appeal that less disciplined minds would be tempted to join and connect fields that simply didn't belong together. "America is a great incubator of synthetic cultures," he wrote, and he named "the Dianetics of Hubbard" as an example of "these cults."[15]

Walter then added a warning that would resonate for an entire generation. He phrased his warning against the cultish in almost poetic language.

This is something to be aware of, for what we need is to preserve and cultivate just these growing-points of science, not to arrange in arbitrary style, like cut flowers, their sterile and exotic efflorescence.[16]

But Wiener's and Walter's warnings were futile. The universal theory of machines that was cybernetics was too powerful and too appealing for too many creative minds. Indeed, it was so seductive that it proved impossible to keep it in the narrow confines of proper science—enter *Psycho-Cybernetics*, a self-help book published in 1960 by Maxwell Maltz, a well-known plastic surgeon. The cover design was tabloid-style, set in black capital letters against red background. The book boasted that "this famous plastic surgeon's remarkable discovery" would help readers "escape life's dull, monotonous routine" and "get more living out of life." That wasn't everything. No, reading Maltz's magical book would even "make you look younger, feel healthier, be more successful!"[17]

The formula must have worked, if sales figures are an indicator. The book became a longtime *New York Times* best seller, up there on the coveted list next to *Heloise's Housekeeping Hints* and the *I Hate to Cook Book*. By the turn of the twenty-first century, the book had sold more than thirty million copies.[18] Maltz's work became an all-time classic in the expansive genre of self-help books. If somebody truly popularized cybernetic theory, it was the doctor working with breasts, not the one with the numbers.

Maltz's breathtaking success illustrates the appeal of cybernetics. He had an epiphany when he read Wiener's *Cybernetics*. Any good plastic surgeon, Maltz reasoned, must also be a psychologist, whether or not he or she wants to be. Cosmetic surgery changes not only a man's face or a woman's tummy. Surgery can change a person's future, behavior, personality, sometimes even basic abilities and talents. Changing the shape of somebody's nose inevitably affects that person's "inner self," as Maltz saw it. The doctor realized that his job came with an "awesome responsibility." Meeting this responsibility meant understanding the self, and how the body interacts with a person's inner self. In his quest to explore this mind-body problem, the cosmetic surgeon traveled widely and researched across disciplines. Finally, he recounted, "I found most of my answers in the new science of Cybernetics."[19]

Thus inspired by the new science of feedback loops and automation, Maltz saw the body as a machine that houses the human mind. By changing the machine through surgery, he would also change the mind. The doctor's newfound insights took him even further. The new science of cybernetics, Maltz realized, had provided ample proof that no such thing as a subconscious mind existed. Maltz didn't see the human soul as a complex and incomprehensible biological system, messed up by traumatic childhood experiences and deep archetypical fears, as psychoanalysts did. Freud and Jung were wrong; missile engineers were right.

"Your brain," as Maltz addressed his readers throughout the three-hundred-page book, is a goal-striving mechanism. That mechanism operates automatically to achieve a certain goal, "very much as a self-aiming torpedo or missile seeks out its target and steers its way to it," the plastic surgeon wrote, relating his concept back to Wiener's original inspiration with the antiaircraft predictor and automatic missile guidance.[20] Nuclear missiles didn't just intimidate; they also liberated—and in rather unexpected ways. The doctor returned to the "awesomeness" of the interceptor missile again and again to illustrate how the human mind works. He saw a goal-striving mechanism that consisted of the brain and the nervous system, all used by and directed by the mind. The consciousness operates the body, which Maltz saw as an automatic and purposeful machine:

This automatic, goal-striving machine functions very similarly to the way that electronic servo-mechanisms function, as far as basic principles are concerned, but it is much more marvelous, much more complex, than any electronic brain or guided missile ever conceived by man.[21]

Comparing humans to machines was a risky move. Few people identify with pistons and crankshafts, let alone with ballistic missiles. Yet most people want to control their fate as if it had a steering wheel. Maltz sensed this tension. "You are not a machine," the surgeon clarified, talking to his readers directly. But at the same time, body and mind functioned like a machine, "like an electronic computer," a mechanical device preprogrammed to work toward achieving a goal. Maltz then suggested that his readers could reprogram their "built-in success mechanism" in order "to automatically steer you in the right direction to achieve certain goals."

This human-machine analogy was powerful. There was no subconscious beyond the control of the individual. No, in Maltz's world everybody could control his or her own "machine." Even making mistakes was okay. The unerring logic of negative feedback would allow for mistakes, ready to correct the course. The most crucial skills in life were learned by trial and error, "mentally correcting aim after an error," until the successful motion or reaction or performance was programmed into the servomechanism that was the human being. Once that had happened, more learning and continued success was accomplished by *forgetting past errors* while *remembering the successful response*, Maltz insisted.

The trick was trusting one's own mechanism to do its preprogrammed work, not to "jam it" by being too nervous and anxious, or by attempting to force something in too much of a conscious effort. "You must 'let it' work, rather than 'make it' work," Maltz wrote.[22] The brain was no gray, gooey mess; it was a whirring machine, ready to be put to work toward a better life. It just made logical sense. Like the guided missile. Maltz convinced tens of millions of readers.

The 1960s oozed optimism and an unshakable belief in technical progress. So Maltz's argument fell on fertile ground. The New York Yankees passed around worn copies of *Psycho-Cybernetics* when it came out in paperback in 1968. Tom Tresh, a Yankees infielder, raved about the book: "There's a passage that even helps me play shortstop." Instead of standing in the field nervously, "trying to outthink the ball and wondering which side of the field it'll be hit on," he would just relax and tell himself that wherever it would be, he had done it before and would do it again this time.[23]

In the foreword to his extraordinary longtime best seller, Maltz added a reflection. It was "rather ironic," he wrote, that an idea that began as a study of machines and mechanical principles "goes far to restore the dignity of man as a unique, creative being."[24] Even the creative Dr. Maltz could not have imagined how far cybernetics had yet to go.

By 1970, cybernetics had already peaked as a serious scholarly undertaking, and it soon began to fade. Its scientific legacy is hard to evaluate. On the one hand, cybernetic ideas and terms were spectacularly successful and shaped other fields: control engineering, artificial intelligence, even game theory. On the other hand, cybernetics as a science entered a creeping demise, with therapists and sociologists increasingly filling the rolls at the American Society for Cybernetics. Kevin Kelly, the *Wired* magazine editor, later observed that "by the late 1970s, cybernetics had died of dry rot."[25]

Yet, to the surprise of the remaining founders, cybernetics lived on—not in Boston's scientific research labs, but in California's counterculture communes. The rising New Age movement found the new discipline's mystic side appealing. The most eccentric expression of this remarkable shift is an ode to cybernetics written during the Summer of Love in 1967 San Francisco by Richard Brautigan, a long-haired hippie poet, "All Watched Over by Machines of Loving Grace":[26]

> *I like to think*
>   *(and the sooner the better!)*
> *of a cybernetic meadow*
> *where mammals and computers*

live together in mutually
programming harmony
like pure water
touching clear sky.

I like to think
       (right now please!)
of a cybernetic forest
filled with pines and electronics
where deer stroll peacefully
past computers
as if they were flowers
with spinning blossoms.

I like to think
       (it has to be!)
of a cybernetic ecology
where we are free of our labors
and joined back to nature,
returned to our mammal
brothers and sisters,
and all watched over
by machines of loving grace.[27]

The idea's journey from the East Coast to the West Coast is extraordinary—the ideology emerged and evolved in a range of military-funded and space-related projects, and then found its way into San Francisco's drug-fueled counterculture, in the short space of two decades. One person played a particularly important role in this metamorphosis: Stewart Brand.

Stewart Brand isn't easily described in one sentence: he has been an influential editor, writer, entrepreneur, socialite, intellectual, community organizer, and futurist.

The Cold War shaped Brand. A nuclear nightmare back from his teenage years in the early 1950s was one of his most vivid memories. His hometown, Rockford, in northern Illinois, was a well-known hub for the production of heavy machinery and tools. The Soviets, young Brand knew, had placed the town high on a list of likely targets for nuclear attack. The dream about the day after the strike was harsh: "There was chaos, and then I looked around and I was the only person left alive in Rockford, Ill., a knee-high creature."[28] He wanted out, to escape the specter of nuclear annihilation. Brand is best known for founding the famous *Whole Earth Catalog*, a publication that itself became an emblem and icon of California's late 1960s counterculture and back-to-the-land movement.

One afternoon, probably in March 1966 in the hills of San Francisco, Brand dropped a bit of LSD and went up on a roof overlooking the city. It was a form of escape. He sat in a blanket, shivering in the cold spring air, overlooking the hills, lost in enhanced thought:

> And so I'm watching the buildings, looking out at San Francisco, thinking of Buckminster Fuller's notion that people think of the earth's resources as unlimited because they think of the earth as flat. I'm looking at San Francisco from 300 feet and 200 micrograms up and thinking that I can see from here that the earth is curved. I had the idea that the higher you go the more you can see earth as round.[29]

Yet no photograph of the whole Earth was publicly available at the time, Brand thought, despite nearly ten years of US space exploration in a Cold War arms race that extended even beyond the planet. Then a skinny twenty-seven-year-old, 6 feet tall, with a big grin, he found this unacceptable. As he stared at the city's high-rises, it seemed to him they were not really parallel, but diverged slightly toward the top—because of Earth's curvature. Brand's mind was racing: "I started scheming within the trip," he remembered. "How can I make this photograph happen?" He had persuaded himself that it would "change everything" if only the people could see this photograph looking at Earth from space.[30] "Why haven't we seen a photograph of the whole Earth yet?" That was the question.

The next morning he started printing buttons and posters asking exactly that question. He got himself a large Day-Glo sandwich board with a little sales shelf at the front, donned a white jumpsuit, strapped on boots and a top hat with a crystal heart, and went off to UC Berkeley. Each button retailed at 25 cents. The dean threw Brand off campus, which earned him an article in the *San Francisco Chronicle* and other papers. He soon branched out to Stanford, then Columbia, Harvard, and MIT. Maybe, just maybe, the picture he sought would change some minds and make people realize how small and precious and fragile Earth was.[31]

Finally, in November 1967, NASA beamed the picture down to Earth from an ATS-3 satellite. Brand was elated and slapped it on the cover of his new publication: the *Whole Earth Catalog*. The first issue came out in the fall of 1968. It had an all-black cover. In the middle was a round and clear image of the whole Earth. Above the pristine globe, the cover said simply, "Whole Earth Catalog: access to tools." Tools, for Brand, had an almost mythical meaning. Anything could be a tool: a hacksaw, a monocular, a pair of Levi's 501 jeans, or the ideas in a book. "Here are the tools to make your life go better. And to make the world go better," he wrote in one of the catalog's introductions. "That they're the same tools is our theory of civilization."[32] The sturdier and the more robust, the better. The catalog was modeled on Sears and L. L. Bean mail-order catalogs. But it was different.

Brand's thinking was simple: if commune dwellers wanted to go back to the basics of self-sustained living and farming, they needed to know the basics first, and find those basics first. That's where his catalog came in. It listed and recommended the basics, or what its maker considered to be the basics: guides on shelter and land use, industry and craft, communications, community, nomadics, and—first and foremost—"understanding whole systems," the first item that the table of contents boasted. It was "purveying the stuff," and it became "a node of a network of people purveying it to each other," Brand recalled later. "And it was designed as a system. I knew about systems. I had studied cybernetics."[33]

Brand had not just studied cybernetics; he oozed it. Naturally, perhaps, the first issue of the *Whole Earth Catalog* offered readers access to that multipurpose tool: cybernetics. Brand didn't review just one cybernetic book in the first and rather slim sixty-two-page catalog; he reviewed seven, in detail, including several classics that had defined the field.

First, of course, was Norbert Wiener's 1948 *Cybernetics*: "Society, from organism to community to civilization to universe, is the domain of cybernetics," Brand wrote, introducing the book to the hippies in bold holistic terms.[34] Wiener had died four years earlier and couldn't object anymore. Brand also reviewed Ross Ashby's 1952 classic, *Design for a Brain*. "This is the learning mechanism," he wrote.[35] Third, he reviewed the MIT professor's second book, from 1954, *The Human Use of Human Beings*, proclaiming it "social, untechnical, ultimate." The fourth classic was more popularly accessible: Maxwell Maltz's self-help bible *Psycho-Cybernetics*. In Brand's words, "This is not a book to read. This is a kit of tools to use in gaining control of your nature for whatever ends you desire." Then there were less well-known tomes—notably, *General Systems Yearbook*, *Industrial Design*, and *Human Biocomputer*—by other authors. On top of all that, Brand's first version of the cult catalog included Brautigan's famous poem "All Watched Over by Machines of Loving Grace," next to a picture of a nude couple with libertarian amounts of pubic hair on display.

The concept worked, and the *Whole Earth Catalog* became a runaway success. Brand and his wife, Lois, had started off selling a print run of a thousand copies out of their Menlo Park home. Version one retailed for $5. They hired staff as their readership, and the subscriptions to the catalog and its supplements, grew exponentially. Brand produced six different editions of the catalog, published every half year, and nine quarterly supplements in total that were much shorter. In 1971 he announced the final issue, which sprawled over 449 pages, listing well more than a thousand items. Two and a half million copies were sold in total. That last issue of the catalog won the National Book Award for Contemporary Affairs in 1972.

But the catalog didn't just proselytize. It would not have been so successful if it didn't have more to offer. Brand practiced what he preached; the opening pages of every single issue reminded its far-flung but growing readership of the catalog's function and purpose: "The WHOLE EARTH CATALOG functions as an evaluation and access device," it said. This device was about access to inspiration, to personal power, to better ways of shaping one's own environment, and to a community for sharing this ongoing adventure.

The ingenious Brand invited his readers to submit suggestions, ideas, and reviews. The supplements played a crucial role in this functional vision, and every supplement made this point as clear as possible: the function of the supplements was to correct and update information with the help of reader reviews, reader comments, and other input from the community. The supplements included new suggestions of items to add or remove. A section called "Other People's Mail" contained letters from "nameless to nameless." Each supplement also carried various announcements. The supplement was, in short, a community forum, complete with spam, trolls, and discussions that must have seemed pointless to many readers.

Each iteration of the full catalog carried forward the most popular and most recommended items, steadily growing in volume. The final issue was almost seven times as thick as the first. New issues of the catalog also included a selection of reader comments, printed alongside the items they recommended for inclusion. In the final issue, for example, one reader, Ron Nigh from Palo Alto, defended the listing of Ross Ashby's work and also strongly recommended including *An Introduction to Cybernetics*, the second book from the former doctor in Barnwood, Gloucestershire, published fifteen years earlier. "An excellent place to start," Nigh recommended, signing off with "Love."[36]

Brand had a vision, a purpose. He stated his vision on the first page of every Whole Earth publication, where he explained the "function" and then the "purpose" of this widely popular publication. According to Brand, the catalog's function was to serve as an evaluation and access device. With it, the "user" could find out what was worth getting and where to get it. The purpose was to promote tools for education, inspiration, and shaping the environment—because, Brand wrote, "We *are* as gods and might as well get good at it."[37]

Brand's vision was to turn the catalog itself into a tool. The CATALOG—he usually spelled it in capital letters—was to form a feedback loop. He wanted it to be a communication device that connected the far-flung community he cared so much about. He wanted the catalog to be part of something that would create an equilibrium. The catalog was part of a whole system, a dynamic and self-regulating system. Brand would collect the crucial negative feedback in the supplement every few months and loop it back to the land by mail, to the readers-turned-cogs of this machine of loving grace. His publications, as he saw it, were part of an adaptive machine, not unlike the magnetic forces that governed the adaptive behavior of Ashby's homeostat. The catalog itself, with its supplements and its community, *was the learning mechanism.*

Learning was a crucial part of counterculture. Learning was perhaps the only way to expand the mind to see the way into a better, more peaceful, and more just future. To those hungry for mind expansion and knowledge, both psychedelic drugs and computers had instant and intuitive appeal. But that appeal, and the connection between the two, was hard to articulate, at least at first.

Michael Rossman was a prominent community organizer, an advocate of open education, and an activist for the Free Speech Movement at UC Berkeley. He was also an avid proponent of the use of psychedelics. By 1969, then in his late twenties and dashingly handsome, Rossman was living the good life in California, praising the educational virtues of the sweet smell of grass in a book on learning and social change. "Psychedelics," he held forth, "are a colorless, tasteless spice that heightens the flavor of whatever is cooking in the personal or social stewpot."[38] To him, taking LSD or smoking pot was strikingly similar to participating in the free learning groups or impromptu therapy sessions that were then popular. Primitive cultures used these substances for collective mythical insights and religious experiences for a good reason, he thought.

Rossman had been an avid reader of the *Whole Earth Catalog*, and he was familiar with Brand's take on tools and thoughts on control and communication, explained in the masthead of every issue and embodied by the catalog itself.[39] Rossman understood that LSD had become a favored chemical "tool" of painters, musicians, and writers. It was under the influence, he wrote, that "neglected or repressed sensory and emotional experiences and memories reassert themselves, often abruptly." Psychedelics would facilitate the connection of diverse elements, so that new patterns could be formed. The effect of psychedelic drugs on society, to Rossman, could just as well be expressed in the language of engineering: "In the cybernetic description of process," he wrote, "the corresponding passage is to a higher order of control—one that makes possible heterarchical rather than hierarchical control systems."[40] What he meant was simple: counterculture was changing established power structures. Top down was the past; bottom up was the future. That's where technology came in.

Rossman understood already in 1969 that computers had a key role to play in the future. As the free-speech activist was considering writing a book, the inventor Douglas Engelbart gave what became known as "the mother of all demos," a now legendary ninety-minute presentation at the Fall Joint Computer Conference in San Francisco. Engelbart introduced the prototype of the first mouse and the vision of a personal computer, a computer that could be owned and operated by everybody, not only IBM and the Pentagon.

To Rossman, that meant technology wasn't on the side of authority any longer. The future was brightening up: the "free use of computer technology" would mean that fifteen years into the future, flat structures would trump centralized power. "By 1984, America could govern itself by a system of totally decentralized authority," Rossman wrote in 1971. By the year 2000, the young idealist predicted, "cybernated" societies would be in even better shape, "They will view the machine as their extension—not vice versa, as is the custom now."[41]

Computer technology would free human labor, the Berkeley activist was sure, and make centralized decisions obsolete, thanks to the advance of cybernetics. "Every other industrial technology is now becoming capable of such re-engineering toward cyber-integrated production and full shared control," young Rossman predicted.[42] Such ideas were becoming common countercultural currency at that time. But to articulate the Zeitgeist in best-seller form required a much more experienced and eloquent thinker.

Nobody was better placed to discover that missing cybernetic thinker than the restless Stewart Brand. Initially he came to cybernetics from biology, from "world saving," as he called it, and from trying to understand mysticism. But he wasn't entirely happy with the phi-losophy that he found and presented so forcefully in the *Whole Earth Catalog*. Three years of scanning innumerable new books as the cata-log's editor did not turn up what he was looking for. "What I found missing was any clear conceptual bonding of cybernetic whole-systems thinking with religious whole-systems thinking," he recalled.

Brand was searching for the meaning of consciousness, for the right for life, for what's sacred. "Tall order," he admitted in *Harper's* maga-zine. But then something remarkable happened. As Brand put it, "In the summer of '72, a book began to fill it in for me: *Steps to an Ecology of Mind*, by Gregory Bateson."[43]

Bateson was a British-born anthropologist and social scientist. His father had been a leading geneticist who hoped his son would follow in his footsteps. But the younger Bateson resisted the family pressure: he went from English public-school boy and fellow at Cambridge Uni-versity's St. John's College to doing anthropological fieldwork in New Guinea and Bali. After an interlude during World War II in Burma working for the US Office of Strategic Services, Bateson became a countercultural icon in the wild San Francisco of the 1960s and '70s. He developed a theory of schizophrenia but never lost his English accent. True to form, Bateson spent the final years of his life at the famous Esalen Institute, a mecca for hippies and dropouts 45 miles south of Monterey at Big Sur.[44]

Bateson had been at the first Macy conference in 1942, at the age of thirty-eight. The discussions in New York inspired him. He helped organize some of the early meetings. Throughout his life, Bateson felt privileged to have been part of the Macy conferences. "My debt," he recalled in the early seventies, "is evident in everything that I have written since World War II."[45]

Bateson put the finishing touches on his most influential book, *Steps to an Ecology of Mind*, when he was at the Oceanic Institute in Hawaii in 1971, working with dolphins. *Steps*, as Bateson and his fans

affectionately called his book, was an instant success among the countercultural intelligentsia. As the then sixty-seven-year-old Bateson
looked back, two historical events stood out. One was the Treaty of
Versailles, which in his reading culminated in Hiroshima and Nagasaki. The other was the discovery of a new idea: "Now I want to talk
about the other significant historical event which has happened in
my lifetime, approximately in 1946–1947," Bateson wrote. It was the
discovery of cybernetics. "I think that cybernetics is the biggest bite
out of the fruit of the Tree of Knowledge that mankind has taken in
the last 2,000 years."[46]

Overstating the profoundness of Bateson's cybernetic ideas is hard.
In the previous twenty-five years, Bateson believed, extraordinary
advances had been made "in our knowledge of what sort of thing the
environment is, what sort of thing an organism is, and, especially,
what sort of thing a *mind* is."[47]

About twenty years before Bateson wrote these lines, he met Ross
Ashby in New York. He had been intrigued by Ashby's ideas and
how Ashby used the homeostat as an inspiration and illustration, as
well as how he insisted that the weird contraption was alive, that it
had mental characteristics. Bateson later read Ashby's books and was
fascinated by them. Ashby had used the examples of a man with an
artificial arm trying to fix an engine, or a sculptor with a chisel shaping a slab of marble. Bateson took the latter example a big step further.
To do that, he needed a stronger illustration with more symbolism. A
chisel wasn't forceful enough.

"Consider a man felling a tree with an axe," Bateson suggested. When
a lumberjack fells a tree, he does so iteratively. The man wielding the
axe strikes the trunk, again and again. Each time, the man will modify
the next stroke, correcting the angle of the blade, adjusting the force of
the axe. The man will do so in response to the cut face in the tree left
behind by the previous stroke. The tree is part of the process, not external to it. The lumberjack's mental corrective process is possible only

because of the tree. For Bateson, the situation had to be understood in cybernetic terms: "This self-corrective (i.e., mental) process is brought about by a total system: tree-eyes-brain-muscles-axe-stroke-tree."

Already in the early 1950s, Ashby and Wiener had suggested that the line between humans and their tools is arbitrary. The chisel is functionally part of the sculptor. The bomber pilot acts like a servo-mechanism. Man and machine were forming one system. This was Cybernetics 101, the very basics. Ashby had then pointed out that the line between system and environment is arbitrary.

Bateson simply took this idea to its logical conclusion: if the axe was an extension of the man's self, so was the tree, for the man could hardly use the axe without the tree. So it was tree-eyes-brain-muscles-axe-stroke-tree—"and it is this total system that has the characteristics of immanent mind," Bateson wrote in *Steps*. One mind resided not in one person's skull; it resided in the whole system: "Mind is immanent in the larger system—man plus environment."[48]

Bateson knew that this would sound wild to most of his American and European readers, who were so used to understanding the world in individualistic terms, not in such a radically holistic way: "This is *not* how the average Occidental sees the event sequence of tree felling," Bateson hastened to add.[49] This was an understatement that, if anything, betrayed the philosopher's British origins.

But to Bateson, the loop of tree-eyes-brain-muscles-axe-stroke-tree was an elementary cybernetic thought. Any sensible behavior had to represent a "total circuit," a completed feedback loop. Therefore, any unit that displayed trial and error could be called a mental system, he believed. And the tree was the part of that unit that enabled the lumberjack to display trial and error in modifying his strokes. Naturally, the tree was part of the same mental system.

"*Any* ensemble" of objects that had the necessary complexity of causal circuits "will surely show mental characteristics," Bateson generalized.[50] Any such system will be self-corrective and automatically strive "toward homeostatic optima," just as Ashby had predicted. The system itself was a transducer, a sense organ.

The word "governor" was a misnomer, Bateson believed, in both engineering and politics. Talking about a governor assumed that the governor was in charge, governing and controlling the rest of the machine or the rest of a political community. But that wasn't just simplistic; it was wrong. The governor was itself part of a larger entity, part of a larger circuit. "The behavior of the governor is determined," Bateson argued, "by the behavior of the other parts of the system," by the environment itself.[51]

Bateson wrote a cybernetic book, and he had met and known all the founding cyberneticists at the Macy conferences. He was a sounding board for their ideas. He mentioned them in the foreword to his most influential book. Yet he never quoted Wiener. He never quoted Bigelow. He never quoted von Neumann. Instead, he was influenced most by the doctor from Gloucestershire. Perhaps Bateson respected Ashby so much because, like Ashby, he worked in a hospital with mentally troubled patients during his formative cybernetic years, at the Veterans Administration hospital in Palo Alto, from 1949 to 1962.

Bateson referred to Ross Ashby in a curious way. He often harked back to the fellow British cybernetic mind without explicitly referencing Ashby's work, as would be customary in academic writing:

Following Ross Ashby, I assume that any biological system (e.g., the ecological environment, the human civilization, and the system which is to be the combination of these two) is describable in terms of interlinked variables such that for any given variable there is an upper and a lower threshold of tolerance beyond which discomfort, pathology, and ultimately death must occur.[52]

Bateson had heard and discussed these ideas long before, when he had quizzed Ashby on his homeostat at the 1952 conference on Park

Avenue in New York sponsored by the Macy Foundation, cocktails included. But he had not immediately grasped the fundamental, even spiritual, implications of Ashby's experiment.

Now Bateson understood: society was a homeostat. All of Ashby's ingredients were there; the dynamic was that of an "ultrastable system," Bateson explained. The system was "self-corrective." The distinction between organism and environment became blurred. The variables could move only within limits. They were interlinked, like the magnets and troughs in the primitive machine from the English west country. Adaptation happened in response to stress. The objective was conservative: to find a new equilibrium. "Again following Ashby," Bateson observed, the distribution of flexibility among a system's different variables mattered most to achieve that equilibrium. Bateson always recommended the fellow Englishman to his students on a legendary reading list that was posted on the door of his corner office at Kresge College in Santa Cruz.[53]

Bateson took Ashby's idea to its logical conclusion: saying that a computer, or a machine, could be "a mental process" was incorrect, he was sure.

The computer is only an arc of a larger circuit which always includes a man and an environment from which information is received and upon which efferent messages from the computer have effect.[54]

The notion of a circuit was important to Bateson. A circuit, to him, was a circular connection or movement. It was something larger than just a loop of information; it implied a hardwired and system-wide connection among various parts. The circuit was the bridge between the feedback loop and the network. It was this larger system, or ensemble, that showed mental characteristics.

The cybernetic epistemology which I have offered you would suggest a new approach. The individual mind is immanent but not only in the body. It is immanent also in pathways and messages outside the body; and there is a larger Mind of which the individual mind is only a sub-system. This larger Mind is comparable to God and is perhaps what some people mean by "God," but it is still immanent in the total interconnected social system and planetary ecology.[55]

To Bateson and to his many disciples, established views were tilted and biased toward the individual. So deeply entrenched were these established views that even he, Bateson, succumbed to them: "If I am cutting down a tree, I still think 'Gregory Bateson' is cutting down the tree." The self, his own mind, was still an "excessively concrete object to him."[56] That was different from the cybernetic epistemology. It was different from the true, correct view of the mind that cybernetics enabled. But there was a huge difference between glimpsing a new way of thinking in exceptional moments, and making that new way of thinking *habitual*.

Bateson needed to get into the habit. So he explored experiences that could help him imagine what it would be like "to have this habit of correct thought," as he called it. He experimented with psychedelic drugs. "Under LSD, I have experienced, as have many others, the disappearance of the division between self and the music to which I was listening," he said at a lecture in early 1970 in New York City. "The perceiver and the thing perceived become strangely united into a single entity."[57] Psychedelics were not an escape into a chemically created artificial reality. The drugs didn't pull down the curtain on the user; they pulled it up. The chemical substance revealed a more accurate, a more correct, and a more wholesome perspective on the world. Psychedelics *liberated* the viewer from an otherwise artificial reality.

Stewart Brand read *Steps* when it came out in 1972. Brand had just stopped editing the *Whole Earth Catalog* and had shifted his attention to similar publications with a different format. Brand was still looking for the book that would explain it all, the whole system. After reading *Steps*, Brand was intrigued. The book, he recalled, provided "the conceptual bonding of cybernetic whole-systems thinking with religious whole-systems thinking."[58] He decided to meet the author who commanded such spiritual persuasion. Brand arranged for an extended interview with Gregory Bateson himself.

Brand spent several days at the anthropologist's home in Big Sur, overlooking the Pacific shimmering in blue and green and gray and reflecting the unique, bright, yet mild light of northern California. He was intrigued by the way Bateson spoke of "circuits." The term appeared more accurate than "feedback loop," more open-system. The network itself began to shimmer like the sea. What kind of networks, Brand wasn't sure. Certainly not just cold computer networks.

Brand's mind was more wholesome, his eyes hovering along the bright surface of the Pacific. Brand imagined a watched porpoise bedeviling its observer, a chilled body shivering until warm, flesh turning to ashes turning to flesh again, ice ages periodically shaping the ecosystem. He even dwelled on indigenous Iatmul culture in Bali, Indonesia, which Bateson had studied in the late 1930s, where all was "in beautiful cybernetic balance," as Brand wrote in a long and winding story about his formative encounter with Bateson in *Harper's* in 1973.[59]

This circuit was cosmic. "Without circuit, without continual self-corrective adjustment, is no life," Brand wrote, reflecting on Bateson's philosophy and the time he had spent with him at Big Sur.[60] The story turned out awkward, Brand thought when it came out. But the ideas it articulated were profound. "Every part of cybernetics research is jumping with fascinating activity," he wrote a year later.[61]

III

For Brand, there were two "cybernetic frontiers," and they were closely linked. The first was Bateson's holistic philosophy. The second was *Spacewar*, the world's first computer video game, released in 1962. The game was hugely influential among the first game developers, and it quickly spread in university campuses and research labs. *Rolling Stone* magazine commissioned Brand to write a long article, with many screenshots of *Spacewar* and extended sections of dialogue. It was 8:00 p.m. on a clear October night in 1972 when Brand came visiting the engineers-turned-gamers in the moonlit and remote foothills above Palo Alto, in Stanford's Artificial Intelligence Laboratory.

The game involved two players, each in control of a spaceship attempting to destroy the other. The two ships were tiny symbols in the screen's vast black space circling around a small sun in the center of the screen that had its own gravitational pull. The shots were lines of small white dots in the screen's dark space. The exchanges among the gamers, chronicled over pages in *Rolling Stone*, went something like this:

"Where am I? Where am I? *click clickclickclickclick*
"Agh!" *clickclickclick clickclick*
"Glitch." *clickclick*
"Awshit."[62]

The game was highly successful. Like so many other successful games, it was addictive. That addictiveness seemed familiar and appealed to Brand. The players in the AI lab used a PDP-10 main-frame, manufactured by the Digital Equipment Corporation, DEC.

The machine, in a full configuration, with separate memory and disk and printer cabinets, could fill an entire room, weighed up to 6 tons, and cost about $200,000. The PDP-10 was the machine that popularized time-sharing, an innovative way of enabling multiple users to share the processing power of a large mainframe computer, each with his or her own "terminal."

When Brand saw the machine in action, he saw the future. Reliably with nightfall in North America, hundreds of computer technicians gathered to play. As soon as the action started, Brand observed, the technicians were

> effectively out of their bodies, computer-projected onto cathode ray tube display screens, locked in life-or-death space combat for hours at a time, ruining their eyes, numbing their fingers in frenzied mashing of control buttons, joyously slaying their friends and wasting their employers' valuable computer time. Something basic was going on.[63]

This ability to zone out reminded him of his own experiments with psychedelics in the 1960s, the oneness with the environment, and the perceptual intensity during the trip. The players were captivated, for hours, intense, frenzied—a spasmic grip on the keyboard and a trance-like gaze firmly fixed on the tiny, low-resolution screen, according to Brand "the most bzz-bzz-busy scene I've been around since the Merry Prankster Acid Tests"[64]—the legendary wild parties that took place at the end of 1965 and throughout 1966 in the San Francisco Bay Area, to experiment with psychedelic drugs. Ecstatic Merry Prankster partygoers were drenched in black lights against fluorescent paint, whipped up by stroboscopes, and entranced by the Grateful Dead's first performances.

The Whole Earth editor instantly saw the cultural appeal of this machine. "Ready or not, computers are coming to the people" was the first line in his *Rolling Stone* piece, "That's good news, maybe the best since psychedelics."[65]

For Brand, *Spacewar* was "a flawless crystal ball of things to come." Indeed, he saw what most engineers were unable to see. He had the bird's-eye view. The game represented a break with established principles of power and authority. It represented the coming revolution. It wasn't about top-down control, it wasn't about batch processing, it wasn't about sending data to a manufacturer to make production of something more efficient, it wasn't about passive consumerism, and it wasn't about the most efficient use of the machines. The game was the opposite of all that. "Spacewar was heresy," Brand recognized, and the harbinger of more heresy to come.[66]

The game was intensely interactive in real time on a computer. It encouraged the users—who were programmers already—to be creative, to tinker, to make something new. It was about communication between humans. It served a deeply human interest, entertainment. And it "bonded human and machine" through graphical interfaces. The "hackers" enabled and created all this, not the managers and the planners. The game was about liberation, subversion, and expanding the mind. It was the electronic version of Bateson's earthy axe-tree-unity—machine-eyes-brain-muscles-keystroke-machine.

Most members of Brand's generation, he recalled, scorned computers as embodiments of centralized control, as tools of a militarized superpower in a senseless war in Vietnam, wielding computer-controlled nuclear weapons that could end all human civilization. Counterculture had tried and tried various ways to overcome war and capitalism and top-down control. Now, a small group of cultural entrepreneurs and hackers embraced computers as a tool of liberation. These machines, not psychedelic drugs or archetypal geodesic domes, "turned out to be the true royal road to the future," Brand recalled later, in the mid-1990s.[67]

The editor of the *Whole Earth Catalog* knew well that a longer historical view was crucial to seeing the bigger picture. The hackers, in Brand's reading of history, came in three waves. The first wave, in the 1960s and early 1970s, came from universities, and from newly minted computer science departments. He had portrayed these tinkerers in his *Rolling Stone* piece on *Spacewar*. They invented time-sharing, against the interest of large corporations, and gave more people access to SAGE-style supercomputers—in effect, turning mainframes into more widely accessible virtual personal computers.

The second wave of hackers, in the late 1970s, overturned mainframes entirely by bringing the personal computer to market. Many of them were hard-core counterculture types—for instance, Steve Jobs and Steve Wozniak, two cofounders of Apple. They had honed their skills by developing, and then selling, so-called blue boxes, illegal phone phreaking devices to make free calls.

Then came the third wave of "hackers," the social hackers of the early 1980s. The personal computer and emerging network technology didn't articulate an entire philosophy and aesthetic just by themselves. Of course, building software tools to connect and educate communities helped, and the then emerging free-software movement offered a promising platform. But writers, intellectuals, artists, and organizers needed to develop and carry these ideas and bring them to life hand in hand with the technology itself. This third wave of hackers was more social than technical, and it had a profound cultural impact—with counterculture as its stage-one boost phase. Technology on its own would have been too narrow, too geeky, too insular. But when combined with counterculture and paired with punk and added art, computers became cool.

High Frontiers was an alternative magazine that started up in 1984, with on-the-cheap layout, illustrated with clipped black-and-white pictures invoking the visual style of 1950s *Life* magazine, with an added dash of Dadaism. The tagline left no doubt about the irony of the publication's title, *High Frontiers*: "the space age newspaper of psychedelics, science, human potential, irreverence & modern art." R. U. Sirius and Somerset Mau-Mau started and edited the periodical, using these pseudonyms as well as their real names, Ken Goffman and Mark Frost. The first issue was so crude it was really more of a pamphlet than a magazine, retailing for $1 on counters in head shops on San Francisco's Haight Street and elsewhere. The hip magazine took advantage of the raw punk aesthetic then popular in the Bay Area underground scene. It declared itself the "official psychedelic magazine" of the 1984 Summer Olympics held in Los Angeles.

The countercultural psychedelic movement had a lot of respect for technology, especially computers. The first editorial of the very first issue of *High Frontiers* observed an "acceleration of our culture" by technology, such as "computers and robotics." This fast clip of change made new perspectives necessary. Cannabis, peyote, psilocybin mushrooms, ergot of rye, LSD—all these substances, Goffman suggested, would "accelerate our minds and cleanse our spirits" in order to be better equipped for fast technological and cultural change. "We are, after all, arriving at the technology which is indistinguishable from magick predicted by Arthur C. Clarke [*sic*]."[68]

Goffman was all too conscious of the paradox he was navigating toward: the same scientific and technical know-how that "can kill us all on any given afternoon" could also be used to create an age of abundance, leisure, personal growth, and space exploration. But to open the world's eyes to the peaceful use of science and technology, perceptual blinders that were shuttered needed to be removed. Goffman called for "flexibility, optimism, and generosity of spirit to choose planetary transformation over oblivion." And this is where peyote and LSD came in: "Psychedelic drugs, used in specific ways, are powerful tools in helping to remove our perceptional blinders."[69]

Counterculture celebrity Terence McKenna donated some money to lift the first issue off the ground, and his wife Kathleen helped with the artwork. McKenna was also one of the "psychedelic heroes" interviewed in the magazine. He was a prominent pioneer of shamanism, alchemy, and plant-based psychedelics at the time, and author of the best-selling *Psilocybin: Magic Mushroom Grower's Guide*. For McKenna, magic was a tool, triggered by plants or drugs or dance or exercise. He had encountered these "shamanic technologies" during his ethnobotanical expeditions to the Amazon basin, while looking for tribes using the hallucinogen aa-koo-he-hey. Shamanism, in McKenna's view, was just a self-consistent method of describing the world. Science was another such self-consistent method. "Voodoo is another one," McKenna was convinced.

Once the perceptual blinders were gone, the new and true technological state of affairs was revealed. Information technology was growing ever smaller and less obtrusive, McKenna emphasized. Nineteen eighty-four was the year Apple introduced the Macintosh, the first successful computer graphical user interface with a mouse. The new Domain Name System was about to be launched, enabling the .gov, .mil, .edu, .org, and .com domains. The system evolved out of ARPANET, originally to make sending e-mails technically easier. By then, about a hundred multiuser domains—often abbreviated as MUDs, or simply called dungeons—were active online.

The early internet was evolving fast. Yet McKenna was ahead of his time. To him, a new form of planetary connection was emerging: "Through electronic circuitry and the building of a global information-system, we are essentially exteriorizing our nervous system, so that it is becoming a patina or skin around the planet," he told *High Frontiers*. "And phenomena like group drug-taking and rock-and-roll concerts and this sort of thing," he said, "these are simply cultural anticipations of this coming age of electronic-pooling-of-identity."[70]

McKenna had been heavily influenced by Stewart Brand's philosophy, the *Whole Earth Catalog*'s infatuation with Wiener and Ashby, and later even edited one issue of the *Whole Earth Review*. McKenna pointed out that this global condition of "informational oneness" had become possible through the "advent of more advanced cybernetic systems and more advanced psychedelic drugs."[71]

To the Amazon-traveling ethnobotanist, the very technology that began its evolution in air defense research—and was then refined in the Cold War—didn't clash with the wholesome peace and oneness of the psychedelic subculture at all. On the contrary, technology and hallucinogens were two sides of the same coin: "I think every time you take a psychedelic drug you are anticipating and experiencing this future state of electronic and pharmacological connectedness," McKenna suspected in 1984. This was not an eccentric view. It represented an entire subculture. To those steeped in countercultural thought, *High Frontiers* was also about *access to tools*. Some thirty-five years earlier, Norbert Wiener and the early cyberneticists had tried to defend their new science against holism in all its forms, from Freud to Hubbard. Now, *High Frontiers* was taking the bogy of holism to a whole new level.

Timothy Leary, a countercultural guru, perhaps best captures this changed meaning of technology for counterculture. In 1959, Leary took an academic position at Harvard University as a lecturer in clinical psychology. The young tweed-donning professor was a straight arrow and narrow-minded person, by his own description. "I was very much against computers at the time," he said. In 1960, when the air force's SAGE network came online, powerful corporations and government agencies owned and used these expensive machines, not private individuals. "So I had this prejudice that computers were things that stapled you and punched you," Leary recalled.[72]

The military-run, prohibitively expensive, all-controlling IBM supercomputer was the epitome of both big business and big government. IBM was "Big Brother," as Leary saw it. This lingering image is what Apple mocked with such ingenuity in its famous one-minute 1984 Super Bowl ad, directed by Ridley Scott. An athletic blonde woman in T-shirt and shorts is seen charging past storm troopers, right into the heart of power, carrying that most iconic of tools, a sledgehammer. Then she throws the hammer, smashing the oppressor's larger-than-life image: "On January 24th, Apple Computer will introduce Macintosh," the video concludes, "and you'll see why 1984 won't be like '1984.'"[73] The personal computer had become the ultimate power tool of liberation.

Leary purchased his first personal computer in early 1983. "I've learned so much about drugs and the brain in the last six months from working with a personal computer," he told the audience at the Julia Morgan Theatre in Berkeley in July 1983. His nine-year-old son, and his grandchildren, ten and eleven, had tutored him. The most notable feature, for Leary, was that the computer needed to be "activated," as he saw it. "There is code," he said. It was just the same thing with the brain—the "human bio-computer"—and drugs. Drugs were the code that enabled you to boot up the biomachine in your head in novel ways: "You can activate it!"[74]

Several months later, early in 1984, Leary hosted a belated coming-out party for his computer, on Wonderland Park Avenue, at his home in Los Angeles. The gig was very 1980s: a flurry of guests in leather and silk, donning oversize glasses and neon-colored hair, sipped white wine and played with Leary's stretch limousine. His new IBM PC was sitting on a red lacquer picnic table, the guests hovering curiously around the magic machine, half-empty glasses in hand. IBM had launched the model two and a half years earlier, at an introductory price of $1,565. The box came from the future, in cream color.

"Tim, it's *beautiful*," one guest said, admiringly.

"The max, Tim, the max."

"Make it talk."[75]

Just days earlier, on January 24, Steve Jobs had introduced the Macintosh at De Anza College's Flint Center near the company's headquarters in Cupertino to an enthusiastic crowd. Jobs, in a baggy double-breasted suit, had made the "insanely great" Macintosh talk, to the roaring applause of the twenty-six-hundred-strong audience.

Leary knew that the young idealists, the early computer adopters, recognized the significance of freeing their perception from conformism and conservatism.[76] The vehicle for this subversion was merging man and machine. Not by implanting chips or "microsofts" into skull bones, as new science fiction proclaimed, although he wouldn't exclude that as a possibility. Leary wasn't a nerd who thrived in isolation; he was a socialite who thrived on flirtation. "Computers are the psychedelic drugs of the '80s—oh, absolutely," the then sixty-three-year-old told his mesmerized guests, "Like psychedelic drugs, they are mind-expanding."

Leary was a gifted writer, speaker, and provocateur. His vision was both sensual and political. He had recognized the liberating power of the machine just after buying that IBM PC on the red lacquer table. "Personal computers and recreational computers, personal drugs and recreational drugs," he was sure, "are simply two ways in which individuals have learned to take these powers back from the state." To achieve this power grab, Leary teamed up with XOR, a Minnesota-based software company then known for its games. LSD, *PC Magazine* joked in an article about Leary's wild party, would now stand for "Leary's Software Development."[77] But Leary wasn't nearly as good at organizing and software engineering. His plans to develop what he then called artificial intelligence with XOR didn't go anywhere.

But it was 1984—the year that wasn't supposed to be like *1984*, as Apple had put it in its Orwell-evoking ad. The Macintosh was here. William Gibson published *Neuromancer*, and an era-defining subculture emerged. Yet something was missing.

IV

It was again Stewart Brand, of Whole Earth fame, who turned a cybernetic vision into reality. A little less than a year after Leary's party, on a day in the late fall of 1984, Brand had a fateful lunch with Larry Brilliant, in a restaurant in La Jolla. Both men were in town for a conference of the Western Behavioral Sciences Institute. Brilliant, a doctor, was a roly-poly man with a goatee and a lot of excess energy. He ran Network Technologies International, a company in Ann Arbor, Michigan, that offered newly designed conferencing software. And he needed a group of people who could bring his new product to life and help him showcase and market its potential. So he pitched an idea to Brand over lunch: take the *Whole Earth Catalog* online, along with its entire hippie community.

Brand was torn. He didn't want to take the old catalog online. He had moved on. He wanted something fresh. He wanted to reach a larger audience and bring in hackers, activists, intellectuals, and journalists. This new community could be really special; ARPANET had been closed off from the public, accessible only to researchers. The nascent internet was growing fast. By 1984, the number of servers connected to "the Net" had exceeded a thousand. The brand-new Domain Name System was introduced that October to simplify its setup. Yet so-called bulletin boards were still geeky hangouts for lone nerds and hacker types, not the online equivalent of a coffeehouse or a countercultural commune where it was fun to hang out and to meet new people. The time had come for something bigger, something more inclusive.

But Brilliant also had a point. The Whole Earth approach had already proved itself in paper form, in slow motion, with people writing in by mail and e-mail, and then waiting for the supplement to ship back to the land. Brand agreed to the deal. Brilliant's Ann Arbor company backed the agreement with what was then a significant investment: $150,000 for a VAX computer, a dishwasher-sized mainframe manufactured by DEC, a rack of half a dozen modems, and initially six telephone lines, plus another $100,000 for the primitive conferencing software, a Unix-based platform called PicoSpan.

Brand needed a name. Why not at least re-create the Whole Earth spirit in name, he thought. After some doodling, he found a quirky acronym so forced that it was also self-deprecating: WELL, which stood for Whole Earth 'Lectronic Link. In the spring of 1985, the WELL's hardware arrived in the ramshackle Sausalito offices of the Whole Earth, tucked between houseboats off the pier. The community platform opened on April 1, 1985.

The WELL's newly hired director was Matthew McClure. McClure had worked as a typesetter for the *Whole Earth Catalog* until 1971. He then spent twelve years on the Farm, a large hippie commune in Lewis County, Tennessee. Brand hired McClure again in 1984 when he returned to the Bay Area. McClure was the ideal man for the WELL: a tech-savvy commune dweller of Whole Earth extraction. This time the two men wanted to make the online commune as accessible as possible to a large number of communards, practically to anyone with a computer and a modem. They pushed the monthly usage fee down to $8, plus $2 per hour. The lower the entry barrier, the better. The user dialed the number with an old-fashioned modem, machine-screaming a whining song of acoustic numbers into the phone. There were no videos, no pictures, no sound. Everything was text, and command based. Thus, denizens of the WELL had to learn to use a clunky system that often broke down. At first the WELL remained a curious and somewhat freakish phenomenon, experienced in low resolution, in rounded, thick glass, and visibly flickering.

Brand advertised the new platform a few times in the *Whole Earth Review*. His instinct was to offer free accounts to reporters, to spread the word faster. By 1986, the online community had grown to about five hundred.[78] Six years later, the number was about six thousand. The WELL was perhaps the first proper online social network with general appeal. It had all the upsides and downsides that come with social media: it was addictive, it was entertaining, it was riveting, and it could be a waste of time. There were great discussions, and there was spamming and trolling. A minority did the majority of the talking.

The social-media pioneers had enough experience to intuitively get the details right. Real names mattered. So, everyone's real name was available on the system, in "finger files." Any user could simply finger another user's screen name—say "hlr" or "rabar"—and pull up the user's real identity. It also mattered that users, not the WELL, took responsibility for what they said. So, Brand and McClure came up with "YOYOW": "You own your own words." The motto greeted every user on the log-in screen. Nobody could reproduce what somebody else had written.

These rules promoted civility and a high quality of discussion. And the initially small number of regionally clustered users meant that the WELL was indeed different from later and larger social networks in one critical aspect: there was an expectation that you would, sooner or later, meet the others in person, face-to-face, perhaps at one of the regular parties at the home of the WELL's mainframe and its administrators in Sausalito, a pleasant half-hour drive across the Golden Gate Bridge from downtown San Francisco. All this made for a real small-town community feel online.

The discussions were organized in so-called "conferences." These threads covered a potpourri of topics: environment, future, books, drugs, sexuality, or "best and worst memories of the sixties." The mideighties version of emoticons was typing "<smiles>" or "<hugs>" in pointy brackets, coding-style. All this appealed to baby boomers, who had come of age in the 1960s and were now in their late thirties and early forties—bright, libertarian, male, with postgraduate degrees. Their demographics and common interests alone made this online community a unique cultural phenomenon, in a "self-absorbed, cabalistic way," observed journalist Katie Hafner in a *Wired* article about the WELL that has become a classic.[79]

One enthusiastic WELL member was Ramon Sender Barayón, a.k.a. "rabar," a San Francisco artist and writer. With the Merry Pranksters, Sender had coproduced the now legendary 1966 Trips Festival, a formative event for the hippie movement. Initially, he had difficulty with PicoSpan, the clunky conferencing software, "but then I felt the energies on the WELL," he recalled.

> It reminded me of the Open Land communes I'd been to in the 1960s. The tribal need is one our culture doesn't recognize; capitalism wants each of us to live in our own little cubicle, consuming as much as possible. The WELL took that need and said, "Hey, let's see what happens if we become a disembodied tribe."[80]

Yet the tribe wasn't completely disembodied. The earliest hippies, like Brand and Sender, understood that the online communication worked so well only because of the face-to-face contact between members of the community. That's why the regular Sausalito meetings were so important, as some more distant members discovered. Jon Lebkowsky started dialing in from Austin, Texas, in 1990. In terms of interests, personality, and culture, he was an ideal fit, and he would later become a well-known activist. But he felt that his posts were ignored until he made the long trip to the Bay Area to attend a WELL party.

The very idea of separating online and offline didn't square with the cybernetic minds of the Whole Earth pioneers. It was like isolating one of Ashby's units of the homeostat, like cutting off Wiener's negative feedback loops, like taking Bateson's axe away from the man felling the tree—the antithesis of balance and whole systems. From the get-go, Brand wanted the Whole Earth network to be self-governing; he wanted the system to be a sociotechnical homeostat, a collective thinking machine and communal learning mechanism.

The experiment succeeded far beyond his expectations. The WELL brought Bateson's metaphorical "circuit" in line with the technical circuits on the mainframe, one powering the other. The line between system and environment, or between offline and online, indeed became arbitrary. The WELL literally linked the "two cybernetic frontiers," Brand had written about ten years earlier in *Rolling Stone* and *Harper's*: the mind and the computer.

To the hippies at the congenial monthly Sausalito parties, the VAX mainframe was the long-awaited machine of loving grace (it had to be!).[81] Cliff Figallo, another former communard from the Farm in Tennessee, and the WELL's nineteenth user, lovingly built the cabinet for the VAX. The WELL's cybernetic origin in Norbert Wiener's air defense research wasn't just palpable in Brand's philosophy, his approach to community organizing, and decades of Whole Earth work. The same type of dishwasher-sized DEC mainframe that powered the escapist hippie machine in Sausalito also powered the Minuteman ICBMs slumbering in the nation's atomic missile silos, ready to wipe out the Soviet Union at the push of a button.[82]

So, it was not a coincidence that the idea of cybernetic space was first brought to life on the WELL's Sausalito server.

6. SPACE

ON MARCH 27, 1984, *CBS EVENING NEWS* REPORTED ON A revolutionary new display technology invented by the US Air Force. An engineer in aviator glasses and a brown leather jacket stood in front of the camera and explained to mesmerized Americans how pilots could now fly in "virtual space," to video game–like images of jets racing along a grid-like electronic globe. Indeed, air force engineers were the first to articulate the idea of virtual space in the early 1970s, and then to realize it in a range of prototypes over an entire decade. This pathbreaking military research became public that March.

It was a potent cybernetic idea: a separate, virtual, computer-generated space distinct from real physical space—what later became known as "cyberspace." For decades, visual artists in advertising and film had articulated and imagined the space inside electronic apparatuses akin to outer space, portraying the atom as a solar system, with electrons orbiting like planets, the whole Earth encapsulated in a lightbulb, or tiny humans working inside giant vacuum tubes.[1] By the early 1980s, the idea had found its moment. The space inside the machine inspired and mesmerized the counter-cultural avant-garde, which already sensed that the technical ground was shifting. The 1980s became a decade of imagination, of "consensual hallucination," in the immortal words of William Gibson.

By the end of the decade, virtual-reality technologies were all the rage: data gloves, data goggles, and data suits triggered eccentric visions of how humans would interact through networked computers and wearable interfaces. The machines of the near future, many readers of the *Whole Earth Review* came to believe by the late eighties, would enable intense immersive experiences that would rival and surpass the most intense emotional experiences available to date: sex, music, and drugs. An entire subculture—cyberpunk—emerged at the curious intersection where technology and networked machines met mind expansion, psychedelics, music, and fashion. By 1988, entrepreneurs and intellectuals, inspired by cybernetics, control, man-machine interfaces, and whole systems, had dubbed this new place "cyberspace."

At first, this novel space was something that could be entered only with fancy hardware interfaces, with goggles and gloves and data suits. It was synonymous with virtual reality. Then, sometime in the spring of 1990, a curious flip happened. Cyberspace became almost synonymous with the entire internet. By the early 1990s, virtual space—discovered by the air force and coveted by hippies—became something that could be accessed with more affordable hardware by millions of legitimate operators; a personal computer and a modem were enough to begin the fantastic voyage to the new frontier. Cyberspace was on its gleaming path to becoming a mythical new realm of freedom and liberty—and of war.

The US Air Force was flying and fighting in cyberspace before Gibson had even coined the term. The same problem that had inspired Norbert

Wiener during the Blitz in World War II had kept air force engineers busy throughout the fifties and sixties: human-machine interaction in the cockpit under stress. The result was the air force's invention of "virtual space" in the early 1970s.[2]

The trigger was Vietnam. The F-4 Phantom and the F-105 aircraft were used extensively in the war. But these ageing planes were at the end of their life cycles. They had limited cockpit space for updating and modernizing the displays.[3] This constraint prompted early innovative work on display technologies. By the early 1970s, McDonnell Douglas was developing the F-15 Eagle, and General Dynamics was working on the F-16.

But instead of improving the cockpit design, these fourth-generation fighter-bombers made the problem worse. An F-15 cockpit at night was a sea of lights and instruments: three hundred switches and seventy-five displays, including the so-called vertical situation display, the horizontal situation display, standby instruments, and radar. The control stick alone had eleven switches, and there were nine more on the throttle. All this was connected to about fifty computers providing stability augmentation, air data on speed and altitude and temperature, propulsion control, electronic warfare systems, various weapons, and so on. By interacting with these computers, an F-15 pilot effectively controlled an energy delivery system hurtling through space beyond the speed of sound.

The complexity was overwhelming. A state-of-the-art cockpit displayed more data to the pilot than a human was able to process, and it did so on tiny displays. Operators complained that their brains would "ooze out of their fingertips."[4] Another problem was the high g-forces during ever-faster high-speed maneuvers. These forces made precise free arm movements difficult. As a result, tracing targets and aiming weapon systems became harder.

Human-machine interaction was ripe for revolution. The air force's Harry G. Armstrong Aerospace Medical Research Laboratory decided to take the idea of the cockpit back to the drawing board, to design the "ideal cockpit," as one of the lead engineers recalled. The question was straightforward, if daunting: "How could we input information to that crewmember"—the pilot—"so he could quickly make decisions?"[5] The air force engineers already had a guiding vision: "The key to solving these problems is to make the interfaces to the machine more 'human-like' rather than requiring the human to be more 'machine-like,'" they understood.[6]

One of the key officers on the team at Wright-Patterson Air Force Base was a freshly commissioned second lieutenant, Thomas Furness. Furness was well qualified, with a degree in electrical engineering from Duke University. In the late 1960s, the lab started off working on helmet-mounted sights, to guide a plane's weapon systems simply by pointing the head at the adversary. The next task was a workable night-flight display that was directly in front of the pilot's face, not down in the crowded cockpit display area. Furness's lab was entirely unaware of earlier, much more primitive work on three-dimensional displays.[7] Flying by a tiny 4-inch display down in the cockpit at arm's length of the pilot's face was clumsy; pilots don't want to "go into the cockpit," in aviator lingo. Head-coupled night vision would be a huge improvement, with an enhanced image projected into the transparent glass visor of a helmet.[8]

Over the next months and years, the air force lab made impressive progress with helmet-mounted sights. By 1976, Furness and his team had come up with the design plan for a "visually coupled airborne systems simulator," or VCASS. Their vision was "visually coupling" man and machine, as the project's lead engineer told a DARPA conference on biocybernetic applications for military systems in Chicago in April 1978.[9]

Inspired by cybernetics, researchers in the Wright-Patterson lab had even more ambitious plans of "deeply coupling" pilot and plane. Their vision was to "input" information into the pilot's brain through the VCASS display—visually, acoustically, and through touch—and then to "output" information from the pilot's cortex back into the computer, by measuring neurologically evoked nerve potentials in the brain through magnetoencephalography. All that, the air force hoped, would work during high-stress air-to-air combat. Such direct "biofeedback" meant that flying machines "would be able to provide superhuman capability."[10] The goal was winning in air-to-air dogfights by simply thinking about maneuvers and firing missiles.

A prototype of the lab's complex helmet display became operational in 1981. The engineers later called it the super cockpit. The visually coupled VCASS helmet was the most sophisticated virtual-reality system ever built. Although it had "simulation" in the name, it was designed to become an in-flight system, to control an actual fighter plane against actual adversaries.

But it looked ugly and clunky. When the first test pilots from Edwards Air Force Base in California came over to the lab in Dayton, Ohio, they were shocked. "You got to be kidding me," one said, incredulously. "Come on, Furness, what is this all about?"[11] One TV report later called the prototype helmet a "ridiculous contraption," and journalists regularly made fun of the design.[12] The prototype was so unwieldy that it needed to be lowered over the pilot's head from the ceiling above the test cockpit.

Thankfully, science fiction came to the rescue. In May 1977, George Lucas's film *Star Wars* was released. Not only did the final showdown feature the hero Luke Skywalker using a targeting computer with helmet-mounted sights attacking the Death Star, but the helmet of the villain, Darth Vader, eerily resembled the air force's own design—so much so that the engineers themselves immediately made the connection: "The VCASS looked like a Darth Vader helmet," Furness recalled.[13]

Other officers, and later the media, loved making the comparison. Oversize helmets with eerie eyes suddenly became cool. But the air force helmet didn't just look like science fiction. The engineers had crammed the system with cutting-edge technologies that were far ahead of anything the private sector had to offer at the time—even Luke Skywalker's computer-controlled sight looked outdated in comparison to the air force's system; the young hero had to pull the clunky thing up from behind his seat as he was trying to obliterate the Death Star.

The lab at Wright-Patterson had spent about $40 million developing the system over the years, with a team of more than a hundred people working on the helmet and its accessories.[14] The display had a high-resolution, high-luminance image source that would paint a panoramic picture in whichever direction the pilot looked, independent of the aircraft's boresight, even when obstructed by the plane's fuselage.[15]

The helmet provided a stereographic view, with one display for the left eye and one for the right, spanning a range of 120 degrees by 60 degrees. It had a partially silvered mirror, took advantage of holographic lenses, and filtered ambient light. The lab had even manufactured 3-D earprints, small models of digital human ears. The engineers had understood that the shape of the ear itself enabled the location of the sound in space. And they needed ears to design truly 3-D sound systems, not just simply stereophonic sound. The system tracked hand movement and even provided tactile touch feedback to the operators. Tiny pneumatic cushions sewn into the gloves meant the pilots could feel that a virtual switch had actually been pushed.[16] The machine was powered by eight room-sized VAX computers and water-cooled electronics.

Once geared up in the oversize prototype helmet, the pilot was able to interact with the aircraft in four different ways: by simply turning the head to look at a target, prompting the plane to aim its weapons accordingly; by voice control, speaking commands into a microphone (such as "Select," "Lock on," "Zoom"); by a touch-sensitive panel, calling up virtual switch panels; and by moving gloved hands, with a magnetic tracker that sensed hand position and orientation.[17]

A fifth control interface was possible but not yet installed in the simulator: control by eye movement, with an eye position tracking system incorporated into the headgear, measuring the orientation of the eye relative to the helmet. Humans usually turn their eyes to orient themselves, not the entire head, so eye tracking was a logical next step. The idea was that the pilot would be able to simply look at switches to toggle them on or off.

Once the helmet was lowered over a pilot's head, Furness and his team would switch on the imaging processor and a crisp, wide virtual image would appear in front of the pilot. The air force didn't want to present a highly complex picture to its aircrews. Complexity had been the problem in the first place. The idea was the opposite: to avoid a highly complex picture. "We fused the information from all these sensors into a cartoon," said Furness.[18]

The simplified map was in one-to-one registration with the real world. The pilot was able to tell how high he was above the ground, which was shown as a rough grid, just by seeing the virtual picture flow by, at low altitude and high speed. A pathway in the sky was automatically computed to avoid danger zones and displayed in the same panorama. Radiating surface-to-air missile batteries were shown as red cones of lethality, like vast and dangerous virtual cathedrals reaching into the sky; nonradiating batteries were shown as simple yellow boxes on the ground. The system displayed friendly airplanes in the sky as white planes, and any potential enemy as a small red plane in a yellow cube. The pilot could see everything at once, in one glance.

The skeptical experimental pilots, with proud "X" badges on the chests of their jumpsuits, were mesmerized. The system was "painting radar into the sky" for them, along with energy management information. They didn't have to go into the cockpit, but could keep their eyes on the adversary instead—eyes, and ears: "They'd be flying along; the test-pilots would hear the enemy coming in from the back," Furness said, because the system was programmed to use sound to indicate the enemy's position to the pilot. Being under the helmet was close enough to the real thing: "You'd hear a few swear words coming out," Furness recalled. After flying and fighting in early 1980s virtual space, the test pilots stepped outside the cockpit with sweat rolling down their faces. They were impressed. "This is fantastic," they would say. "When can we have it? It is the only way to go!"[19]

"The super cockpit was a cockpit that you wear," the program's chief engineer explained later. "You put on a magic flight suit, you put on a magic helmet, you put on magic gloves. You plug into the computer, and you create this panorama of three-dimensional information that you see, hear, touch."[20]

The Darth Vader helmet was a revolution—not just practically, but conceptually. Air force designers slowly came to understand the significance of computer-generated space. "The notion of virtual space really evolved in the early seventies for us, in my lab," recalled Furness, the lead engineer.[21] The air force already understood that to surround operators with three-dimensional information was not sufficient. Just watching wasn't good enough. Pilots needed to be able to interact with the displayed information and manipulate it—say, by flicking a virtual switch. This interaction is what created the perception of being in a separate space.

The engineers had long anticipated virtual space. On May 17 and 18, 1977, one week before *Star Wars* opened in theaters, Dean Kocian, one of the Wright-Patterson lab's lead engineers, had gone down to Phoenix, Arizona, to a conference at Williams Air Force Base. Kocian predicted that their project would enable his team to create "virtual space." State-of-the-art computer images could create synthesized hemispherical visual spaces. This approach was cost-effective and flexible. Once the hardware was created, the software designers would have full creative freedom: "Visual display configurations in virtual space can be developed and altered by simply changing the related software," Kocian told an audience of engineers and air force personnel in Arizona.[22]

Developing the display presented many problems. Some of them were unexpected. Already in 1978, the Wright-Patterson team reported discovering what it called "display fascination" to a DARPA conference on biocybernetics. Extensive testing and a body of anecdotal evidence showed that "crew members often become enthralled or 'drawn into' their display," so that it becomes difficult for them to interrupt or change the focus of their attention.[23] The lure of the display could potentially present problems during operations. The air force was worried that it took test pilots consistently longer to redirect their attention from the display to the real world than from the real world back to the display. It was as if the operators would default into the machine.

Despite these early clues, the air force engineers initially didn't realize the full power of what they were developing. That conception changed dramatically in 1981, when the full prototype system with all components was ready for testing in the lab—by the engineers themselves, not yet by test pilots. "When we turned the lights on the first time, it was like somebody reached out of that computer and pulled us inside. Now, we were not looking at a picture anymore, we were in a place, a place that was generated by that machine,"[24] said Furness, recalling the reaction in the lab at Wright-Patterson. "It was like you left your seat and went to another place."[25]

The air force leadership began to recognize the technology's potential, including its public affairs potential. Sometime in early 1984, Furness got a call from the Pentagon. A high-ranking officer asked whether the lab at Wright-Patterson could put together a news release. CBS Evening News with Dan Rather called first to schedule a slot. David Martin, a CBS reporter, spent an entire day on the base for filming. On the evening of March 27, 1984, Americans saw test pilots flying in virtual space in what the reporter called a "Darth Vader helmet."

Furness, now a civilian but still in aviator leather jacket, told CBS how the pilots were flying "in what we call virtual space." Lieutenant Colonel Arthur Bianco, an F-16 pilot and program manager, had an even better description: "The simplest way to think about it, conceptually, is *Star Wars* and R2-D2," Bianco said, against CBS's background displaying the final battle scenes of George Lucas's science fiction film, with Luke Skywalker flying his four-winged attack glider against the Death Star, the iconic R2-D2 robot as his copilot, and the Empire's attack gliders screaming by. "We're taking the first steps on a very long path of getting a real R2-D2 in our fighters," the F-16 pilot said.[26]

McDonnell Douglas and Kaiser Electronics had already started developing an operational prototype, in cooperation with Furness's lab in Ohio. The helmet was called the "Agile Eye." And to improve the clumsy original VCASS design, the air force engineers turned to Lucasfilm, among others. "We actually used the same industrial designers that had designed Darth Vader's helmet to design the Agile Eye," Furness said.[27]

The group had nearly a hundred designs. To select the best helmet, they took the designs to a wind tunnel. Aerodynamic properties were critical. During an emergency ejection at a speed of about 560 miles per hour, the lift forces on the pilot's neck can be as high as 500 pounds, enough to kill him. The shape that the Lucasfilm designers came up with, by accident, acted like a spoiler with superb aerodynamic properties, reducing the lift in case of an emergency ejection by half. "Uh, this is really cool," the engineers said. By early 1987, the Agile Eye was ready for testing and soon thereafter for operations. It doubled the kill ratio of pilots during air-to-air combat training.[28]

Meanwhile, the CBS story triggered an avalanche of news coverage, and the coverage, in turn, sparked general interest in virtual-display technology. For a few more years, the lab at Wright-

Patterson Air Force Base was the sole pioneer pushing out this technical frontier. But the huge public interest from journalists, from academics in other fields, and from the general public led to a change of mind-set.

Two of the engineers from Wright-Patterson took part in an aerospace conference on simulation in January 1986 in San Diego. It was a civilian conference, not held at a military base. Furness and Kocian gave a talk titled "Putting Humans into Virtual Space." They summarized their research and presented the virtual cockpit. They explained how the display could be stabilized against different features of the environment to give the pilot a steady focus: against the pilot's head for selecting virtual switches without shaking; against the cockpit to render instruments; against Earth to navigate way points; or against another plane for aiming fire.

But soon the air force team had applications in mind that went beyond dogfights and bombing runs. "Using this system," two of the military engineers told the audience in San Diego, "the operator becomes part of a 'designer's world' created in virtual space."[29] They began to understand that there was no reason why this should be limited to an F-15 Eagle. The operator could be a welder, with the mask displaying temperature and gas mixture in real time; or a shopper trying on virtual clothes; or a surgeon traveling into the virtual space that is the patient. Furness spelled out this vision for *Popular Mechanics* in 1986:

As he [the surgeon] makes his incredible journey inside this human being, he sees a whole new world from inside the blood vessel. He "pilots" the catheter probe, navigating toward the heart, while hearing the gurgle of blood around a defective heart valve. As he approaches the heart valve, he reaches out with his hand to remotely control a miniature suturing machine which corrects that valve malfunction.[30]

The story was illustrated with an image of a doctor in a white lab coat, sticking a green-glowing catheter into a patient's chest while wearing the Darth Vader helmet.

By now, Furness foresaw that developing the hardware would be straightforward. "We know what to do," he said. "But the same cannot be said of the mindware."[31] The air force engineers felt that humans and machines were getting so closely coupled that "software" wasn't an appropriate term anymore, so they suggested "mindware" instead. Either way, developing the correct code would be the real challenge: "It is the mindware which provides the virtual workstation or super cockpit environment for the pilot," Furness told a symposium in May 1988, in his thick southern accent.[32] A few months later he left the air force to focus on nonmilitary uses of what would soon be called "cyberspace" (although nobody had yet used the term outside of Gibson's science fiction stories).

While the air force's Wright-Patterson lab was busy playing with virtual space, the idea began to capture the imagination of science fiction writers. Perhaps the single most influential book on the imaginary space inside the machines isn't Gibson's first and famous novel, *Neuromancer*, but rather Vernor Vinge's novella *True Names*,[33] published in 1981, the same year the VCASS was switched on in secret. Vinge is often credited with being the first to articulate the then futuristic vision of computer-generated parallel worlds as new domains of human interaction.[34] The author was well placed to be a pioneer: he then worked as a professor in computer science and mathematics at San Diego State University.

True Names plays out in an imagined future. It tells the story of a group of computer hackers who lead a double life—one in the "real world" and one on the "Other Plane," a virtual world inside computers, processors, and switches. Roger Pollack, Vinge's main protagonist, would enter this "data space" through so-called Portals. These entry points were installed in the protagonists' homes, hooked up to a network operated by familiar-sounding service providers such as Bell, Boeing, or Nippon Electric. Vinge's description of Pollack's delicate ascent to the processor-generated Other Plane has become an iconic image:

> Then [Pollack] sat down before his equipment and prepared to ascend to the Other Plane.... He powered up his processors, settled back in his favorite chair, and carefully attached the Portal's five sucker electrodes to his scalp.... For long minutes nothing happened: a certain amount of self-denial—or at least self-hypnosis—was necessary to make the ascent. Some experts recommended drugs or sensory isolation to heighten the user's sensitivity to the faint, ambiguous signals that could be read from the Portal.[35]

Vinge's vision was not unlike that of the Wright-Patterson engineers: man and machine were deeply coupled, through the brain's electrical signals, linked by sucker electrodes. While their real bodies were left sitting in the real world, in their favorite chair, "users" made the mental ascent, their representations finding themselves in a bizarre, *Alice in Wonderland*–style world of talking frogs, magma moats, and icy mountains. Vinge's protagonists would wear asbestos T-shirts and use fantasy pseudonyms, like Erythrina or Don.Mac, not their true names. Those true names remained secret on the Other Plane. Pollack's nickname was Mr. Slippery.

Vinge's story played with the relationship between the real world and the virtual:

Its moats and walls were part of that logical structure, and though they had no physical reality outside of the varying potentials in whatever processors were running the program, they were proof against the movement of the equally "unreal" perceptions of the inhabitants of the plane. Erythrina and Mr. Slippery could have escaped the deep room simply by falling back into the real world, but in doing so, they would have left a chain of unclosed processor links.[36]

In Vinge's story, the various landscapes and features of the Other Plane depicted real-world phenomena in a dreamlike way. A swamp, for instance, "represented commercial and government data space," while a 2,000-ton satellite in static orbit over the Indian Ocean, which created a 900-millisecond time delay in communication, was "represented as a five-meter wide ledge" near the top of a mountain rising from those swamps. That ledge was a safe meeting space to conspire against an all-controlling government. *True Names* articulated one of cyberpunk's dominant themes: escaping into computer networks. "Some experts" in Vinge's story even "recommended drugs" to smooth the entry into the portal. Yet cyberpunk's postapocalyptic aesthetic was largely absent.

In the heady days of the early 1980s, writers did not have to look hard for inspiration. Counterculture was ripe with excitement about the rise of computers, yet cowed by the omnipresent and very real possibility of nuclear annihilation. In November 1983, a NATO crisis simulation exercise dubbed "Able Archer" was nearly misread by the Soviet Union as an impending nuclear attack, marking perhaps the single most dangerous moment of the Cold War.[37]

The tension between hippie utopia and nuclear dystopia was palpable.[38] It expressed itself in an escapist counterculture characterized by an intoxicating blend of punk, futurism, surreal collages, psychedelic visual art, a virtual-reality technology fetish, industrial and electronic music, and drugs. Cyberpunk had its own distinct fashion, with ponytails on men, tattoos on women, all-black leather jackets with mirrored shades on everybody, and hats brimmed with irony. The subgenre's symbolic father figure was William Gibson, an American-Canadian novelist with more links to counterculture than to computer science.

Gibson explained his muse in an interview with the *Whole Earth Review*. He wanted to write a novel, so he was looking for a place to set the story. The science fiction of Gibson's own childhood was space travel. The vehicle was the rocket ship. But outer space didn't resonate with Gibson. Inspiration came to him one day as he walked past the windows of a video arcade. Gibson recounted watching the kids play with the primitive purpose-built gaming machines, in rows, emitting bleeping sounds and flashing lights, like slot machines in Vegas:

> I could see the physical intensity of their postures, how rapt these kids were.... You had this feedback loop, with photons coming off the screen into the kids' eyes, the neurons moving through their bodies, electrons moving through the computer. And these kids clearly believed in the space these games projected.[39]

In Gibson's mind, a new level of human-machine interaction had been reached. "The body language of just like intense longing and concentration," he thought. "It felt to me like they wanted to, like, go right through the glass at the back of the machine. They wanted to be inside there with—the pong or whatever."[40]

Now he had a space: inside the machines. Gibson just needed a name. He tried different ones. "Dataspace" didn't work. "Infospace" didn't work. "But cyberspace!" Gibson wrote the word down on a notepad. It sounded like it meant something, or as if it might mean something, he thought. But then maybe not. "As I stared at it in red

Sharpie on a yellow legal pad," the author recalled, "my whole delight was that it meant absolutely nothing."[41]

It was exactly what a science fiction author wanted, a hot but meaningless idea. Gibson was able to charge it with meaning, to specify the rules for the arena. He had derived it from cybernetics—another word that to him sounded evocative, spiritual, computer-related, deep and dangerous. Gibson first used the word "cyberspace" in passing in his 1982 science fiction story "Burning Chrome," published in one of the hottest literary and cultural magazines of its day, *Omni*. In the story there is a "chic bar for computer cowboys, rustlers, cybernetic second-story men," and Gibson mentions "cybernetic virus analogs, self-replicating and voracious."[42]

Two years later, in *Neuromancer*, Gibson introduced the new space inside the machines in the trademark language that made his work so popular. The segment describes the fictional thoughts of Henry Dorsett Case, a low-level drug dealer in the dystopian underworld of Chiba City, Japan:

> A year here and he [Case] still dreamed of cyberspace, hope fading nightly. All the speed he took, all the turns he'd taken and the corners he'd cut in Night City, and still he'd see the matrix in his sleep, bright lattices of logic unfolding across the colorless void. ... The Sprawl was a long strange way home over the Pacific now, and he was no console man, no cyberspace cowboy. Just another hustler, trying to make it through.[43]

"Cyberspace," for Gibson, was meant to evoke a virtual, disembodied world of computer networks that users would be able to "jack into" through consoles and portals. By far the most quoted paragraph comes later in the book. These few lines of science fiction have become the canonical description of cyberspace, to be repeated many times in countless scholarly and military publications to come:

Cyberspace. A consensual hallucination experienced daily by billions of legitimate operators, in every nation, by children being taught mathematical concepts.... A graphic representation of data abstracted from the banks of every computer in the human system. Unthinkable complexity. Lines of light ranged in the nonspace of the mind, clusters and constellations of data. Like city lights, receding.[44]

Gibson's novels had their own aesthetic. People lived in sprawling cityscapes, crammed and gritty and dark. Washed-up computer cowboys and hustlers with alien tattoos, shades, and neural implants that were blurring the line between human body and machine part. The stories had an apocalyptic feel, like Ridley Scott's 1982 cult film *Blade Runner* or *The Matrix*, a 1999 film directed by the Wachowskis.

Gibson romanticized the technology. When he shaped the language and the aesthetic of cyberpunk, he didn't even know that hard drives had spinning disks. "Fortunately I knew absolutely nothing about computers," he recalled.[45] Until late in 1985, the fêted science fiction author and creator of cyberspace didn't even own a personal computer. And people talking about computers bored him. For Christmas that year, Gibson finally bought an Apple II at a discount. The machine's successor model, the Macintosh, had been launched so effectively nearly one year earlier with the legendary cyberpunk ad "1984," but the older Apple II was still a best-selling device.

When Gibson booted up the machine at home and got ready to use it, he was shocked by the computer's mundane mechanical makeup. "Here I'd been expecting some exotic crystalline thing, a cyberspace deck or something, and what I'd gotten was something with this tiny piece of a Victorian engine in it, like an old record player."[46] The science fiction writer called up the store to complain. What was making this noise? The operator told him it was normal; the hard drive was simply spinning in the box that was the Apple II. Gibson's ignorance about computers, he recounted, had allowed him to romanticize technology. And romanticize he did:

> She slid the trodes on over the orange silk headscarf and smoothed the contacts against her forehead.
>
> "Let's go," she said.
>
> Now and ever was, fast forward, Jammer's deck jacked up so high above the neon hotcores, a topography of data he didn't know. Big stuff, mountain-high, sharp and corporate in the non-place that was cyberspace.[47]

Those two science fiction visions of computer-generated spaces would appeal to different yet overlapping communities: Gibson's *Neuromancer* appealed to a wider and, in the short term, more influential community passionate about counterculture, aesthetics, virtual reality, and drugs. Vinge's *True Names* appealed to a narrower group that became influential only in the long term: those passionate about engineering, gaming, encryption, and privacy.

But for now there was a problem. The air force had developed the hardware in secret. Vinge and Gibson had developed the vision in novels without even knowing of the air force's first steps in virtual space. Vision and prototype needed to be connected.

Jaron Lanier embodied what the *Whole Earth Catalog* stood for: offbeat, dreadlocked, bohemian, raised under a geodesic dome in Mesilla, New Mexico. Lanier went from performing on the streets of Santa Cruz to writing software for Atari, an arcade game company. At Atari, Lanier had created *Moondust*, a primitive art-music game. The game confused many players because it was so different, not a first-person shooter but peaceful, "trippy," as one called it.[48] In 1984, the year *Neuromancer* was published, Atari's business started to sour and Lanier lost his job.

But Lanier had already started working on a "postsymbolic" visual programming language. Programming by coding seemed archaic and unnecessarily complicated and exclusive to young Lanier. His vision was manipulating objects in three-dimensional space. He had been working on a visual programming language that he called Mandala. Then the popular magazine *Scientific American* devoted its entire September 1984 issue to software. The editors had heard of Lanier's project and chose one of his visual programming experiments as a cover illustration. It showed a kangaroo, an ice cube, a score, colored swallows, and a trumpet. One day in August, Lanier received a panicked call from an editor. "Sir," the editor told Lanier on the phone, "at *Scientific American* we have a strict rule that states that an affiliation must be indicated after a contributor's name." But Lanier didn't have one at the time, so he made something up on the spot. "VPL Research," he blurted out, for "visual programming language." After the issue came out in September, investors started calling, and he founded the company for real.[49]

Soon Lanier discovered that having the traditional combination of screen and keyboard and mouse was a limiting factor for visual programming. The right interface was missing. There was no hardware to move things around. One possibility, Lanier thought, was that data gloves would do away with the mouse. Elegant hands-on screens would replace clunky arrow pointers and old-fashioned cursors. That way, even unskilled users could simply grab an object in a screen, twist it and turn it and reposition it, and interact with the machine. It was a bit like playing the drums or like conducting. Lanier wanted to be able to wave his hands and arms and make electronic music through motions.

In a stroke of good fortune, Lanier met Thomas Zimmerman at an electronic music concert in Stanford. They had both worked at Atari but never met there. Coincidentally, several years earlier Zimmerman had started developing a "data glove." In 1982 he had even filed for a patent on the glove.[50] Zimmerman's invention mounted optical flex sensors on the hand's individual fingers to measure how the fingers bent; not even the US Air Force engineers at Wright-Patterson Air Force Base came up with that idea. Zimmerman's passion for developing the glove came from a long-frustrated desire to play air guitar and create actual sounds by touching strings that weren't there.[51] Like Lanier, Zimmerman had envisioned conducting an entire orchestra playing electronic music by hand waving. He had also studied ballet at MIT, so he immediately thought his input device could extend over the entire body.[52] Zimmerman joined VPL in 1985.

The idea of a virtual concert was especially appealing. "Why not have it in a space that you're actually in?" the founders reasoned. They had an input device with the glove. But to create the perception of virtual space, they needed an output device—some sort of head-mounted display to show the machine's output to the human. So VPL started working on what the young company called the "eye phone," little screens that sat on the eyes like the speakers of an ear phone sat on the ears. The key feature of the new device was a pair of color LCD displays that looked like two small TV monitors that had seen better days.[53]

Unfortunately, the prototype helmet was rather uncomfortable. The screens were heavy, so the contraption had lead weights to keep what looked like a black cyclist's helmet balanced on Lanier's or Zimmerman's head. Just like the air force, VPL was discovering that the device's center of gravity mattered a great deal. Putting the eye phone on wasn't easy: two helpers were needed to strap the machine onto the wearer, with one person pulling apart the headphones and another one lowering the bulky front end. A finger-thick tube wound around the neck to feed input data to the screens. The goggles were heavy and left indented red lines on the forehead.

The glove was less of a problem. It was made of thin Lycra fiber, with optical sensors out of fiberglass. When the wearer flexed a finger, or turned a wrist, the fibers would be bent, thus transmitting less light. The computer would measure the loss of light and translate these values into commands. Over time, VPL improved the finger-bending measurements, by scratching and scraping the fibers at the right joints, so that the glove would ignore overstretched open palms but precisely measure a finger pulling a trigger.[54] Another sensor located that hand's position in three-dimensional space. A computer collated the data from the sensors, drawing an image of a moving hand on a screen. A tangle of wires dangling from the wearer's wrist and neck connected to an expensive state-of-the-art computer, which at that time was a custom Macintosh IIx design/control workstation, a machine with up to 128 MB of memory and a 16-MHz processor.

The logical next step was computerized clothing, the full-body data suit. The suit would make a foray into cyberspace a truly immersive experience. It looked like a diver's outfit. When the wearer lifted an arm, a crude animated figure on a small screen lifted an arm. If the diver stepped forward, the avatar stepped forward, in real time. It was primitive, but exciting. VPL hired more Bay Area engineers, and soon sixteen people were developing hardware and software on "Virtual Reality," as the trend quickly became known, usually capitalized.

The machines were custom-made and labor-intensive, and production runs for the eye sets remained small. A full set of equipment cost an unaffordable quarter million dollars. Still, many potential clients and reporters came to VPL's offices to try out the machine. Again the glove was different. The company sold 1.3 million data gloves to Mattel as a gaming device and joystick replacement, and a smaller number of more sophisticated and more expensive versions to IBM and NASA. "It really sold itself 'cause it was so cool," one engineer recalled.[55]

Naturally, other entrepreneurs jumped on the bandwagon. Their companies had fancy names: Autodesk, Inc., Sense8 Corporation, Virtual Research Systems, Pop-Optix Labs, and TiNi Alloy Company for tactile feedback systems; Polhemus, Inc., for a widely used head tracker.[56] Even Apple considered joining NASA's Ames Research Center and Autodesk to further refine the gloves-and-goggle gadgets.

Yet the two dozen VR companies that sprouted up like mushrooms lacked four things that the air force had benefited from during the previous decade: a significant budget, skilled engineers, time, and—perhaps most important—a clear purpose.

The Bay Area pioneers had a vague idea for a product, but not a clear problem to solve—no defining purpose, such as improving the kill ratio of an air force pilot under stress in an ill-designed F-15 cockpit. Pilot-plane interaction had occupied some of the world's brightest engineers for two generations, but in the San Francisco of the electrifying 1980s, counterculture trumped airpower. VPL even abandoned its original purpose: "The programming language fell by the wayside," one of Lanier's engineers recalled.[57] The developers were driven by a hazy vision of virtual worlds. They took Gibson's phrase "consensual hallucination" quite literally, it soon turned out.

"The idea is that by wearing computerized clothing right over your sense organs, you transport your sensory system into a reality that could be of any description," Lanier told one interviewer, sitting on the grass in front of a houseboat. He found the research "deliriously exciting."[58]

One evening in the spring of 1989, Adam Heilbrun drove down from Sausalito to Lanier's home in Redwood City to report a story for the *Whole Earth Review*. He arrived at 8:30. Lanier was busy fine-tuning a virtual Pacific Bell logo for an upcoming gig, but he immediately started creating a more impressive virtual world for the visiting journalist. Three hours later, by 11:30, they were ready for action.

Heilbrun recounted in awe how the blue figure of Lanier's girlfriend, wearing a fully-body data suit, twisted strangely on the floor of Lanier's apartment, slowly trying to find the "right place," as if her movements were controlled by a "distant, internal logic." Lying next to her was Lanier, his Rastafarian dreadlocks spread out on the floor, twirling along. Then came Heilbrun's turn. The entire setup seemed eerily familiar: "The room had the leftover aura of psychedelics," Heilbrun thought of this dreamland. "Well, I'm addicted," he suggested to Lanier. "Please don't use that word with this," Lanier interrupted softly. "Look what happened to mushrooms."

Lanier was referring to the ban on psychoactive drugs. Several of his friends had seen their academic careers coming to a screeching halt in response to psilocybin becoming illegal. Lanier mentioned Terence McKenna. "I'm really worried that virtual realities may become illegal," the twenty-eight-year-old entrepreneur sighed.[59]

In this pioneer's vision at least, virtual-reality gear was indeed a psychoactive substance. He envisioned that the user of his wearable technology could choose to be anything in virtual reality. A user could choose to be a cat, for instance: when the user smiled in the real world, the gear would read the facial expression, so that the cat in virtual reality also smiled. The movement and gestures and facial expressions of the real body, however subtle, would control the virtual body. That virtual body might as well be a mountain range or a pebble on the beach or a galaxy or a piano. "I've considered being a piano," Lanier told his guest. "I'm interested in being musical instruments quite a lot." This wouldn't even be that remarkable, he suggested:

> You could become a comet in the sky one moment and then gradually unfold into a spider that's bigger than the planet that looks down on all your friends from high above.[60]

Clarke's 1961 vision of becoming a spaceship or a TV network didn't seem too far off any longer, and the psychedelic scenes in *2001* suddenly made complete sense.

Again and again, Lanier emphasized the social aspect of his new technology. Strapping on goggles and gloves was more like using the telephone than watching TV; it was a two-way street, a shared experience, a way to socialize. Only far more intense. Switching on the virtual-reality engine would be, he imagined, like having a collaborative lucid dream. "It's like having shared hallucinations, except that you can compose them like works of art," Lanier told his guest.[61]

VR was even better than LSD because it was, by definition, a social experience, not chemically induced isolation. The effect would be wholly positive. Lanier believed his new technology would "bring back a sense of the shared mythical altered sense of reality that is so important in basically every other civilization and culture prior to big patriarchal power." The goggles-and-gloves technology had originally been developed to make F-16 bomber pilots more lethal. But now Lanier was convinced that it "has a tendency to bring up empathy and reduce violence."[62]

Indeed, the psychoactive trope came to dominate the debate on virtual reality. Ken Goffman, a.k.a. R. U. Sirius, the psychedelic editor and enfant terrible of San Francisco's publishing scene, was also friends with Lanier. Sirius published the very first nonscholarly article on the countercultural version of virtual space, which came out in the summer of 1988 in the psychedelics-and-tech underground magazine *High Frontiers*, which had been renamed *Reality Hackers*.

The influence of Stewart Brand's access-to-tools philosophy was palpable: the hipster magazine was all about psychedelics, mind machines, and "artificial reality technology." If that link to mind expansion wasn't clear enough, then the person reporting about the new technology left no room for doubt: the infamous Timothy Leary, illustrated with a psychedelic image of a man sitting at a personal computer, with the machine as the access portal to a trippy spiritual world, a Buddha looking down on the "cybernaut," as Leary referred to the user.[63]

The technology was brand-new and had never been covered in any magazine or newspaper before. Leary introduced the readers of *Reality Hackers* to helmet-mounted liquid crystal displays, head tracking, three-dimensional sound equipment, speech and gesture input, and VPL's prototype glove. Humankind, the psychedelic guru pointed out, had already entered a "post-industrial cyber-era." The real world was losing relevance, the magazine reported. A few months later, the *New York Times* was the first big national paper to cover the trend. "You and your lover trade eyes so that you're responsible for each other's point of view," Lanier told the nation's newspaper of record. "It's an amazingly profound thing."[64]

But so far, nobody had spelled out how cyberpunk, Gibson's stories, and virtual space were connected. This changed in September 1988 at Autodesk, then a six-year-old Sausalito-based company specializing in computer-aided design, a technology for displaying 3-D objects on screens. John Walker was Autodesk's founder and still one of the company's leading minds. The programmer was frustrated by people who referred to computers as "electronic brains." In Walker's view, this expression imputed characteristics to machines that they didn't have. "When you're interacting with a computer, you are not conversing with another person," he wrote in an internal memo. "You are exploring another world."[65]

But the user interface was the problem; how the user interacted with the machine was far more important than the computer itself. And the history of man-machine interfaces was ripe for revolution, in Walker's mind: yes, there was progress through the years from plug-boards to punch cards, from time-sharing to menus, and eventually to graphical controls and windows. But a mouse couldn't "transport the user through the screen into the computer," Walker lamented. Just two weeks earlier he had read about the air force's brand-new helmet-mounted display, the *Star Wars*–inspired Agile Eye, and was impressed by it.[66]

Like Gibson, Walker now needed a name for that synthetic place. He dismissed virtual reality as an oxymoron and then suggested "cyberspace." Gibson misused the root "cyber," Walker pointed out, by talking about computers rather than control. So he suggested bringing the word back to its Greek origin, even using the Greek spelling in the Autodesk white paper. "Since I'm talking about means of man/machine interaction," wrote Walker, "I can make the case that 'cyberspace' means a three dimensional domain in which cybernetic feedback and control occur."[67] Autodesk tried unsuccessfully to register the term as a trademark. William Gibson was nonplussed and jokingly threatened to retaliate by filing a trademark application to register the name "Eric Gullichsen," then the lead programmer on the Autodesk project.[68]

By the spring of 1989, science fiction, drugs, and emerging computer technologies were all the rage in the Bay Area. So Stewart Brand, cybernetic pioneer and Gregory Bateson devotee, decided to get ahead of the game by lifting this shrewd phrase from science fiction and obscure start-up memos: "Cyberspace" was how he titled a lead story in the summer issue of the *Whole Earth Review*. To report the story, Brand took a few friends who worked at Autodesk down to NASA's famous Ames Research Center in Mountain View, a good half-hour drive south of San Francisco. The Ames press officers took the hippie researchers into the lab. There they strapped on the most advanced publicly visible virtual-reality gear (the air force's work, far ahead of NASA's, was still mostly classified then). And Brand was simply amazed:

> The first thing most of us did when we arrived in virtual reality was study our own hand, looking for all the world like stoned kids: "Have you ever really, REALLY looked at your hand!!?"[69]

It was spectacular. Wearable devices, and the virtual reality they would simulate so perfectly, would enable human operators to break free of desks, screens, mice, and entangling wires. The new technology promised new ways to move inside a human body, in atoms, or inside buildings that existed only as design plans. Virtual reality would revolutionize industrial design, medicine, architecture, space exploration, entertainment, education, games—even sex and drugs. "I have seen the future, and flew in it!" Brand reported in his own magazine. "I lost my body almost instantly, except as a command device (ultimate mouse), and thoroughly enjoyed life as an angel. Oh wings of desire," he raved.[70]

But all that was theory. All those bold claims about social interaction in virtual reality were vision alone. It had never been done. The actual technology wasn't there yet. In practice, Lanier's cyberspace remained a solitary experience. Limitations in computing power and bandwidth did not allow for a truly interactive experience through glove and eye phone (at least not for another quarter century). That practical limitation, of course, didn't hold back the enthusiasts. Even the *Wall Street Journal* was enamored of the idea and breathlessly repeated Lanier's claim to its reporter that his gear and goggles were as good as "electronic LSD."[71] The young entrepreneur, it seemed, wasn't afraid any longer that the new psychoactive gear could be outlawed. The possibility of escaping, of escaping into a synthetic realm of the mind, was too appealing to resist.

Lanier met Timothy Leary in a getaway car. The VPL founder's girlfriend at the time knew Leary. One day Leary was running an all-week workshop down at the Esalen Institute. But for some reason Leary wanted to get out, and fast. He called Lanier's girlfriend to ask if they could come get him. He had a Tim Leary impostor on standby and wanted to smuggle the fake Leary in while they smuggled the real Leary out. Lanier and his girlfriend drove down from the Bay Area with two cars to make the exchange less suspicious. Lanier, then in his twenties, drove the flight car and ended up smuggling Leary out. Back at Esalen, nobody noticed. "They were all way too stoned," recalled Lanier.[72]

Leary had famously pioneered the use and study of psychedelic drugs. Yet in hindsight, until the late 1970s he found that he had no language to express how the brain actually worked. But the psychologist had an epiphany when he got his first computer. He suddenly understood that cybernetic terminology was ideal to describe the workings of the mind and the brain, that "human biocomputer" he had spoken about already in Berkeley in 1983, covered in the drugs-and-technology magazine *High Frontiers*. Comparing the body to manufactured artifacts had been common for a long time: hydraulic engineering helped us understand the heart as a pump with valves and pipes and the blood as a circulation system, Leary believed. Now cybernetics enabled the understanding of the brain's piping. Net-worked computers, for Leary, were the operating system of the mind itself. Millions of young Americans understood that "the best model for understanding and operating the mind came from the mix of the psychedelic and cybernetic cultures."[73]

Leary's lectures at universities were packed. For one event at Sonoma State University, the flyer was an actual LSD packet, a plastic bag with an ominous sugar cube to swallow.[74] It was a PR stunt. By now the LSD guru was turned on by wearable technology. He gave talks like "From LSD to Virtual Reality." "Can this computer screen create altered states?" he would ask the students. "Is there a digitally induced 'high'?"[75] These were rhetorical questions; of course the answer was yes—or would be in the near future.

The main function of a computer was not merely to be a personal computer, but to be networked: to offer *interpersonal* communication, Leary was convinced. Computer networks were liberation: "individuals and small groups that go off to start learning how to program, reprogram, boot up, activate, and format their own brains."[76] Leary's prose captivated droves of students: "We are creatures crawling to the center of the cybernetic world," he told them, "But cybernetics are the stuff of which the world is made. Matter is simply frozen information."[77]

Leary's language resonated with an emerging subculture. The Grateful Dead's songwriter John Perry Barlow once described him as a reverse canary in the coal mine, meaning that whatever Tim Leary was interested in, mass culture would discover a few years later.[78] Several of

US Army Air Force B-17G Flying Fortress, with ventral ball turret, dropping its bombs in 1944/45.

The merging of human and machine, made more urgent by warfare on an industrial scale, was best captured by Alfred Crimi, an Italian-born modernist painter working for the Sperry Corporation, here in a turret drawing, 1943. *Courtesy of Northrop Grumman Corporation.*

The trailer version of the SCR-584 automatic tracking radar, a system designed for fire control in air defence and used by the British Anti-Aircraft Command against German V-1 attacks at the end of World War II, as conceived by a US Army artist.

The world's first cruise missile was the German buzz bomb, or V-1. To contemporaries, it was a robot. In 1944, a "war of the robots"—in the words of the head of the British air defence—ensued over the English Channel.

The SCR-584 had 140 vacuum tubes, weighed 10 tons, and cost about $100,000. Combined with the new proximity "VT" fuse, the SCR-584 offered an effective defence against the V-1.

One of the secret "sea forts" positioned by the Royal Navy in the Thames Estuary during World War II to protect London. The forts were an effective defence against German attacks.

Norbert Wiener's initial cybernetic research was a $2,325 defence contract approved in December 1940. Here, Wiener (center) is pictured with two senior army officers: Brigadier General Leonard Greely (left) and Colonel Donald B. Diehl (right).

In the summer of 1946, Wiener coined the term "cybernetics" from the Greek *kubernetes*, for "steersman," inspired by observing man-machine servomechanical systems at war.

W. Ross Ashby was an early British cyberneticist and innovator. He built the homeostat in 1946, a contraption then touted as the world's first "thinking machine," based on Royal Air Force bomb switches.

Ashby's homeostat was not just self-adaptive, but both system and environment at the same time. At a 1953 meeting, the machine deeply impressed Gregory Bateson, who became a key figure in counterculture.

Ashby with his homeostat. The machine's purpose was to do nothing, to remain in balance when disturbed.

Ashby's laboratory at Barnwood House, a mental hospital near Gloucester, in the English West Country.

Wiener in a classroom at MIT in May 1949 with the primitive robot known as Palomilla. The photographer was Alfred Eisenstaedt, famous for the image of the sailor kissing a nurse in Times Square on V-J Day.

Julian Bigelow (far left) and John von Neumann (far right) closely collaborated with Wiener. Also pictured are Herman Goldstine (second left), one of the original ENIAC developers; and Robert Oppenheimer (second right), "father" of the atomic bomb. They stand in front of the IAS computer that was used in the American hydrogen bomb project.

J. C. R. Licklider, a pivotal computer science pioneer, took part in Wiener's early cybernetic discussion circles in Boston.

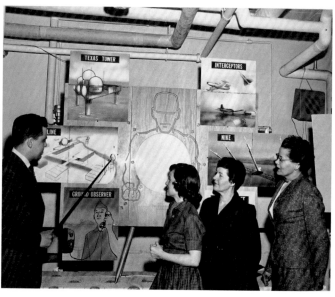

The US Air Force's response to the Soviet bomber threat was the Semi-Automatic Ground Environment (SAGE), which logged the course, speed, altitude, and location of all aircraft flying over North America at any given moment. The first sector at McGuire Air Force Base near Trenton, New Jersey, became operational in July 1958.

One of SAGE's many innovations was the light gun, a new input device and ancestor of the touch screen. SAGE was an inspiration for science fiction; the war room in Stanley Kubrick's *Dr. Strangelove* is one example.

The interior of SAGE Combat Center CC-01 at Hancock Field, New York.

Fear of push-button war and of losing labor to machines was ripe
in the 1950s, as Leslie Illingworth's "Friend or Foe?" cartoon in the
June 29, 1955, issue of *Punch* magazine intimates. *Copyright Punch Limited.*

The world's first cyborg was a rat in a lab at Rockland State Hospital, New York, in 1960.

Manfred Clynes, the researcher who came up with the cyborg idea, had this framed image in his office for years. The picture, by space-age illustrator Fred Freeman, a World War II navy veteran, was published in *Life* magazine on July 11, 1960.

Alice Mary Hilton—an author, organizer, and acolyte of Norbert Wiener—was one of the most passionate and eloquent proponents of automation. By 1963, Hilton worked hard to bring about the "age of cyberculture."

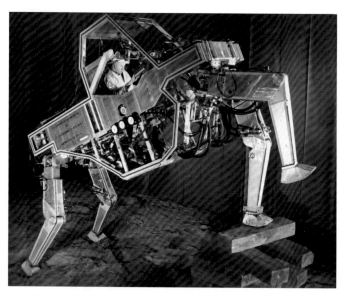

By late 1969, the US Army hoped to use the four-legged walking truck to carry loads and weapons into the jungles of Vietnam, through ditches and slopes impassable by wheeled vehicles.

An unidentified model poses in the gentle grips of the 80-ton Electric Beetle, one of the feedback-driven cybernetic anthropomorphous machines (or CAMs) built in 1962 by General Electric for the US Air Force to handle radioactive material for nuclear aircraft propulsion.

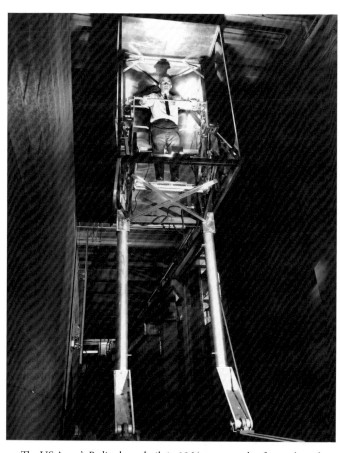

The US Army's Pedipulator, built in 1964, was a study of a two-legged
walking machine manufactured by General Electric.

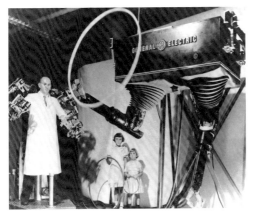

Handyman—"man" stood for "manipulator"—tried to combine man and machine into "an intimate, symbiotic unit that will perform essentially as one wedded system," wrote GE's Ralph Mosher (left) in 1967.

Hardiman, a man-augmentation system built for the US Navy around 1970, had thirty powered joints and could lift its own weight plus 1,500 pounds.

Anthropologist and social theorist Gregory Bateson was part of the original cybernetic Macy conferences and later applied cybernetics on a higher level, articulating his theory in the 1972 cult book *Steps to an Ecology of Mind*.
Barry Schwartz Photography.

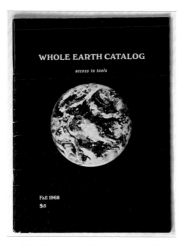

The first issue of Stewart Brand's *Whole Earth Catalog* was published in 1968. It was meant to be a printed feedback loop for back-to-the-land communards, and it reviewed six books on cybernetics.

Stewart Brand (left) and company play with the Earth Ball at the New Games, an event that Brand organized in California, October 1, 1973. © *Ted Streshinsky/CORBIS.*

Stewart Brand holds a copy of the *Whole Earth Catalog* on July 6, 1984. That same year, he launched the Whole Earth 'Lectronic Link, or WELL, the first real computerized social network. © *Roger Ressmeyer/CORBIS.*

The US Air Force pioneered the concept of "virtual space" in the late 1970s. Staff Sergeant Vernon Wells is shown here with the visually coupled airborne systems simulator (VCASS) at the Armstrong Aerospace Medical Research Laboratory at Wright-Patterson Air Force Base, June 1, 1985.

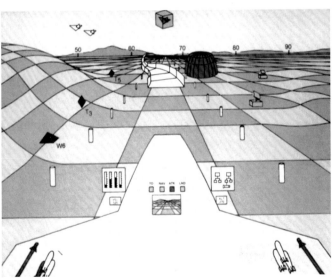

Typical computer-generated imagery projected inside the pilot's helmet of the VCASS.

Timothy Leary speaking at the Barbizon Hotel in Amsterdam on May 10, 1987. The psychedelics pioneer and counterculture guru of the 1960s had discovered the computer as a mind-expanding device in the early 1980s.

These psychedelic Buddha screens illustrated Timothy Leary's 1988 announcement in *Reality Hackers* magazine: reality no longer had a monopoly on stimulation in the "post-industrial cyber-era." It was the first text on the new virtual-reality trend.

Science fiction author William Gibson. In a 1982 short story, he coined the term "cyberspace" to describe the virtual space inside machines.

Mondo 2000, a San Francisco underground magazine, shaped the cyberpunk aesthetic between 1989 and 1993. It linked psychedelic drugs, virtual reality, and the rise of computer networks, as this typical illustration shows.

The group attending Michael Benedikt's Cyberconf, May 1990. Of note in this photo are John Perry Barlow (tall in the first row); to Barlow's left, Sandy Stone; to Barlow's right, Howard Rheingold, and then gaming theorist Brenda Laurel, followed by Michael and Amelie Benedikt. Behind the Benedikts stands Nicole Stenger, and behind her are *Habitat* pioneers Chip Morningstar and Randy Farmer.

French virtual-reality artist Nicole Stenger, in VPL gear. Stenger gave one of the most widely read presentations at Cyberconf: "Mind Is a Leaking Rainbow."

Jaron Lanier, founder of the virtual-reality company VPL, wearing one of the company's prototypes, the head-mounted display as an output device.

Virtual-reality gloves and full-body data suit prototypes developed by VPL. Lanier envisioned goggles, gloves, and suits as the future of human-machine interaction, enabling users to "enter" cyberspace.

A full-body data suit by VPL.

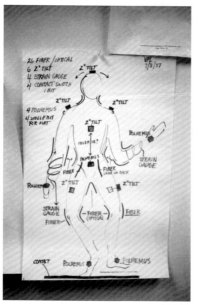

A data suit diagram by VPL.

An early virtual-reality machine, similar to the devices on display at the
Whole Earth Institute's Cyberthon in October 1990.

A *Mondo 2000* illustration of "cyberspace," the "new frontier" that could be "colonized." The virtual open range inside the machine was a mythical dimension where physics, laws, and identities would have new meanings, with its walls held up by cryptography.

One of the most persistent topics of human-machine interaction in cyberspace was sex. In the summer of 1990, colorful writer Howard Rheingold coined the memorable term "teledildonics" to describe futuristic forms of bodily interaction through machine interfaces.

Mondo 2000's guide to cyberpunk, laced with irony.

Former Wyoming cattle rancher, Grateful Dead lyricist, and gifted writer John Perry Barlow was one of the most charismatic figures "in cyberspace" in the early 1990s. He covered and shaped the early tech counterculture and cofounded the Electronic Frontier Foundation.

In a military library in Virginia, EFF cofounder John Gilmore discovered a coveted classified document that the NSA had refused to release to him, and he confronted the government. Here he celebrates his victory in late 1992.

Timothy C. May, an early Intel employee, was the most eloquent cypherpunk and one of the most radical. He wrote the "Crypto Anarchist Manifesto," came up with BlackNet in August 1993, and coined phrases like "crypto = guns."

Tim May (left), John Gilmore (right), and Eric Hughes, who appeared wearing these masks on the cover of *Wired* magazine's second issue, in May 1993, their PGP fingerprints written on their foreheads.

Ryan Lackey, inspired by the cypherpunks, ran HavenCo, a desolate anarchist server platform in the North Sea, starting in late 1999.

HavenCo operated from Roughs Tower, a repurposed old World War II sea fort located about 7 nautical miles off the Suffolk coast.

HavenCo lacked both bandwith and demand from customers. Its server racks were never full, and the business venture failed. By September 2001, Lackey operated a cypherpunk remailer from the air defence platform.

Owen Davies's article "Robotic Warriors Clash in Cyberwars,"
published in *Omni* magazine in January 1987, was illustrated
with this painting by Paul Lehr.

Businessman, author, and organizer Winn Schwartau coined the phrase "electronic Pearl Harbor" in January 1991. This illustration by Harry Whitver conveys Schwartau's conception of the new security threats.

This photo of an aviator's night vision system illustrated a 1992 *Bulletin of the Atomic Scientists* article that predicted "cyberwar," described as war fought through "robots" and autonomous weapons.

Deputy Secretary of Defense John Hamre, announcing on August 11, 1999, the creation of the Pentagon's Joint Task Force-Computer Network Defense (JTC-CND), the unit that would come to lead the Moonlight Maze investigation.

The trove of files taken by the Moonlight Maze attackers was vast—"as high as the Washington Monument," if piled up, said an internal government report in 1999.

By late 1998, the Moonlight Maze investigation was so labor-intensive and stressful that members of the FBI's coordination group made their own T-shirt as a souvenir, with "Byte Back!" printed on the back.

the Bay Area firms saw marketing potential in Leary. Autodesk asked him whether he would be willing to pitch their cyberspace vision in a promotional video. "We believe that this is inevitable technology," said Eric Lyons,[79] technology director at the Sausalito-based company. Leary knew this already. He agreed.

"The concept of cyberspace, creating realities on the other side of computer screens, opens up a new and very thrilling chapter in the human adventure," Leary said in his typically soft-spoken narrator voice, sitting stylishly in front of a gray wall in a gray suit. He was hypnotic. Somewhere within our brains, he said, there was a universe of wonder and novelty that needed to be "accessed, booted up, turned on, activated." He sounded like a prophet, and even looked like one. For thousands of years, he argued, such activation was done through yogis, meditation, music, dance, drugs, or other mystical experiences. Men and women used to come back from the experience unable to express what they had just witnessed. People came back from exploring their brains saying simply, "WOW—," searching for words. Leary's "wow" sounded very convincing. Sometimes brilliant artists could capture the experience in a picture, he added, or a still frame.[80]

Now technology was changing all that. Leary understood two things: that the brain *is* a computer. The brain can't just be compared to a computer. It doesn't just work like a computer. It is a computer. And he understood that the brain, because it is a computer, can be connected through computers: "Now in the late 20th century, here at Autodesk, a band of explorers has assembled, given us the hardware and software, to move around in the cybernetic universe." Previously, all that computer users could do was put their noses up against the computer screen, looking into another universe, like looking into an aquarium. But now, with new equipment and software, users could actually go to the other side of the glass and swim around, he said, and meet other people. "It's cyberspace," he concluded, "and it's a nice place to be."[81]

For Leary, the possibilities were boundless. "There are no limits on virtual reality," he told one interviewer in 1990. Data glove and data suits were the future. "The donning of computer clothing will be as significant in human history as the donning of outer clothing was in the Paleo-lithic."[82] Leary knew of Wiener and Bateson but probably found their writings too impenetrable. Leary was a man of action, a sensual man. The numerous speeches and articles of the psychedelics guru are playful and inspiring, but undisciplined and rambling. Perhaps without knowing it, Leary had fully absorbed the cybernetic myths of the previous decades:

> Telephone, telegraph, teletype, cars, jet planes.... Today, at the end of the industrial age, at the dawning of the cybernetic age, most digital engineers and most managers of the computer industry are not aware that we live in a cyber-culture surrounded by limitless deposits of information which can be digitized and tapped by the individual equipped with cybergear.[83]

Leary even deployed the same image that Clynes had used in his landmark 1960 paper that presented the cyborg. "Cyberwear," for Leary and early 1990s counterculture, was a "mutational technology." Individual brains could now have out-of-body experiences, just as "landware" like legs and lungs had permitted fish to escape the water and have out-of-water experiences.

Intrigued by the hype, and an admirer of Leary, Grateful Dead lyricist John Perry Barlow wanted to see for himself. The problem, he understood, had long been the interface. During the twenty years leading up to 1990, human relations with "these magic boxes" had become ever more intimate, at a fast clip. First came austere batch-processing punch cards as input devices—with simple printers as output devices, of the kind Wiener had used when he first interacted with a computer. Then came the light gun in the air force's SAGE system, the keyboard, the mouse, and Lanier's gloves to get informa-tion into the machine—and graphical operating systems, screens in ever-higher resolution, and even the air force's helmet-mounted displays to get information back out.

But the link between user and computer was still slow and far too clumsy. The interface, to Barlow, was "the mind-machine information barrier." The solution seemed simple enough: that barrier needed to be eliminated.

> The thin alphanumeric stream which drips from our fingertips and into the computer is a pale reflection of the thoughts which produce it, arriving before the CPU at a pace absurdly mismatched to its chewing/spitting capacities.[84]

The human CPU has always been fast, and the machine's CPU was getting faster—but the connection between the two was still only at trickle speed. To make matters worse, the speed of human-machine interaction wasn't the only issue. Another problem was that the interaction was only two-dimensional. No actual space was involved. Yet humans remember stuff in three dimensions, Barlow was sure. Something needed to be done.

So Barlow went to Autodesk to find out. There, John Walker took him "through the looking glass," as he had written in his 1988 internal Autodesk memo. "If cyberspace truly represents the next generation of human interaction with computers," Walker told Barlow that day, "it will represent the most profound change since the development of the personal computer."[85]

Walker had Barlow wired up with state-of-the-art equipment. The machine that generated the illusions was a Compaq 386, running Microsoft Windows 2.1, with a pair of Matrox graphic processors. The output device was a pair of VPL eye phones, with two video screens that were parallax-corrected, for the right viewing angles. When Barlow moved his head to look down from hovering hundreds of feet up, a Polhemus magnetic sensor tracked and measured this movement.

The graphics engine then used the movement data to adjust the image that Barlow was seeing, with as little lag as possible. For input, he was wearing a VPL DataGlove on his right hand, also linked to a Polhemus positioning sensor.

The two tracking sensors forwarded the position of Barlow's head and the position of his right hand in real time to the Compaq processor. If the software worked properly, Barlow could then see the position of his right hand. When he made a fist or pointed his index finger, the fiber-optic cables sewn into each of the glove's fingers would bend (or not bend), letting light escape at the knuckles and joints, thus relaying the position of the single fingers to the CPU. The computer could then both use the hand's position data as input commands and display the gesture back to Barlow.

"Suddenly I don't have a body anymore," the former Wyoming cowboy reported from his first trek into the new frontier of cyberspace.

All that remains of the aging shambles which usually constitutes my corporeal self is a glowing, golden hand floating before me like Macbeth's dagger. I point my finger and drift down its length to the bookshelf on the office wall.[86]

Barlow tries to grab a book, but his hand passes through it as if it doesn't exist. "Make a fist inside the book and you'll have it," a female voice says, seemingly coming out of nowhere. It is one of Autodesk's employees. Barlow does as he's told, making a fist in the real glove and moving his virtual hand again, and the book remains embedded in it. "I open my hand and withdraw it." Magically, the book remains suspended above the shelf, hovering in the emptiness of cyberspace.

Barlow is still standing in a small room, equipped with goggles and gloves. But in his virtual reality, he is exploring an office. He points up with his virtual hand. He ascends as if flying, passing right through an overhead beam as he zooms up. Several hundred feet up, he turns and looks down. It's a confusing situation, hovering up there in some undefined space. He feels a wave of loneliness and decides to head back down to the office building. He points down with his glove, but he's falling too fast. "I plunge right through an office floor and into the bottomless indigo below." In a panic, he tries to recall how to stop the fall. But he can't point his body. "I flip into brain fugue."[87]

It was nothing if not exciting. Barlow needed to see how others reacted, people he knew already. He called Jerry Garcia, singer, songwriter, and lead guitarist for the Grateful Dead. Was he game for a cyberspace demo? "When?" responded Garcia. Barlow had him try the goggles and the glove and delve right into virtual space. When Garcia was back in the reality he was used to, Barlow wanted to know how it was. "Well," Garcia said, "they outlawed LSD. It'll be interesting to see what they do with this."[88]

This tech-drug comparison at the time kept an entire subculture busy, complete with its own journals. "The closest analogue to Virtual Reality in my experience is psychedelic," Barlow wrote in *Mondo 2000* that summer. The hot new cyberpunk magazine, successor of *High Frontiers* and *Reality Hackers*, was still managed by the same crew from a villa in the lush Berkeley Hills. "Cyberspace is already crawling with delighted acid heads," Barlow reported.[89]

In fact, the comparison was so common that it had its own term, the "cyberdelic." "It's a Disneyland for epistemologists," Barlow wrote. The experience was so surreal that he might as well have described an acid trip. He had read Bateson's *Steps* when it came out in the early 1970s. Barlow began to see "an underlying grammar of nature," he wrote obliquely.[90] When he was strapped into the VR gear, the British cybernetic philosopher again came to mind: "Gregory Bateson would have loved this," Barlow thought.

It wasn't just dazzling. It was revolutionary. The "colonization of cyberspace was beginning," the would-be frontiersman wrote in *Mondo 2000*.[91]

III

Ironically, the actual colonization of cyberspace began on a rather more primitive device: a Commodore 64. It wasn't real people in virtual space that did the colonizing. It was avatars. In fact, the creators of the early game *Habitat*—Chip Morningstar and Randall Farmer—came up with the idea and with the word "avatar" for the game.

The first version of *Habitat* was built for the C-64, a common home computer in the mid-1980s that already appeared "ludicrous" to the developers themselves.[92] The game was ambitious. Lucasfilm Games, the company that designed the virtual world, envisioned a population of twenty thousand users, with plans to expand to as many as fifty thousand. All these people, the equivalent of a small city, would meet "in a single shared cyberspace," as two of the lead developers outlined in 1990. Interaction would happen in real time. Users could "play games, go on adventures, fall in love, get married, get divorced, start businesses, found religions, wage wars, protest against them, and experiment with self-government," Morningstar and Farmer wrote in a famous paper on the lessons of their pioneering game.[93]

Morningstar and Farmer found much of the work that was done on cyberspace technology in these heady days exciting and promising, but also misplaced. They didn't appreciate the "almost mystical euphoria" about data gloves and head-mounted displays. Hardware, in the view of the game developers, was a distraction. Cyberspace was not about experiencing hardware; it was about experiencing people. This is also the reason why Vinge was so popular with engineers. He was one of the few science fiction authors who understood that the input-output devices would remain slow and bulky. "The IO devices were very low bit rate," Vinge said later, explaining how he had constructed *True Names*. "They depended on the viewer's imagination to fill the gaps, which is exactly what happens when you read a book."[94] The *Habitat* engineers had the same approach: the game depended on the player's imagination to fill the gaps.

Habitat was meant to represent the real world—at least to a degree. The game designers liberally added childhood memories of games of make-believe, "a dash of silliness, a touch of cyberpunk," and of course, their remarkable technical skills in what was then called "object-oriented programming." The objects were the furniture of the *Habitat* world: houses, trees, gardens, mailboxes, books, doors, compasses, but also more controversial objects like clubs, knives, and guns. The game's little cartoon characters, controlled by the gamer, could buy and sell these items with in-game money that they had in their in-game bank accounts. Tokens were the currency in the land of *Habitat*, commonly abbreviated simply as T. For each new player joining *Habitat*, an avatar was created, or "hatched," and a starting amount of 2,000T was placed in the player's personal account. Each day the player logged in to the game, the money grew by 100T.

The game was inspired by science fiction, "notably Vernor Vinge's novel, *True Names*," the game designers explained.[95] ATMs, which in *Habitat* stood for "automatic token machines," gave avatars access to their money. One token was a twenty-three-sided plastic coin, slightly larger than a US quarter. Remarkably, the game coins had a portrait of Vinge on their face, adorned with the motto "Fiat Lucre" and the line "Good for one fare" on the back. But this was the 1980s. Such details could only be explained in the handbook and were lost on the bulky, curved, low-resolution screens of C-64s that started up with a painfully high-pitched beeping sound and never really stopped flickering.

The idea for the avatars came to Morningstar and Farmer out of cybernetic disappointment, out of a lost hope, or perhaps a lost fear: "Nobody knows how to produce an automaton that even approaches the complexity of a real human being, let alone a society," the two wrote at the time. So they decided they shouldn't even try. "Our approach, then, is not even to attempt this, but instead to use the computational medium to augment the communications channels between real people."[96] The way to do this was to have small cartoon figures represent the players in the virtual world.

Morningstar and Farmer often walked around the offices of Lucasfilm Games, chatting and discussing how to name things—for instance, the "ATM," the token machine. Morningstar, who had been a bookish kid, had read about the Hindu concept of *avatāra*, which stood for the descent, or the appearance, of a deity on Earth. When the cord to the heavenly world snapped in the myth, as he recalled it, the avatar was pulled back.

The image worked very well for the developers at Lucasfilm Games: "There's this sense of being attached to the divine," Morningstar recalled. "I was in love with that."[97] The deity was the gamer in front of the Commodore 64, the avatar represented the user, and the mystical silver cord that connected the two was the telephone line hooked up to the computer's modem. "You were reaching out into this game quite literally through a silver strand," added Farmer. "The avatar was the incarnation of a deity, the player, in the online world."[98]

The computer game became a classic. The approach that Lucasfilm's developers took with cyberspace and representation was very different from that of VPL and the virtual space pioneers: they prioritized low bandwidth over high bandwidth. *Habitat*'s avatars coupled human and machine loosely, not tightly. Morningstar and Farmer wanted to move complicated hardware and software out of the way, thus allowing for a richer social interaction. That, they thought, was what cyberspace was all about: "The defining characteristic of cyberspace is the sharedness of the virtual environment and not the display technology used to transport users into that environment," they believed.[99]

The promise of cyberspace was palpable. And virtual worlds needed to be built; the cyberspace environment, after all, was a man-made environment. Naturally, architects became interested in building these new worlds. Michael Benedikt was a professor of architecture at the University of Texas at Austin who had been inspired by William Gibson's work. "Cyberspace was an idea that was only just beginning to flower," he thought then.[100]

It was time for a large academic conference to explore the possibilities. So Benedikt reached out to Gibson and asked him to join the four-person conference committee. The science fiction author accepted. Over the summer of 1989, Benedikt started pulling together the first large meeting of minds. His e-mail announcement emphasized that this conference was "not about the enabling technology of 3-D interfaces." Instead, Benedikt wrote, "It is about the nature of cyberspace conceived of as an independent realm."[101]

The subject came natural to Benedikt; he was an architect. "The design of cyberspace is, after all, the design of *a new world*," he wrote.[102] And Benedikt understood that cyberspace was not something that is just discovered; it needed to be built, entity by entity. "That was a huge opportunity for architecture of an amazing kind; there'd be no weather, there'd be no rain, there'd be no gravity," he recalled. Even the laws of physics were adjustable. There were no building regulations. The possibilities were limited only by psychology, by how much disorientation users were able to handle.

Benedikt's First Steps conference was historic. In hindsight, it marked a symbolic shift: cyberspace transcended virtual reality in Austin, Texas, on May 4–5, 1990.

It took the better part of a year to organize the meeting. Eventually, in the first week of May 1990, the university's School of Architecture and Department of Computer Science jointly sponsored the conference; the fee was $115. One of Benedikt's students had suggested advertising the meeting only by e-mail, which had never been done before. The stunned organizers received sixty abstracts from artists, technologists, computer scientists, entrepreneurs, architects, and sociologists, many of them WELL members, some from as far as Sweden and Italy. And the organizers were amazed again when these people actually arrived in Austin by plane, fresh out of cyberspace. Several of the fifty attendees would go on and shape the emerging debate, including the colorful author Howard Rheingold and science fiction legend Bruce Sterling, best known for *Mirrorshades*, an anthology that defined the cyberpunk genre.

John Perry Barlow was one of the first to respond. He mailed in an abstract titled "Music in Cyberspace." The cattle rancher and Grateful Dead lyricist pointed out that his band had long been trying to blur the line between audience and performer. Jaron Lanier's data glove, Barlow remarked, had been developed as a means for the guitar player wannabe to fulfill that desire, and then he asked, "Could we develop a system of shared cyberspace in which the band and members of the audience could get together and 'jam' in real time?" When Benedikt and the conference committee read Barlow's abstract, they were confused. Benedikt didn't know the rancher, and Barlow's only academic credential was being a songwriter. "We very nearly didn't invite him because he seemed crazy," Benedikt recalled—but crazy in a good way. They quickly agreed: "This guy's a live-wire; we have to have him."[103]

At nine in the morning on the fourth of May, a beautiful Friday, the audience had coffee and then gathered in the Flawn Academic Center, a vast modernist white cube with ornamented windows, hovering over the landscape on pillars, itself a symbol of the 1960s space age. The architectural setup was inspiring: "It seems that the political arena of the sixties is slowly drifting to the existential arena of computers in the nineties," one speaker noted after the conference.[104]

One of the first presentations was "Mind Is a Leaking Rainbow," by Nicole Stenger, a French-born American artist and MIT-based virtual-reality pioneer. Stenger was a glamorous woman with intense eyes, eccentric earrings, and long curly blond hair. Stenger visibly stands out in the group photo of thirty-six. "The exploration of Cyberspace will

become the first massive trip of humanity into hallucination," she said
in a thick French accent, striking a chord with many of the WELL
members in the audience. "Without exaggeration, cyberspace can be
seen as the new bomb, a pacific blaze that will project the imprint of
our disembodied selves on the walls of eternity," Stenger told the audience.[105] Her talk was more performance art than scholarship. It was
very well crafted, and later very widely read and very frequently cited.

"On the other side of our data gloves," she said, "we become creatures of colored light in motion, pulsing with golden particles." By
now, some of the technologists had incredulous expressions on their
faces. Stenger was unfazed. "We will all become angels, and for eternity! Highly unstable, hermaphrodite angels, unforgettable in terms
of computer memory. In this cubic fortress of pixels that is cyberspace,
we will be, as in dreams, everything: the Dragon, the Princess, and the
Sword."[106]

Stenger compared the rise of cyberspace with the first days of
creation:

> In this primeval garden where a synthetic sun will rise, inner
> voices will whisper, immaterial kisses hover in the air, you will
> lie in the reconstructed sense of fur. For blind bards as for near-
> sighted whiz kids, cyberspace will feel like Paradise![107]

Her artful performance contrasted sharply with some of the drier
scholarly presentations: "Borne of Disneyland like luscious candy
cotton. Atoll of grace between the West and the Orient. Soufflé of
desires revolving in the light, whispering the names of the world's
fiancés: Laure, Beatrice, Peter Pan, John Lennon," she explained.
"Cyberspace the dessert of humanity!"[108]

Many of the computer scientists and the practically minded engineers could not believe what they were hearing. This blond French
artist wasn't even sure sex would survive in its natural form. "How

will your boyfriend know that you've been in your pajamas for weeks if you only meet in a cyberspace exchange?" she wondered. "You won't need condoms any more. Cyberspace will be the condom."[109]

This idea resonated with many in the audience. Allucquère Rosanne Stone, an early transgender performance artist and media scholar, later organized the second conference of cyberspace, "2Cyberconf." The paper she presented at the 1990 conference in Austin would become one of the most read and most cited. "To become a cyborg, to put on the seductive and dangerous cybernetic space like a garment, is to put on the *female*," Stone said. Cyberspace both disembodied, she argued, and it reembodied.

> As the charged, multigendered, hallucinatory space collapses onto the personal physicality of the console cowboy, the intense tactility associated with such a reconceived and refigured body constitutes the seductive quality of what one might call the *cybernetic act*.[110]

Morningstar and Farmer, the engineers of the first virtual world that was actually used, were also in the audience. To the two game developers it seemed as if these people were tripping. They could not understand a word. "Holy crap," Morningstar said to Farmer. "I could not keep up."[111]

Morningstar and Farmer presented in the afternoon. They explained the difficulties of building a virtual environment on limited processing power and low bandwidth. Their talk was hands-on in the way early settlers talked about the practicalities of homesteading the frontier. The audience was thrilled. "They clapped and stood up," Farmer recalled later in Palo Alto. "They loved it," Morningstar added.

These were the days when connecting through a computer was done by phone. So Barlow, who didn't present, interjected from the floor that cyberspace is "where you are when you are on the telephone."[112]

The line got great laughs in Austin, and people started using it in the conference. The participants agreed that "how you plug yourself in" was secondary; helmet and goggles or keyboard and mouse—it didn't really matter. In Austin the entire idea became a more abstract concept, a metaphor. Henceforth, Barlow and Rheingold and others would talk about "cyberspace" in metaphoric terms, just as the scholars at the conference had done. The conference volume was one of MIT's best sellers for several years.

Tightly coupling man and machine, of course, retained a nearly irresistible appeal. The most bizarre articulation of cyberspace must be Howard Rheingold's vision of teledildonics. After hearing Stenger's wild presentation in Austin, Rheingold articulated his own vision of future sex in the summer of 1990 in *Mondo 2000*. The first fully functional teledildonics system, Rheingold clarified at the outset, would *not* be "a fucking machine." Users did not want to have intercourse with a cold piece of technology; they wanted to make love to *other people*. But "at a distance," hooked up to each other via modem or other data links, and "in combinations and configurations undreamt of by precybernetic voluptuaries."[113]

Rheingold's vision went something like this: a night out in the virtual village meant getting dressed first, of course. But getting dressed now was somewhat different: it involved entering a suitably padded chamber loaded with sensors, putting on a head-mounted display with high-resolution goggles, and then slipping into the second skin of a lightweight body suit, hopefully diaphanous by then, he added. The electronic stocking would fit tightly, "with all the intimate snugness of a condom." The machine magic was embedded inside the suit, on the inside of its surface; tiny actuators and smart effectors would replicate the feeling of touch, and of course also the feeling of being touched. The miniature actuators would be ultrathin vibrators, hundreds of vibrators per square inch.

Once dressed, a user could pick up virtual objects, run fingers over a virtual surface, and feel texture and edges through the gloves and the body suit, Rheingold explained. There, alone in the virtual chamber, it was suddenly possible to feel virtual satin at the cheek, or to encounter virtual human flesh. Rheingold was intrigued by the possibilities: "You can gently squeeze something soft and pliable and feel it stiffen and rigidify under your touch."

Once set up and ready to go, punters dialed a phone number to meet their partners (in the 1990s, modems required dialing a phone number to get online). "Your partner(s) can move independently in the cyberspace," Rheingold outlined, matter-of-factly. The avatars are able to touch each other, "even though your physical bodies might be continents apart."

Nineteen ninety technology wasn't quite ready for action yet. Rheingold acknowledged this limiting detail. Every nook, every protuberance, every knob or plane or valley of a body's surface would require its own processor. Computers weren't powerful enough yet to control hundreds of thousands of sensors and actuators. But bandwidth wasn't a problem any longer; fiber-optic networks would already be able to handle the flood of dildonic data, Rheingold believed. Actuators were getting better fast. "Today's vibrators are in the ENIAC era," he was sure, referring to a famous computing machine initially designed during World War II to calculate artillery firing tables for the US Army. Nevertheless, the future was slowly coming into glorious shape. "Part of the infrastructure for a dildonic system exists already," the enthusiastic WELL member wrote.

To Rheingold, such a web would enable entirely new forms of human interaction—rather stimulating forms of interaction. "There is no reason to believe you won't be able to marry your genital effectors to your manual sensors and have direct genital contact by shaking hands," he predicted. A handshake was no longer just a handshake. The applications were titillating, the possibilities of gadgetry limitless. There was no room for doubt; his pioneering decade was merely foreplay. "Teledildonics is inevitable," Rheingold wrote. Twenty years from 1990, remote sex will be "ubiquitous."

IV

The emerging internet had a dark and ugly side. A few months before the conference in Austin, in December 1989, *Harper's* magazine hosted an online conference on the WELL. The editors invited forty participants to discuss privacy, hacking, and the computer underground. The group was a mix of the who's who of tech-hippies and hackers, including Stewart Brand and Kevin Kelly of Whole Earth fame, and Lee Felsenstein, who had founded the Homebrew Computer Club. John Draper (a.k.a. Cap'n Crunch) was there, along with two actual hackers from a shady group called Masters of Deception who did not reveal their names at the time: Mark Abene (a.k.a. Phiber Optik) and Eli Ladopoulos, who hacked as Acid Phreak. John Perry Barlow, one of the more prolific WELL members, was included as well.

At one point in the discussion about hacking, Barlow became agitated. The hackers implied that open systems deserved to be exploited. "You seem to argue that if a system is dumb enough to be open, it is your moral duty to violate it," Barlow interjected. "Does the fact that I've never locked my house—even when I was away for months at a time—mean that somebody should come in and teach me a lesson?"[114] A heated exchange ensued.

"Barlow, you leave the door open to your house? Where do you live?" Acid Phreak wrote. The debaters were at home, in their familiar environment. Barlow felt emboldened: "Acid, My house is at 372 North Franklin Street in Pinedale, Wyoming. Heading north on Franklin, go about two blocks off the main drag before you run into a hay meadow on the left. I'm the last house before the field. The computer is always on." It was late in the evening, and Barlow had had enough. "But do you really mean to imply what you did with that question? Are you merely a sneak that goes around looking for easy places to violate? You disappoint me, pal. For all your James Dean-On-Silicon rhetoric, you're not a cyberpunk. You're just a punk."

Forty-eight hours later the hackers took revenge. Barlow had implied that Acid Phreak and Phiber Optik were simply alienated kids who were playing with modems instead of skateboards. The two didn't appreciate that comparison. "You have some pair of balls comparing my talent with that of a skateboarder," Optik shot back, and he pro- ceeded to post a full copy of Barlow's credit history right there on the WELL, open for everybody to see. He had hacked TRW, a company that logged credit histories. (Many years later, pranks involving public revelations of private details would become known as "doxing.") "I'm not showing off," Optik added. "I just find your high-and-mighty atti- tude annoying and, yes, infantile."[115]

Barlow was shocked. "I've been in redneck bars wearing shoulder- length curls, police custody while on acid, and Harlem after midnight, but no one has ever put the spook in me quite as Phiber Optik did at that moment."[116]

Barlow wasn't the only one spooked by computer hackers. By coincidence, the federal government was about to launch a massive crackdown on illegal hacking that had been nearly two years in the making. Operation Sundevil responded to complaints by businesses and organizations to federal law enforcement agencies. Hackers alleg- edly abused long-distance telephone services and voice-mail systems, and stole credit card numbers. On May 8, 1990, the US Secret Service deployed more than 150 special agents in fourteen cities across the United States, swooping down on the homes of twenty-eight hack- ers with search warrants. The raids hauled in forty-two computers, more than twenty-three thousand floppy disks, telephone testing equipment, and documents. The Feds arrested two suspects for crimes related to the investigation: one nineteen-year-old woman from Pittsburgh and a twenty-year-old man from Tucson, Arizona.[117]

Around that time, the *Harper's* magazine editor took Barlow and the two hackers who had spooked him so badly to dinner at a Chinese restaurant in Manhattan. The former farmer was surprised to find the "cyberpunks," as he called them, to be harmless, spotless, and fashionable—"dangerous as ducks." Barlow asked Acid Phreak why they had chosen such a threatening name for themselves, *Legion of Doom*. "You wouldn't want a fairy kind of thing like Legion of Flower Pickers or something," Acid replied. "The media ate it up."[118]

Over dinner, Ladopoulos told Barlow what had happened on January 24, 1990, a few weeks after their heated encounter on the WELL: a platoon of Secret Service agents, guns drawn, had burst through the doors of Ladopoulos's apartment. The only person at home at the time was his terrified twelve-year-old sister. The agents packed up Acid's computer, notes, books, answering machine, ghetto blaster, and entire collection of audiotapes. The place was a mess when his mother returned from work.

The raid wasn't part of Operation Sundevil, but to Barlow that made no difference. He saw injustice at work. Worse, he saw the federal government trying to encroach on the free and wide-open spaces of the virtual world. Barlow sat down in Pinedale, Wyoming, and wrote a call to action.

In the fall of 1990, Stewart Brand published the text as "Crime and Puzzlement: In Advance of the Law on the Electronic Frontier."[119] The story recounted the episode of the *Harper's* discussion on the WELL, of Barlow getting doxed by Acid, of a clueless FBI agent visiting Barlow in Pinedale, and of the subsequent government crackdown that was Operation Sundevil.

The story was liberally illustrated with Wild West imagery. There was a faceless silhouette of six desperado hackers in cowboy hats, just after sunset on a dirt road that stretched out into the horizon, throwing long shadows, all looking at a computer screen. There were cactuses and shot-out screens, still smoking, and a fence with a "No Trespassing" sign nailed against the planks. The clueless FBI agent was portrayed as the hapless sheriff, star pinned to his chest, with a clown's hat. The WELL was a "frontier village" and "like open range," Barlow wrote, with boundaries "hard to stake and harder still to defend." Lines that used to be clear in the brick-and-mortar world were gone: between free speech and data, between land ringed by barbed wire and property infinitely reproducible, between trespassing and simple access.

"Cyberspace, in its present condition, has a lot in common with the 19th Century West," Barlow wrote. He squeezed every last drop out of the corny Wild West comparison: "It is vast, unmapped, culturally and legally ambiguous, verbally terse . . . , hard to get around in, and up for grabs."[120]

And it was time for action. In May 1990, just after returning from the rousing Austin conference, Barlow decided to establish the Electronic Frontier Foundation. He got crucial support from Mitch Kapor, a wealthy Silicon Valley entrepreneur. Their goal: "the extension of the Constitution into Cyberspace."[121]

In the heady days of 1990, the frontier spirit was strong. A cyber gold rush was on. With Woodstock on their minds, Kelly and Brand from the Whole Earth Institute decided it was time for a similar type of mythical event.[122] They called it Cyberthon.

Cyberthon was a twenty-four-hour marathon conference, fair, and exposition on virtual-reality culture, held the first weekend of October 1990. The venue was a vast warehouse studio of Colossal Pictures, at the edge of San Francisco's Bayview District near Hunters Point, next to a junkyard. The organizers built a disorienting wooden maze of 10-foot-high corridors, hung with black curtains to hide snake dens of cables, "with calculated irreverence toward day, night, right, left, and the entire history of empiricist philosophy," in the words of Peggy Orenstein, a reporter for *Mother Jones* at the time. (Orenstein was making an inside joke about the then popular postmodernist critique of empiricism, not unlike the views that had surfaced at Benedikt's Austin conference several months earlier.)[123]

The technology was presented in the small spaces between the plywood dividers. The maze was hot from the machines and the bodies of the excited guests. There wasn't enough virtual-reality hardware for the four hundred visitors, so the organizers held hourly VR lotteries whose winners got to enter the promised land. Wavy Gravy, veteran political activist and hippie clown, announced the winners with a megaphone. About three hundred managed to catch a glimpse.

Three companies provided the main gear: Sense8, Autodesk, and Lanier's VPL. There was a lot to see: Tahiti, Mars, or the human brain in 3-D. At the Autodesk stand, visitors could strap on tightly fitting goggles and gloves. The picture they saw was grainy, as if they were sitting too close to an old-fashioned television set. By tilting back their heads and looking up inside the warehouse, they could see a virtual bland sky, with a fish swimming by in the air. TiNi, the company providing the tactile feedback, attached electrodes to the fingers of curious visitors, sending tiny pulses into their fingertips at the push of a button. There was a game room, to play handheld versions of British arcade video games. *Habitat* designers Farmer and Morningstar presented their social game.

When Farmer explained the game to Orenstein, two figures appeared on the live screen, with real people playing in real time. A "man" with a standard-issue nondescript body said he was from Nebraska. The "woman" was from North Carolina, with equally standard-issue big breasts and a slim waist. The man extended his arms and grabbed the woman's breasts. "Nice boobs," he said.

Orenstein didn't like the game's built-in sexism. Farmer told Orenstein that the gesture was bad form. But he pointed out that the woman was walking around headless.

"I'm a man, you idiot," she said.

"What's with the boobs, then?" typed the man from Nebraska, confused.

"They're muscles. Now get off 'em."[124]

The entire event seemed strange. Comedian and actor Robin Williams was there, muttering to himself as he tried to fly his avatar through cyberspace in a Sense8 machine. "What was it like?" asked Orenstein when he was done. "Try it," Williams said; then, lowering his voice to a whisper, "Don't be scared."

"Was it fun?"

"Yeah, in a vertigo kind of way," Williams said.

"Cyberthon's flavor was pure Woodstock," the science fiction author Gregg Keizer reported from the VR gig. "I haven't seen so many tie-dyed shirts since high school 20 years ago." Everything was new and exciting. Keizer found the enthusiasm, the idealism, and the naïveté "infectious."[125] Everybody was there. "Cyberthon brought together Berkeley acid-heads, techie nerds, art damage punks, and the people who are actually building cyberspace," the *Whole Earth Review* reported.[126]

Science fiction legends William Gibson and Bruce Sterling were there. "It was supposed to be ironic," Gibson said of *Neuromancer*. "I didn't expect anyone to actually go out and build one of these things."[127] Barlow was there, and Rheingold of teledildonics fame, and of course the crew from *Mondo 2000*. Mind expansion gurus Terence McKenna and Timothy Leary were there, telling guests at three in the morning that all reality is virtual.

"It was that whole San Francisco mixture of psychedelia and computer technology," remembered Erik Davis, who attended as the *Village Voice* rock critic.[128] Davis said he'd gone because he had always been fascinated by psychedelics, the occult, and weird religious ideas. Some started calling the event the "Acid Test of the Nineties," in reference to the Merry Pranksters of the sixties.

VR was powerful. Seeing the new technology with their own

goggles convinced many skeptics. "Maybe all the mystical talk about the technology and how it will change communication, play, and work isn't so mystical after all," Keizer mused.[129] Orenstein also recalled arriving at the Cyberthon maze "curious, but terribly smug," indeed with what she called a neo-Luddite feeling of superiority. But by daybreak she had changed her mind. Helmet-mounted displays and glove control weren't just a fad for geeks with weird hair; this was real.

Stewart Brand, with his trademark combination of tech enthusiasm and skepticism, issued a warning to his guests at the nightly San Francisco cyber show: "We have to lower our expectations," he said about virtual-reality technology. "It'll take ten years to live up to the expectations of today."[130] Even that prediction turned out to be far too optimistic.

The same year, not long after the Cyberthon, Jaron Lanier's VPL, the virtual-reality pioneer, filed for bankruptcy. The company's fall from grace was vertigo inducing. But this unexpected development turned out to be prophetic. In these years, every issue of *Mondo 2000* was sprinkled with ads from virtual-reality companies. "It's a really remarkable institution," said Leary, the technology prophet, about the magazine: "A beautiful merger of the psychedelic, the cybernetic, the cultural, the literary, and artistic. It shouldn't last a long time."[131]

The same was true for the short-lived virtual-reality craze of the early 1990s. It didn't last a long time. By 1993, the hype was in overdrive. "Enthusiasm for VR has reached a critical level such that an overabundance of media hype threatens further advances," observed the NSA in an internal and unpublished text in *Cryptologic Quarterly*. "Laymen are beginning to expect *science fiction* capabilities and not *scientific* capabilities," America's vast electronic agency concluded.[132] The subculture was subsiding. By 1996, all three iconic formative magazines of the cyberpunk era had fallen: the *Whole Earth Review*, *Omni* magazine, and *Mondo 2000* were no longer. The world everywhere was moving on.

Meanwhile, cyberspace was going mainstream. The internet was growing fast, with the dot-com boom taking off in 1995. The world's fewer than forty million internet users were able to view just over twenty thousand websites.[133] Netscape, builder of one of the first browsers, went public that year; Yahoo!, Amazon, and eBay opened their gates too. On February 8, Bill Clinton signed into law the Telecommunications Act of 1996. The day was historic. The United States would update its telecommunication law for the first time in more than sixty years.

The law included a highly controversial provision, the so-called Communications Decency Act. The law was notably restrictive. It went beyond prohibiting the distribution of pornography to children. For one, it attempted to limit the public debate on abortion, at least in the eyes of many activists. Most controversial, however, was criminal-izing "indecent" expressions. The law defines indecency as "any com-ment, request, suggestion, proposal, image or other communication that, in context, depicts or describes, in terms patently offensive as measured by contemporary community standards, sexual or excretory activities or organs." Numerous websites on the still young internet protested by going dark for forty-eight hours. The American Civil Liberties Union argued that the law placed unconstitutional restric-tions on free speech online.

Barlow was outraged. The law made it "punishable by $250,000 to say 'shit' online," as he saw it. He decided it was time to "dump some tea in the virtual harbor." With characteristic grandiosity and pomp, as Barlow himself said, he gave the world a "Declaration of the Indepen-dence of Cyberspace." The text's opening paragraph has become iconic:

> Governments of the Industrial World, you weary giants of flesh and steel, I come from Cyberspace, the new home of Mind. On behalf of the future, I ask you of the past to leave us alone. You are not welcome among us. You have no sovereignty where we gather.[134]

The wide and free global social space would be "naturally independent" of the tyrannies of government. "You have no moral right to rule us, nor do you possess any methods of enforcement we have true reason to fear," Barlow wrote in his influential pamphlet, which he posted to the WELL. The lyricist laid out an oft-repeated yet deeply flawed and naïve vision of the future. As the NSA had observed, Barlow was one of the laymen who expected a reality based more in science fiction than in technical and political actualities. Nevertheless, his "Declaration of the Independence of Cyberspace," addressed to the governments of the world's industrial nations, became a landmark document in the history of the internet:

> Cyberspace does not lie within your borders. Do not think that you can build it, as though it were a public construction project. You cannot. . . .
>
> Cyberspace consists of transactions, relationships, and thought itself, arrayed like a standing wave in the web of our communications. Ours is a world that is both everywhere and nowhere, but it is not where bodies live.
>
> Your legal concepts of property, expression, identity, movement, and context do not apply to us. They are based on matter. There is no matter here.
>
> Our identities have no bodies, so, unlike you, we cannot obtain order by physical coercion.[135]

Two sharply different communities seized on the myth of cyberspace as the new frontier: West Coast privacy activists and the East Coast defense establishment. For the former, cyberspace was a space of freedom; for the latter, it was a space of war.

7. ANARCHY

PUNK IS RESISTANCE. DURING THE 1980S AND '90S, THE subculture was resistance of a special kind: heavy on fashion and light on politics. Punk generated eccentric hairstyles, tattoos, boots and leather outfits, drug habits, and hard-core music that oozed being against stuff. Yet fashion trumped direct action. Punk was aesthetic anarchy.

When computers and networks were added to the mix, cyberpunk was born. The 1990s were a time of extraordinary hope. The decade came barging right through Brandenburg Gate, with the Berlin Wall crashing down in the background. The end of the Cold War and the peaceful collapse of the Soviet Union released an intoxicating sense of optimism, at least in the West. Washington debated the "end of history," with liberal market economies coming out triumphant. In the Persian Gulf War of 1991, perhaps America's shortest and most successful ground war operation to date, the Pentagon overcame the mighty Iraqi army—and with it the lingering Vietnam hangover.

Silicon Valley and America's technology start-up scene, still bathing in the crisp utopian afterglow of the 1980s, watched the rise of the New Economy, with vertigo-inducing growth rates. Entrepreneurs rubbed their hands in anticipation. Intellectuals were inebriated by the simultaneous emergence of two revolutionary forces: personal computers and the internet. More and more PC owners connected their machines to the fast-growing global computer network, first with clunky, screeching modems, then with faster and faster broadband connections.

But amid the hype and a slowly but steadily growing economic bubble, it dawned on a number of users that something was missing: privacy and secure communications. History, thankfully, was gracious. Even more than that: nature itself was generous to humans in front of plastic keyboards. Unrelated to either PCs or the internet, cryptographers had made a third and no less far-reaching discovery in the 1970s. They didn't just invent a technology; more like explorers than innovators, they *discovered* an algorithm based on a beautiful mathematical truth. That truly revolutionary technology was finally unleashed for widespread public use in June 1991: asymmetric encryption, also known as public-key cryptography.

When free crypto was added to the computer underground, "crypto anarchy" emerged. Now people with mirror shades, modems, and PCs could be against stuff. And even better, despite the decade's spirit for unrestrained optimism, they had found something concrete to be against: the government's attempts to regulate ciphers. And so cypherpunk was born, a pun on "cyberpunk." The ideology was powerful—far more powerful and durable than those whimsical and short-lived names implied.

Cryptography is the art of secret communication. Diplomats and military commanders began using secret keys to encrypt their missives thousands of years ago, long before the invention of computers or even the telegraph. To establish secret communication, participants must first have the secret key. Thus arises the problem of key distribution—how to share a secret key with all participants of a secure conversation before the conversation starts. For centuries, key distribution gave large organizations a big advantage. The more resourceful a state's military and intelligence establishment, the more easily it could manage the logistics of key distribution.

Perhaps the single most significant invention in the history of cryptography came to be in 1973: public-key encryption, or "non-secret" encryption, as its inventors called it. It is probably the only mathematical algorithm that spurned its own political philosophy. Ironically, "non-secret" encryption was first discovered in secret at the British eavesdropping agency GCHQ. And it was kept secret for many years.

Public-key encryption was revolutionary for a simple reason. It solved the age-old security problem of key distribution. Sharing a secret key had previously required a secure communication channel. If Alice wanted to send Bob a secret message, she would first need to share the secret key with him. But a secret could not be shared on an insecure channel. Suppose Alice sent Bob a letter containing the secret key and asking him to use it to scramble their subsequent correspondence—say, by replacing every letter with a specified alternative letter. Eve (cryptographers like to call the supposed evil eavesdropper "Eve") could simply intercept the letter and make a copy of Alice's secret key en route to Bob. Eve would then be able to read all future messages encrypted with this key.

By the 1960s, the British military had started worrying. Tactical radio had become more widespread, along with computers and telecommunication technology, making the problem of key distribution worse. "The management of vast quantities of key material needed for secure communication was a headache for the armed forces," recalled one of the British government's leading cryptographers at the time,

James Ellis.[1] Ellis first believed, as was generally assumed then, that no secret communication was possible without a secret key first being shared. His view changed with the random discovery of a World War II report, *Final Report on Project C-43*, by a Bell technician, Walter Koenig, also prepared under an NDRC contract.[2]

Back in October 1944, Koenig had suggested a theoretical way of securing a telephone call by having the *recipient* of a call add noise to the signal and then subtract it afterward. Only Bob could subtract the noise, because only he knew what he had added in the first place. An eavesdropper, Eve, simply would not know how to modify the noise, because she wouldn't have access to the noise that had been added to the phone conversation in the first place.

The system was impractical at the time. But Ellis got the decisive and entirely counterintuitive cue: there was no need to assume that only the sender could modify the message; the recipient could have a role as well. "The noise which had been added," Ellis wrote in 1970, "had been generated by the recipient and is not known to the sender or anyone else." The recipient, therefore, "takes an active part in the encipherment process." In theory, at least, Ellis seems close to solving the age-old key distribution problem.[3]

Now the secret British cryptographers needed to find a mathematical way to enable the recipient to take part in ordinary encryption and decryption. "The unthinkable was actually possible," Ellis recalled. But because he was not a mathematician, he could not solve the underlying challenge of finding a suitable one-way function, a mathematical operation that could be performed in only one direction—something that could be done but not undone.

Three years after Ellis's thought experiment, in 1973, a young Cambridge mathematician, Clifford Cocks, joined the spy agency in Cheltenham. Six weeks into his job, a supervisor casually told Cocks about Ellis's "really whacky idea." Cocks understood that finding a suitable one-way function had been the problem. The twenty-two-year-old had worked on number theory before, and the problem of factoring—finding how a number could be divided into other numbers—was familiar to him. "If you wanted a function that couldn't be inverted," he remembered, "it seemed very natural to me to think of the concept of multiplying quite large prime numbers together."[4]

Multiplying two large primes is easy, even if they are more than a hundred digits long. Factoring the two numbers from the much larger product is hard—very hard. It took the freshly recruited spy about thirty minutes to come up with this prime solution. "From start to finish, it took me no more than half an hour. I was quite pleased with myself. I thought, 'Ooh, that's nice. I've been given a problem, and I've solved it.'"[5]

Cocks didn't grasp the implications of what he had just done. But soon colleagues started approaching the wunderkind from Cheltenham in admiration. The young mathematician's discovery seemed immediately applicable to military communications, and it would become one of GCHQ's most prized secrets. The only person Cocks could tell was Gill, his wife, who also worked for the spy agency. GCHQ called its discovery "non-secret encryption."[6]

But there was a problem. In the mid-1970s, room-sized mainframe computers were not yet sufficiently powerful to crunch large primes into a secure one-way function fast enough. Neither GCHQ nor the NSA turned the theoretical possibility of non-secret encryption into a practical algorithm or crypto product that could actually be used to secure communications.[7] Computers, ironically, were one of the main reasons why cryptographers in the shadows neglected the magic of public keys—and why those in the open discovered this magic.

Meanwhile, a few public academics kept working hard on solving the puzzle of how to exchange a shared secret on a nonprivate channel. Unsurprisingly, the breakthrough happened in the San Francisco Bay Area of the mid-1970s, with its inspiring mix of counterculture and tech entrepreneurship. These pioneers were Whitfield Diffie and Martin Hellman of Stanford University, and Ralph Merkle of UC

Berkeley. Their discovery resembled what the British spy agency had already found in secret.

In November 1976, a history-changing article appeared in an obscure journal, *IEEE Transactions on Information Theory*. It was titled "New Directions in Cryptography." Diffie and Hellman knew that computers would be coming to the people, as Stewart Brand had just reported from the mesmerized *Spacewar* players on their own campus. And they knew that these computers would be networked. "The development of computer controlled communication networks promises effortless and inexpensive contact between people or computers on opposite sides of the world," they wrote in the introduction to their landmark paper. Computer networks, the two cryptographers believed, would be "replacing most mail and many excursions." Going digital posed a new security problem.[8]

A good old-fashioned paper contract could be signed, sealed, and mailed reasonably securely. Anyone could easily recognize a handwritten signature as authentic, but no one other than the legitimate signer could easily produce it. "This paper instrument," they wrote, needed to be digitally reproduced. The task was hard. The simple paper system didn't just work; it worked on a very large scale, and it worked cheaply. Their answer was a public-key cryptosystem, in which "enciphering and deciphering are governed by distinct keys."

Diffie and Hellman had now suggested a theoretical solution, but much like Ellis at GCHQ four years earlier, they had not found a practical mathematical function that actually implemented this cunning scheme. Yet they inspired dozens of other cryptographers to try. It took about four months.

The solution emerged on April 3, 1977—on Passover. Ron Rivest, Adi Shamir, and Leonard Adleman, three MIT academics, discovered an actual and elegant method for public-key cryptosystems, after experimenting with more than forty mathematical functions. Rivest, then a twenty-nine-year-old assistant professor, was the driving force.

That evening, after returning home from a seder with friends that included Adleman and Shamir, he had the eureka moment while sitting on a sofa after midnight, eyes closed: the desired one-way function could be based on very large, randomly chosen prime numbers, over a hundred digits long.

Rivest's idea exploited the same curious one-way property of large primes that Cocks had discovered in secret. But the academic trio moved right to implementation. The numbers are easily multiplied, but it is nearly impossible to reverse the step and find the two primes that were used to generate the product. Multiplication took seconds; factoring would take millions of years, even with the most powerful computers. The algorithm that Rivest, Shamir, and Adleman suggested took advantage of this asymmetric factorization problem. The public encryption key would contain the product; the private decryption key would contain the two primes. It was safe to share the public key on an insecure channel because the factorization problem was so hard that it was, in effect, already encrypted, scrambled by a one-way function that was easy to perform but nearly impossible to reverse.[9] It was magic.

In April 1977, the trio drafted a technical memo that would soon send shivers down the spine of the NSA. The memo remained obscure at first. After typing it up, Rivest mailed it out for informal review to colleagues, addressed simply as the Computer Science Lab at MIT, 545 Technology Square, Cambridge, MA. One of the recipients of this first draft was Martin Gardner, a columnist at *Scientific American*. Gardner saw the idea's potential and mentioned Rivest's work in the popular "Mathematical Games" column in August 1977.

Gardner announced a "new kind of cipher that would take millions of years to break." He didn't have space for technical details, so the column referred to the memo that was "free to anyone who writes Rivest at the above address enclosing a self-addressed, 9-by-12-inch clasp envelope with 35 cents in postage." The column also mentioned that the National Science Foundation and the Pentagon, more specifically the Office of Naval Research, had funded this remarkable crypto work.[10]

The response was overwhelming. Seven thousand letters came flooding in.[11] They came from all over the world. "Some were from foreign governments," Rivest recalled. They all wanted to get their hands on Rivest's revolutionary encryption algorithm in Technical Memo Number 82.

NSA employees also read Gardner's column in *Scientific American*. Cocks's innocent secret discovery in Cheltenham now made the NSA become defensive, and the agency overreacted. Inside the Triple Fence the story sounded not as if somebody was simply making an important discovery, but as if academics were stealing a secret that America's spies already possessed and guarded closely. That was a problem. The powerful Fort Meade machinery sprang into action.

The public spread of cryptographic knowledge needed to stop, so the spies looked into changing legislation. The agency put pressure on academic publishers. NSA employees warned cryptographers that presenting and publishing their research could have legal consequences. They issued gag orders. Fort Meade tried to censor the National Science Foundation and to take over funding crypto research directly. Vice Admiral Bobby Inman, then the director of America's most secretive agency, even tried a softer approach: he gave the first public interview ever in the NSA's history to *Science* magazine.

"One motive I have in this first public interview is to find a way into some thoughtful discussion of what can be done between the two extremes of 'that's classified' and 'that's academic freedom,'" Inman told the magazine.[12] The Texas-born naval career officer said he was deeply concerned about the "burgeoning" academic interest in this field, although he did not explicitly mention public-key encryption. He doubled down and gave a speech on the growing interest in "public cryptography" five months later, in March 1979, to the Armed Forces Communications and Electronics Association:

There is a very real and critical danger that unrestrained public discussion of cryptologic matters will seriously damage the ability of this government to conduct signals intelligence and the ability of this government to protect national security information from hostile exploitation.[13]

The academic discovery meant that sources would soon go dark, the NSA feared.

GCHQ's tightly guarded secret was no longer a secret. The NSA would try everything it could to stop strong crypto from going public, and it would continue to try over the next two decades: cutting government funding of cryptographic research, or taking over the funding; vetting papers before publication; threatening scholars with criminal proceedings, or trying to convince them that publication damaged the national interest. The agency's attempts to stop crypto were clumsy and ham-handed. Even their most potent tool, classifying encryption as a weapon under the International Traffic in Arms Regulations, would ultimately fail.

The NSA's attempts to reign in crypto in the late 1970s foreshadowed a trend: the government's endeavors to counter the rise of strong encryption confirmed the worldview of those who were inclined to distrust Washington's secret machinations. The leaked Pentagon Papers and the ensuing Watergate affair earlier in the decade had eroded trust in the federal government, especially on the libertarian left. Resistance was brewing.

Rivest, Shamir, and Adleman's motivation was a conservative one. They wanted to preserve the status quo, not topple it: "The era of 'electronic mail' may soon be upon us," the trio suspected, correctly. It was therefore the task of cryptographers to "ensure that two important properties of the current 'paper mail' system are preserved": privacy and authentication—that messages remained confidential and they could be signed.[14]

Public-key cryptography made it possible to keep a message *private*:

The sender would scramble the clear text with a key that the *recipient* had "publicly revealed." Then the recipient, and only the recipient, could use the matching private key to unscramble the message's cipher-text. But the new technique could do even more. Public-key cryptography made it possible to "sign" a message electronically, by doing exactly the opposite: having the sender encipher a signature with a *privately* held encryption key, thus enabling the recipient to verify the message's origin by deciphering that signature with the sender's publicly revealed key, thereby proving that only one party, the legitimate sender, could have scrambled the message's signature. Everybody could decipher and read the signature, but in only one way: with the sender's public key.

This form of authentication was like a handwritten signature on steroids: signatures could be verified by everybody and forged by nobody. Electronic mail could now be even better than old-fashioned snail mail, with sealed envelopes that only the intended recipient could open and signatures that were impossible to fake, guaranteeing confidentiality and authenticity.

Perhaps the best part was that the "public" in public-key encryption really had two meanings: the key was public and, equally important, the method was simple enough for widespread public use. That was because the cryptographic breakthrough came just at the right time. It coincided with the advent of the mass-market personal computer, the PC, and soon the spread of the internet. Strong crypto was becoming a public good, no longer a privilege of governments and companies. And what GCHQ had called non-secret encryption was about to inspire an entire set of ideas—some realistic, some utopian—that would come to shape the twenty-first century.

The mix was potent: computers, networks, and public keys clearly would have a huge impact. But exactly what kind of impact it would be wasn't obvious. A few scholars who were tracking the pulse of recent technical developments started exploring these possibilities. One of them was David Chaum.

■

Throughout the 1980s, Chaum was torn between fear and hope. The Berkeley graduate looked like a cliché: gray beard, full mane of hair tied to a ponytail, and Birkenstocks. Chaum was concerned that "automation of the way we pay for goods and services" was advancing in large strides. He shuddered at the prospect of somebody else connecting the dots of his life. Chaum knew that an irritatingly detailed picture could be pieced together from hotel bookings, transportation, restaurant visits, movie rentals, theater visits, lectures, dues, and purchases of food, pharmaceuticals, alcohol, books, news, religious and political material. "Computerization," he lamented in 1985, "is robbing individuals of the ability to monitor and control the ways information about them is used."[15]

Individuals in both the private and public sectors would routinely exchange such personal information about consumers and citizens. The individual user, Chaum was concerned, would lose control and visibility; there was no way to tell whether the information collected in bulk was accurate, obsolete, or inappropriate. "The foundation is being laid for a dossier society, in which computers could be used to infer individuals' life-styles, habits, whereabouts, and associations from data collected in ordinary consumer transactions."[16] Such an outcome, Chaum suspected, would be unacceptable to many.

Thankfully, public-key encryption had emerged just in time to save privacy from automation, computerization, and data-hungry corporations and governments. So Chaum started working on concrete solutions: untraceable electronic mail, digital pseudonyms, anonymous credentials, and general protection of privacy. Chaum is best known for yet another revolutionary cryptographic discovery: blind signatures.

The nondigital equivalent to a blind signature would be using carbon paper to sign a letter that is already in an envelope, without having read the letter first. A signature, in short, is blind when the content of a message is disguised before the signature is added. This signature can then be used to verify the undisguised message.

Chaum had two situations in mind where blind signatures could be put to use: one was digital voting. Alice might want to prove that she cast a vote in an election while keeping her actual vote anonymous. Chaum's sophisticated digital blind signature scheme made this possible. A voter could sign the ballot without revealing the cast vote. It became possible to confirm all this electronically: Alice votes anonymously, Bob sends her a blind receipt, and Eve doesn't see any of it.

But Chaum's true passion was another purpose for using blind signatures: digital cash. In 1983, he suggested a "fundamentally new kind of cryptography" that would enable a better form of money: third parties could not determine payee or the time or amount of the payment. Individual privacy and anonymity was guaranteed, as when paying cash at a gas station or in a drugstore. At the same time, individuals could provide proof of payment, and they could invalidate payments if someone stole their medium of payment, as when using old-fashioned credit cards.

Chaum combined the best of both worlds: the anonymity of cash and the security of plastic. The article that spelled out the idea became one of his most influential papers, "Numbers Can Be a Better Form of Cash Than Paper." But using this improved form of cash was not only about convenience and security. If crypto cash would *not* be adopted widely, Chaum feared, "invisible mass surveillance" would be inevitable, "perhaps irreversible."[17]

Chaum's idea was magically simple and powerful. Steven Levy, a perceptive chronicler of the grand cryptography debate of the 1990s, called him the "Houdini of crypto."[18] So powerful were Chaum's ideas that an entire movement arose. That movement believed crypto was en route to making the state as we know it obsolete.

Many of these early cryptographers had been exposed to a powerful streak of American culture: civil libertarianism with its deep-seated distrust of the federal government—or of any government. Counter-culture, with its focus on free speech, drugs, and sexual liberation, was constantly pushing the boundary of what was legal. Meanwhile, the NSA's hysterical reaction to basic crypto scholarship amplified this hostility toward government in the emerging computer underground of the 1980s. So it was no coincidence that Bay Area cryptographers unearthed what would become one of the most potent political ideas of the early twenty-first century.

One of the intellectual founding fathers of the nascent crypto movement was Timothy May. The son of a naval officer, May grew up in a suburb of San Diego.[19] When Tim was twelve, his father was posted to Washington, DC, and the family made the move to the East Coast. Young Tim, not even a teenager, joined a local gun club. A fascination with firearms would stay with him. He later owned a .22 revolver, a .357 Magnum, an AR-15 assault rifle, a Ruger, a pair of SIG Sauers, and other weapons.[20] Holding a pleasantly heavy, cold metallic firearm felt liberating and empowering. So did reading Ayn Rand, the queen of youthfully aggressive libertarianism.

May was a ferocious reader of fiction as well as nonfiction. Crypto was so new and so radical in its implications that inspiration simply couldn't come from science, he thought; it could only come from science fiction. Vernor Vinge's novella *True Names* came to May's attention in 1986. "You need to read this," a friend told him, giving him a dog-eared Xerox copy of the entire short story. Vinge feared total identification and transparency: "It occurred to me that a true name is like a serial number in a large database," the science fiction writer recalled later. The names could serve as identifiers, connecting otherwise disparate information, as what intelligence officers call "selectors." Whoever had access to a true-names database would have power over the objects in the database.[21]

Vinge's 1981 novella spelled out the very same tension that was driving Chaum's fear of the "dossier society" at the very same moment. May was "riveted," he said later. He thought the story articulated a number of themes that were swirling around in "computer circles" at the time—notably, the role of digital money, anonymity, pseudonyms and reputations, and countering the government's interest to impose control "in cyberspace."[22]

Cyberspace was a familiar notion to May, even before it was articulated under that name. May keenly followed science and technology trends, including Jaron Lanier's early work on virtual reality. The September 1984 issue of *Scientific American*, the software issue, had on its cover a visualization of Mandala, Lanier's visual programming language. May, then working at Intel, had also contributed an illustration to that issue: a blue-and-green scan of an electron micrograph showing a small part of an Intel 80186 microprocessor. One day that September, May ran into Lanier at Printers Inc., an independent bookstore in Palo Alto and a gathering spot for San Francisco Peninsula intellectuals, not far from Stanford University. Lanier was sitting two stools over, and they struck up a conversation about cyberspace.[23]

"Encryption makes it easy and even safe to ignore most local laws about what can be done in cyberspace," May later argued.[24] For May and many other crypto anarchy pioneers, this change was a first-order opportunity. In true cyberpunk fashion, May took the space idea literally. "There is no reason to expect that this capability won't be a major reason to at least partly move into cyberspace," May wrote at the time. The nostalgic frontiersman expected that the World Wide Web's explosive growth, secure communication, and the coming availability of digital money would accelerate the "long-awaited colonization of cyberspace."[25]

In mid-1988, ten years after Rivest's pathbreaking discovery and two years after reading Vinge's *True Names*, May penned the "Crypto

Anarchist Manifesto," whimsically modeled on another famous manifesto with revolutionary ambitions: "The technology for this revolution—and it surely will be both a social and economical revolution—has existed in theory for the past decade," May wrote, "but only recently have computer networks and personal computers attained sufficient speed to make the ideas practically realizable."[26]

The possibilities were extraordinary. "Two persons may exchange messages, conduct business, and negotiate electronic contracts without ever knowing the True Name, or legal identity, of the other," May wrote in his manifesto, using capital letters in honor of his favorite science fiction author. Then May mobilized that most powerful American myth, the Frontier. Barbed wire, a seemingly minor technical invention, had enabled the fencing off of vast ranches and farms in the open rangeland of the West. Barbs on wire had altered forever the concepts of land and property rights in the frontier states, and it had caused the Fence Cutting Wars a century earlier. May sided with the wire-clipping cattlemen and cowboys. On the electronic open range, the barbed wire need not be accepted as immutable fact.

The comparison was odd, but it sounded powerful: Crypto was a game changer. It also emerged as a seemingly minor technical invention at first, from some obscure branch of mathematics. But this time, technology worked for freedom and liberty, and against those who wanted to build fences around their property. For May, crypto was like "the wire clippers" that would dismantle the illegitimate fences around intellectual property. The federal government, he observed with horror, wanted to slow or halt the spread of this technology, and Washington justified the clampdown with vague references to national security. And yes, just as in *True Names*, criminals would abuse it and take advantage of the renewed liberties. But none of this would stop the rise of crypto anarchy, May knew. He ended his pamphlet with this battle cry: "Arise, you have nothing to lose but your barbed wire fences."

The "Crypto Anarchist Manifesto" already contained the seeds of what would become a potent political ideology: technology itself, not humans, would make violence obsolete. May distributed his pamphlet electronically and in print among like-minded activists at the Crypto '88 conference in Santa Barbara and again at the Hackers Conference that year. But something was missing. The message didn't quite get out.

III

By early 1992, Timothy May and a friend, Eric Hughes, were becoming annoyed with the glacial progress of actual cryptographic technologies that could be used by normal people. Yes, Phil Zimmermann had just released his home-brewed PGP (for "Pretty Good Privacy") 1.0 to the public. This was a significant step. Zimmermann, in violation of export control regulations as well as patent law, gave public-key encryption to the people. And the uptake of his rogue app, as well as the uproar, was big. But the first version of PGP was buggy and clunky to use. Much more was possible, and May and Hughes knew it.

May had a taste for rugged frontier individualism, often sporting a wide-brimmed Stetson hat—a true crypto cowboy. The curious physicist had recently retired from Intel as a self-made man at forty, now independently wealthy and living on a self-sufficient ranch in the Santa Cruz Mountains. Back in 1970, young May had read the *Whole Earth Catalog*, and he later subscribed to the *Whole Earth Review*. He also was a former member of the Homebrew Computer Club.

Hughes was just under thirty, with blond hair halfway down his back and a long wispy beard. He had studied math at Berkeley. In May 1992, Hughes came down to Santa Cruz from Berkeley to hunt for a house. But the two men were drawn into their common passion:

crypto anarchy. "We spent three intense days talking about math, protocols, domain specific languages, secure anonymous systems," May said later. "Man, it was fun." The crypto rebels had been inspired by Martin Gardner's famous 1977 column from *Scientific American*, fifteen years after it came out. "Wow, this is really mind-blowing,"[27] May had thought when he read the piece.

May and Hughes began to rope in others. A group of sixteen people started meeting every Saturday in an office building near Palo Alto full of small tech start-ups. The room had a conference table and corporate-gray carpeting. Stewart Brand was at one of the first meetings, as were Kevin Kelly and Steven Levy, the two *Wired* writers. They were all united by that unique Bay Area blend: passionate about technology, steeped in counterculture, and unswervingly libertarian.

The crypto group also shared one other thing: a frustration with the slow pace of cryptographic progress. Chaum's ideas were ten years old, yet there was still no digital cash, no anonymity by remailer, no privacy, and no security built into the emerging cyberspace. They played games, simulating encrypted messages with envelopes, sometimes for up to four hours, with role playing to see how cryptographically guaranteed anonymity would play out in the marketplace. Simply by using envelopes, the group simulated signatures, trust and reputation systems, and even online black markets.

But already at these first meetings, some were concerned that anonymity could be abused. "Seems like the perfect thing for ransom notes, extortion threats, bribes, blackmail, insider trading and terrorism," Kelly said to May during an interview in Santa Cruz in the fall of 1992, referring to these early ideas for black markets. Brand shared this skeptical view and had already banned anonymity on the WELL for similar reasons. May was unfazed. "Well, what about selling information that isn't viewed as legal, say about pot-growing, do-it-yourself abortion?" the self-described anarchist responded to Kelly. "What about the anonymity wanted for whistleblowers, confessionals, and dating personals?"[28]

Most activists sided with May. In September 1992, a few crypto pioneers decided to do the obvious: take their meetings from the Palo Alto office into cyberspace and organize themselves in a mailing list. People were just beginning to use e-mail accounts, so an e-mail list seemed the best and most open way to network the group. Unlike on the WELL up in Sausalito, no membership fees were required and, more important, everyone could sign up anonymously. The list was open; no finger files wanted. Everybody could subscribe by simply e-mailing cypherpunks-request@toad.com. The list was hosted on a machine owned by John Gilmore, an early San Francisco–based crypto activist with a long mane, a flimsy beard, and a keen interest in recreational drugs. Gilmore was one of five original employees at Sun Microsystems, and like May, he was independently wealthy at a relatively young age.

Cypherpunk hailed from science fiction. The name was a give-away, of course. "You guys are just a bunch of cypherpunks," Jude Milhon exclaimed at one of the group's first meetings.[29] Then an editor at *Mondo 2000*, Milhon was better known as St. Jude, a bois-terous feminist hacker remembered for demanding that "girls *need* modems." St. Jude's play on words combined the then piping-hot sci-ence fiction trend with the British spelling of "cypher." The would-be anarchists loved their new hipster nickname. "The 'cyberpunk' genre of science fiction often deals with issues of cyberspace and computer security," May explained later, "so the link is natural."[30] May went up to Berkeley to a couple of *Mondo* parties hosted by Ken Goffman and his successor as the magazine's editor in chief, Alison Bailey Kennedy (a.k.a. Queen Mu).

In November 1992, St. Jude, herself a WELL member, ran one of the first stories on the nascent crypto movement in her ultimate cyber-punk magazine. Milhon related the appropriately cryptic anecdote about meeting two masked cypherpunks at a Screaming Meemees concert in the Black Hole, a Bay Area club.

"Actually, unmasking your real identity could be the ultimate collateral," one of the masked punks told her. "Your killable, *torturable* body. Even without kids, you've got a hostage to fortune—your own meat."

"AAIEEeeee," Milhon responded. "That's great *covert gear* you got there, guys."

"The revolutionists can be contacted at cypherpunks@toad.com," she added.[31]

Crypto anarchy spread. Soon, local chapters popped up in London, Boston, and Washington, DC. Like any reputable subculture phenomenon, the cypherpunks had their own jargon: pseudonyms and anonymous handles simply became "nyms," for instance, and they called themselves simply "c-punks."

Steven Levy and Kevin Kelly attended some of the first Palo Alto c-punk meetings. Levy portrayed the new movement in a famous cover story for *Wired*, in the magazine's second issue, which came out in May 1993. On the cover were Eric Hughes, Tim May, John Gilmore, holding up an American flag, their faces hidden behind white plastic masks, Gilmore even sporting an EFF T-shirt complete with the internet address of the then newly founded Electronic Frontier Foundation. The geeky rebels had their PGP fingerprints written on the foreheads of the masks.[32]

The same year, in the summer of 1993, Kelly published a long story about the crypto anarchists in the anniversary issue of the *Whole Earth Review*, guest-edited by its founder, Stewart Brand. Earlier that year, Mosaic 1.0 had been released, the world's first browser that could display graphics and text on the same page. The software, distributed for free, brought the web to life with color and images. Traffic exploded. The *Whole Earth Review* pointed out that the blooming network made encryption ever more necessary.

By November 1992, when *Mondo* first mentioned the list, it had about a hundred members, including journalists and even a few people with .mil addresses. Radical libertarians dominated the list, along with "some anarcho-capitalists and even a few socialists."[33] Many had a technical background from working with computers; some were political scientists, classical scholars, or lawyers. Two years into cypherpunk, the list had about five hundred people on it, after outages on the host machine had knocked the list back from more than seven hundred subscribers. The opinions didn't easily separate according to the political left and right, so the founding cypherpunks advised members not to rant on hot-button issues, like abortion or guns.

The list, and indeed the group, had no formal leadership. "No ruler = no head = an arch = anarchy," May clarified,[34] and he recommended looking up the etymology of the word "anarchy," just to be sure. Eric Hughes administered the list for the first years.[35] The emerging movement had no budget, no voting, no leaders. Yet the community remained active for many years. John Gilmore did the math: from December 1, 1996, to March 1, 1999, the list processed 24,575 messages. That's approximately thirty messages each day for more than eight hundred days.[36]

But science fiction wasn't just a namesake. Fiction stoked the cypherpunk movement's utopian ideas. On the list, as well as in articles and FAQs, May recommended to "read the sources." Those sources were not scientific articles on encryption or possibly pamphlets of libertarian or anarchist political thought. No, the recommended sources were novels—namely, George Orwell's *1984*, John Brunner's *The Shockwave Rider*, Ayn Rand's *Atlas Shrugged*, and especially Vernor Vinge's *True Names*. In fact, Vinge's work is referenced about twenty times in the *Cyphernomicon*, a sprawling three-hundred-page log that is perhaps the closest thing the movement has to a canonical document, organized as an appropriately messy and never-ending list of frequently asked questions. The only nonfiction source recommended in the document was Chaum's classic 1985 article "Security without Identification."

Inspired by fiction, the activists debated what was "holding up the walls of cyberspace," as May put it.[37] Gibson had famously described the new virtual realm of cyberspace as a consensual hallucination. The crypto activists didn't like that phrase. Something merely consensual wouldn't cut it. The new frontier couldn't just be hallucination. Their idea of cyberspace was rock solid, sturdy enough to withstand the assembled might of the US government, FBI agents and NSA spooks included. This explains why *True Names* holds much more appeal for the cypherpunks than *Neuromancer* does.

Vinge's heroic protagonist—Roger Pollack, a.k.a. Mr. Slippery— was a successful middle-class professional with a suburban house, garden, and car. But on the Other Plane, the Feds suddenly appeared powerless, at the mercy of anonymous hackers equipped with superior power, in need of Pollack's help. Gibson's drug-ravaged and addicted hustler Henry Dorsett Case was the most unappealing contrast. Cyberspace couldn't just be a collectively imagined illusion; the cypherpunks preferred to see the vast online frontier as a new territory that could be as rugged and dangerous to outsiders as the high plains of the Rocky Mountains. The question that hardened the walls of cyberspace was therefore a pertinent one. May tackled it in one of his longer essays:

> What keeps these worlds from collapsing, from crumbling into cyberdust as users poke around, as hackers try to penetrate systems? The virtual gates and doors and stone walls described in *True Names* are persistent, robust data structures, not flimsy constructs ready to collapse.[38]

The answer, by now, was obvious. Cryptography provided the "ontological support for these cyberspatial worlds," he understood.

The astounding mathematical power of quite large primes guaranteed enduring structures in the vastness of a new space that could now be safely "colonized," May believed. Owning a particular "chunk of cyberspace," he explained, meant running software on specific machines and networks. And the owners of such virtual properties made the rules. They set access policies and determined the structure of whatever was to happen there: "My house, my rules." Anybody who didn't like the rules in a particular virtual world would be welcome to stay away. And, May was convinced, anybody who wanted to call in old-fashioned governments to force a change of the rules would face an uphill battle.

For the libertarian minded, crypto anarchy meant that "men with guns" could not be brought in to interfere with transactions that all participants mutually agreed on. Taking violence out of the equation had two wide-reaching consequences. Two types of men with guns would find crypto hard to cope with. The first were the police and agents of federal law enforcement. No longer would they be able to trace and find those who refused to declare income or deal in illegal goods. The state, in short, would lose a good deal of its coercive power. If financial transactions became untraceable, enforcing taxation would be impossible. And that, of course, was a good thing. "One thing is for sure," May told Kevin Kelly of the *Whole Earth Review* already in late 1992, "long-term, this stuff nukes tax collection."[39]

But crypto wouldn't affect only the government and the rule of law. The other kinds of men with guns were criminals. And the same applied to them. Criminals would also lose their power to coerce others with threats of physical violence. If the buyers of drugs, for instance, would be untraceable not just for the Feds but also for gangs, then markets that were chronically plagued by violence and abuse would become nonviolent and abuse would stop. Anonymously ordering LSD online was much less risky than going to dodgy street corners and talking up shady pushers.

Strong crypto, made widely available, enabled totally anonymous, unlinkable, and untraceable exchanges between parties who had never met and who would never meet.[40] The anarchists saw it as a logical consequence that these interactions would always be voluntary: since communications were untraceable and unknown, nobody could be coerced into involuntary behavior. "This has profound implications for the conventional approach of using the threat of force," May argued in the *Cyphernomicon*. It didn't matter if the threat of force would come from governments or from criminals or even from companies: "Threats of force will fail."[41]

Crypto anarchy would not just take force out of chronically violent markets; it would also be a shot in the arm of dysfunctional markets. One of May's favorite analogies was guilds. Medieval guilds had monopolized information—for instance, how to make leather or silver. When independent entrepreneurs would try to produce these goods outside the guilds, "the King's men came in and pounded on them because the guild paid a levy to the King."[42] The police, tax collectors, and corporate interests joined forces.

Printing broke the oppressive system, May argued. Suddenly someone could publish and distribute a treatise on tanning leather, and the king would be unable to stop the knowledge from spreading like wildfire. But even in the age of printing, May lamented, some firms retained a firm grip on specialized technologies, such as gunsmithing. This era would now come to an end. Encryption, he reasoned, would liberate expertise and proprietary knowledge. "Corporations won't be able to keep secrets because of how easy it will be to sell information on the nets," he said, in the dated language of the 1990s.[43] All kinds of transactions would become possible, without restrictions.

The *Cyphernomicon* felt raw, unedited, and self-absorbed in style. The pamphlet was clear: there was a great divide; it was either privacy or compliance with laws. Both at the same time, it implied, was impossible. The gun debate offered the template of self-protection versus protection by law and police—"crypto = guns," as May put it. Both enabled the individual to have "preemptive protection." Some of the most potent cypherpunk slogans were simply copied and pasted from America's great gun debate: "If crypto is outlawed, only outlaws will have crypto." One of the movement's most popular slogans became an adaptation of those famous five words coined so dramatically by the NRA's Charlton Heston: now it was not guns but crypto being pried "from my cold, dead hands."[44]

From the start, cypherpunk was about getting stuff done, not just debate and organization for debate and organization's sake. In the summer of 1992, John Gilmore, one of the three original crypto rebels on the *Wired* cover, had made a bold move against the NSA that catapulted the movement into the national spotlight.

Gilmore had come across two cryptography books that piqued his interest: The first, *Military Cryptanalysis*, by the NSA cryptographer William Friedman, was a four-volume text published while World War II was still raging. The second, *Cryptanalytics*, in six volumes published between 1956 and 1977, was coauthored by Friedman and one of his students, Lambros Callimahos. In each case, the first two volumes had been declassified, which is why Gilmore knew about them.

In early July 1992, Gilmore filed a Freedom of Information Act request with the NSA, asking the agency to declassify the remaining Friedman volumes. Gilmore wasn't requesting some coffee-table crypto book. Friedman was a legendary cryptographer: he had cofounded the US Army's Signal Intelligence Service in the 1930s, a direct predecessor to the NSA. The National Cryptologic Museum idolizes him as the "Dean of American Cryptology." Friedman himself had coined the term "cryptanalysis," and he even had a five-hundred-person lecture theater at Fort Meade named after him.[45]

Without knowing it, Gilmore had requested one of the NSA's founding documents. The agency, as is common, simply dragged out its response. Gilmore has a soft voice and the appearance of a peaceful hippie, with John Lennon glasses and a serene smile. Not one to be easily deterred, however, he considered filing an appeal.

Then a cypherpunk from the East Coast got in touch with Gilmore. "You know, I think I saw something like that at a library," he told him.[46] After Friedman's death in 1969, his personal papers had gone to a public library on the campus of the Virginia Military Institute in Lexington, Virginia. The papers included the unpublished manuscript of the coveted book. The cypherpunk simply went there, Xeroxed the book's page proofs, complete with Friedman's annotations, and sent a thick packet to Gilmore in California. In early October, a week after the packet landed with a thud on Gilmore's doorstep, a second letter from the East Coast arrived in his mailbox. It was from the NSA.

The agency had written to let Gilmore know that it would not release the books to him. In its response, the NSA referred to a statute—18 U.S.C., Section 798. That statute made it a federal crime to publish classified cryptologic information. That was when it dawned on Gilmore that he had a problem. The documents in his possession were classified, and disseminating them—even showing them to experts—would be a crime.

But the NSA didn't know he had the documents already. Gilmore carefully considered his options. He decided to submit copies of the classified documents to a federal district court under seal. The activists had limited trust in the judicial process. They trusted math, not the law. So, before Gilmore filed the sealed document with the judge, he made several copies of the classified book and hid them in extremely unlikely places.[47]

For a while it looked as if the situation could escalate. A Justice Department lawyer representing the NSA demanded that Gilmore surrender his illegal copies and threatened that the NSA might send its own operatives, or FBI agents, to seize Friedman's proofs from Gilmore. The cypherpunks were worried—especially Gilmore's lawyer, Lee Tien. They were worried not just for their own personal freedom, but for the freedom of the country. At academic crypto conferences, rumors were making the rounds that the NSA had already raided the New York Public Library and reclassified documents that used to be public, and that in 1983 they had already removed Friedman's personal correspondence from public access. An early court case seemed to affirm the government's right to snatch any given document. "If they could do a black-bag job on everyone who had it," Gilmore recalled in a café in Haight-Ashbury, "then they could classify anything."[48]

By now the anarchists were beginning to understand what the NSA feared. The c-punks had seen that the spooks at Fort Meade could tweak the law and play politics. But the activists also knew the secret agency hated publicity. So Gilmore started calling some of the technology reporters he knew through the cypherpunks list. One of the best-known journalists in San Francisco at the time was John Markoff, from the *New York Times*. Gilmore reached out to him. The *Times* later ran the story. "In Retreat, U.S. Spy Agency Shrugs at Found Secret Data," the headline read.[49]

Gilmore's plan worked as predicted: the NSA shunned the light. The agency declassified the Friedman documents in response to the high-profile publicity. The NSA's lawyers didn't call Gilmore's lawyer to tell him that they had finally yielded and would declassify the Friedman books; they told Markoff straightaway. "I heard it from Markoff," Gilmore recalled. "They wanted to kill the story."[50]

Friedman's actual book was useless for the cypherpunks. There was nothing in its pages that made any difference for the kinds of cryptographic tools the activists were developing. But the episode was a psychological success. It reaffirmed two views: that "the NSA was the enemy," as Gilmore put it—and that this enemy wasn't almighty. Even a few long-haired hippie activists could score a win against the mighty military machine. And it was only late 1992. The cypherpunks were just getting started.

"Cypherpunks write code," was the mantra. Naturally, that was a statement of principle and a bit of an exaggeration. Not everybody on the list wrote code. In fact, only 10 percent of the cypherpunks wrote code, and only 5 percent worked on encryption-related projects. Early on, the activists had set their sights on a fundamental privacy problem in the digital age that was, at least then, unrelated to encryption: anonymity.

Making strong encryption available to John Doe was a big step forward for privacy. But it didn't even begin to solve a fundamental problem: scrambling plaintext to ciphertext beautifully protected the letter inside an envelope. Whoever opened the envelope could not read what was inside. That was great. But the encryption available at the time didn't conceal what was *on* the envelope: the sender's address, the receiver's address, and some other information about when and how the letter was sent. The correspondents' identity was openly revealed. Encryption protected the *content* of packets but not the *headers*—what later would be called "metadata." The now publicly available PGP protocol left metadata unprotected. PGP on its own, in short, created confidentiality, not anonymity. The cypherpunks wanted a solution to this problem.

Remailers were the solution. These were dedicated machines programmed to take scissors to the envelopes, encrypted or not, to physically cut out the sender's address and then forward the e-mail to the recipient. Remailers, in other words, were servers that automatically stripped e-mails of information that could identify the sender: the code running on the remailer would cut out the metadata, removing the sender's address, replace it with a nonexistent placeholder such as nobody@shell.portal.com, and forward it to the intended recipient. It was like writing a letter with no sender's address, or like calling somebody from a public phone with a distorted voice. Remailers could also be chained, to increase security, just in case one remailer kept a log file that could identify the sender. Court orders or lawsuits were ineffective against machines that automatically forgot data. But integrating PGP into remailers was a problem, at least initially.

Eric Hughes and Hal Finney wrote the first such remailers in 1992, in the programming languages Perl and C. By 1996, several dozen remailing machines would be operational. They had many uses. It became possible, for instance, to publish sensitive information simply by e-mailing it to a publicly accessible e-mail list, because nobody could trace the e-mail back to its sender. In this way the remailers were used to "liberate" ciphers that had not been published before, to spill a few government secrets, and to reveal secrets of the Church of Scientology.

By late 1992, things had started to move. The use of PGP was on the rise, and the first remailers were coming online. On December 1, 1992, two days after Gilmore had scored his symbolic victory against the NSA, John Perry Barlow addressed a meeting of national-security and intelligence officials in McLean, Virginia: "I believe you folks in the Intelligence Community are going to [be] challenged by these issues as directly as anyone," he told the spooks. The EFF cofounder knew that intelligence agencies were working under strict guidelines separating the domestic from the foreign. "You're not supposed to be conducting domestic surveillance," Barlow lectured the gathered officials. "Well, in Cyberspace, the difference between domestic and foreign, in fact the difference between any country and any other country, the difference between us and them, is extremely blurry. If it exists at all."[51]

For Barlow and the cypherpunks, this Vingean prophecy of a borderless cyberspace secured through the power of large primes was a shimmering pacific dream slowly inching into reality; for national security–minded government officials, it was a sinister, threatening nightmare. So in April 1993, the White House under Bill Clinton tried a new approach: if they couldn't stop the spread of crypto, they could perhaps control it. The government proposed a new federal standard for encryption.

The proposal was officially named the Escrowed Encryption Standard, or EES. It was designed to enable encrypted telecommunication, especially voice transmission on mobile phones—but with a twist. The standard encompassed an entire family of cryptographic processors, collectively and popularly known as "Clipper chips." The government expertise for designing such a system, naturally, resided in the country's mighty signal intelligence agency, the NSA. The proposal was then to be implemented through NIST, the National Institute of Standards and Technology.

The system's basic feature was simple in theory: when two devices establish a secure connection, law enforcement agencies will still be able to access the key that was used to encrypt the data. In short, communication was protected, but the FBI could read the mail or listen in when needed. The technical implementation of that simple idea was more difficult than expected.

NSA engineers came up with what they thought was a neat trick. To make a secure phone call, two phones would first establish a so-called session key to encrypt the conversation. That much was a given. The session key would unlock the ciphertext and reveal the plaintext. So the NSA needed to find a way to make the session key accessible to law enforcement without compromising the phone's security. To do that, they created a so-called Law Enforcement Access Field, abbreviated LEAF. The LEAF would retain a copy of the session key. That retained session key, of course, was sensitive, and was itself encrypted with a device-specific key, called the "unit key." This unit key was assigned at the time the Clipper chip was manufactured and hardwired into the device. Unit keys were held in "escrow" by two government agencies. The Feds, in short, had a spare key for encrypted traffic. The White House argued that Clipper would achieve twin goals: the chip would provide Americans with secure telecommunications, and it would not compromise law enforcement agencies in their ability to do legal, warranted wiretaps.

The cypherpunks, predictably, called BS on Clipper. The chip wasn't just controversial; it was a bombshell. The tiny chip was the big cause the movement was waiting for. The very idea that a government, whatever its constitutional form, should be allowed to hold a copy of all secret keys was simply absurd to the growing number of crypto activists. "Crypto = guns" now meant that the Clinton administration faced the combined rage of First Amendment and Second Amendment activists, of those in favor of free speech and armed self-defense: Berkeley academics = NRA types. "Would Hitler and Himmler have used 'key recovery' to determine who the Jews were communicating with so they could all be rounded up and killed?" May asked on the list, rhetorically.[52]

Graffiti sprayers took up the theme: "Stop Clipper—Fuck the NSA" appeared on a garage door at the corner of Sixteenth and Harrison Streets in San Francisco, in March and April 1994. Realist-minded privacy advocates pointed out that any key escrow database would be a juicy target for aggressive intelligence agencies. The activists saw the LEAF and Clipper as a government-mandated "back door" into secure systems. Key escrow effectively meant "key surrender," the EFF argued.

One famous anti-NSA slogan of the crypto wars of the early 1990s was the chip-mocking "Big Brother Inside," a play on the famous tagline of one of the world's leading chip makers: "Intel Inside." Others designed a logo for T-shirts and pins, "Fight the Clipper," with the old tagline from May's "Crypto Anarchist Manifesto": "Arise, you have nothing to lose but your barbed wire fences!" John Perry Barlow had one of the most powerful lines: "You can have my encryption algorithm," he thundered, yet again using that favorite line, "when you pry my cold dead fingers from my private key."[53]

Graffiti and punch lines didn't kill the chip. A hack did. This final blow was expertly administered by one of the cypherpunks, Matt Blaze, who then worked for AT&T. The Clipper's cipher algorithm, known as Skipjack, remained classified. The government would provide the cipher algorithm only in preimplemented, tamper-resistant modules. And it would provide the hardware only through vetted vendors. AT&T was one of those vendors, and it was there that Blaze was able to test the chip.

Blaze found that the LEAF, the spare key, was flawed and indeed vulnerable to tampering. He published his results in a now-famous paper in August 1994.[54] Blaze's finding meant, ironically, that the Clipper could be fixed by breaking it—at least from the point of view of privacy activists: by subverting the law enforcement access field, the encryption remained intact but law enforcement no longer had access. Soon the Clipper was axed, and the cypherpunks bagged another victory against the government.

Several crypto activists also battled the government in court, with varying degrees of success. Three notable cases concerned the status of computer code and machine language as free speech. In the first case, filed in 1994, Phil Karn, a former Bell Labs engineer and cypherpunk, unsuccessfully brought a suit against the Department of State claiming that restricting the export of a diskette with Bruce Schneier's book *Applied Cryptography* on it infringed on his First Amendment rights. In 1996, cypherpunk list member Peter Junger, a professor at the Case Western University School of Law, filed a suit against the Commerce Department for enforcing export regulations against him because he was teaching a computer law class in the United States—ultimately also without success.

Most influential would be a bundle of court cases brought by Daniel Bernstein, known as *Bernstein v. United States*. The young mathematician, represented by the EFF, challenged the State Department for requiring him to get a license to be an arms dealer before he could publish a small encryption program. Then, by the end of the century, on May 6, 1999, the US Court of Appeals for the Ninth Circuit would rule, in a first, that software code was constitutionally protected speech, eventually ushering in the end of the hated cryptographic export control regime.

Emboldened by success and mainstream media coverage, crypto anarchy became not less extreme but more so. Stewart Brand, for one, remained highly skeptical of anonymity online. On April 1, 1994, when the Clipper debate reached its fever pitch, a "nobody" reported on the newsgroup that Phil Zimmermann had been arrested and that the crypto pioneer was being held on bail of $1 million.[55] Brand was about to sit on a panel with Zimmermann and was not amused. He responded to the cypherpunks two days later: "The Zimmerman[n] prank," he wrote, "hardens my line further against anonymity online. At its best, as here, it is an unholy nuisance."[56] Brand commanded tremendous authority among subscribers, so the founders didn't appreciate his snide comment.

"You can't get rid of anonymity," Hughes responded the next morning. He pointed out that there is no clear difference between saying something anonymously and saying it by using a pseudonym. "The first use of a pseudonym is as good as anonymous, because it has no past history," he wrote.[57] This was a clever move. The activists were fond of pointing out that cypherpunks had even been among the nation's founding fathers. Alexander Hamilton and James Madison had chosen the handle Publius to publish the *Federalist Papers*. The cypherpunks even compared themselves to the founding fathers: Gilmore explained that May was the Thomas Jefferson, "the essayist," while Hughes was more the Benjamin Franklin, "the coder."[58] Basically, Hughes responded to Brand that opposing anonymity was un-American.[59]

But Brand was right. Several influential ideas emerged that polarized crypto anarchy and drove the uncompromising and self-defined founding fathers of online anonymity further to the fringe. One such idea, and perhaps the most prophetic one, was BlackNet.

BlackNet was the anti-WELL. It was the machine of loathing and disgrace: completely anonymous, without physical location, and amoral by design. An anonymous voice out of the emptiness of cyberspace (it had to be!) introduced the idea on August 18, 1993. The "Introduction to BlackNet" came as an e-mail to the cypherpunks list through a remailer. The message started ominously, with a hint of irony and faux exaggeration that was obvious to most members on the list:

> Your name has come to our attention. We have reason to believe you may be interested in the products and services our new organization, BlackNet, has to offer. BlackNet is in the business of buying, selling, trading, and otherwise dealing with *information* in all its many forms.[60]

The anonymous author made clear that BlackNet would use public-key cryptosystems to guarantee total and perfect security for customers. The marketplace in cyberspace, the anonymous author wrote, would have no way of identifying its own customers, "unless you tell us who you are (please don't!)."

BlackNet was also elusive. "Our location in physical space is unimportant," the message read: "Our location in cyberspace is all that matters." The mysterious anonymous voice gave a fictional address: "BlackNet <nowhere@cyberspace.nil>." That was obvious geek irony: .nil didn't exist as a top-level domain. But the e-mail then ominously added, "We can be contacted (preferably through a chain of anonymous remailers) by encrypting a message to our public key (contained below) and depositing this message in one of the several locations in cyberspace we monitor." These locations were two Usenet groups—alt.extropians and alt.fan.david-sternlight—and of course the cypherpunks list itself. This didn't look like irony.

The idea for BlackNet was to remain "nominally" nonideological. But the author of the ominous passage made clear that he considered nation-states, export laws, patent laws, and national security "relics of the pre-cyberspace era." These things simply served one nefarious purpose: to expand the state's power and to further what the BlackNet pioneers called "imperialist, colonialist state fascism."[61]

The curious e-mail pamphlet then made clear that BlackNet would currently build up its "information inventory," and that it was interested in acquiring a range of commercial secrets. "Any other juicy stuff is always welcome," the anonymous voice added. BlackNet was specifically interested in buying trade secrets ("semiconductors"); production methods in nanotechnology ("the Merkle sleeve bearing"); chemical manufacturing ("fullerines and protein folding"); and design plans for things ranging from children's toys to cruise missiles ("3DO"). Oh, and BlackNet was also interested in general business intelligence—for instance, on mergers and buyouts. "Join us in this revolutionary—and profitable—venture," the message concluded.[62]

May soon claimed credit for the letter. He had come up with Black-Net in the summer of 1993, as an example of what could be done, "an exercise in guerrilla ontology," as he called it.[63] The goal was to find a way to ensure fully anonymous, untraceable, two-way exchanges of information. The core idea was to use a public message pool as the main channel. The sender would use a chain of remailers to deposit a note in the pool anonymously and untraceably. The sender would have encrypted this message with the intended recipient's public key. The intended recipient, and only the intended recipient, could then simply download, decrypt, and read the public message—and respond in the same way.

The hoax was too good not to be true. May claimed that he had come up with BlackNet just for the purposes of education and research. But the interests he listed—in particular, the commercial secrets—seemed a bit too detailed for a prank. It was credible. What was even more remarkable is that May's overall idea for BlackNet sounded like a business plan. It could actually work.

Al Billings, then an anthropology student at the University of Washington, responded on the list just a few hours after the invitation went online: "It had to happen," he said. "Even if it isn't real, it will happen soon enough. I'm all for it."[64]

"I think it's not real, or at least wasn't intended to be," another cypherpunk responded, picking up on the subtle irony. "My best guess is that it's all a joke, but that the author will soon start receiving genuine replies; it may yet turn into the real thing."[65] They were both spot-on.

The master cypherpunk had created a "proof-of-concept implementation of an information trading business with cryptographically protected anonymity of the traders," in the words of Paul Leyland, an Oxford University number theorist and cryptographer.[66]

The editors at *Wired* apparently didn't spot the bluff. The November 1993 issue announced BlackNet as if it were real: "Sent to us anonymously (of course)," *Wired* wrote and then quoted at length from May's faux pamphlet. A few months later, in February 1994, Lance Detweiler, a former computer science student from Colorado who had become a cypherpunk troll, posted the BlackNet announcement to more than twenty newsgroups and lists, pushing it out to many thousands of recipients.[67]

Some copies made their way into sensitive networks. Oak Ridge National Laboratory issued an advisory to employees and recommended reporting any contacts with BlackNet to supervisors. Indeed, May soon did receive genuine replies and a number of strange propositions, including an offer to sell information about how the CIA was blackmailing diplomats of certain African nations in Washington and New York.[68] May says he decrypted the message with BlackNet's private key and then put it away and never responded.[69]

One of the main motivations for BlackNet, as May had outlined, was economic espionage. By the spring of 1994, the idea had made the rounds on forums, newsgroups, and the then novel World Wide Web, a browsable interface with clickable links that has become synonymous with the internet for most users. The FBI began taking the threat of BlackNet-organized industrial espionage seriously and started investigating. One anonymous post on the list claimed that two federal agents had interrogated Detweiler in Denver about Black-Net, and even correctly identified one of the agents.[70]

The Feds also contacted May and other cypherpunks. The crypto cowboy was so concerned about the unwanted attention, and about his idea's obvious potential for abuse, that he emphasized again and again that he wasn't the person who had posted the initial message on the cypherpunks list. But BlackNet was never actually used as a message pool to sell information. One of the main ingredients was missing: digital cash, which existed only as an idea then.

Soon, an even nastier suggestion appeared on the list: a public hit list for politicians, set up as a portal for contract killing. It also came from a former Intel engineer, Jim Bell. He was one of the most radical minds on the cypherpunks list (Bell and May met only once, in the early 1980s).[71] In August 1992, Bell had read an article by David Chaum in *Scientific American* titled "Achieving Electronic Privacy." Chaum, then the head of a cryptographic research group at Amsterdam University, painted a dark picture: making a phone call, using a credit card, subscribing to a magazine, paying taxes, all these bits and pieces of information could be collected and combined into "a single dossier on your life—not only your medical and financial history but also what you buy, where you travel and whom you communicate with."[72]

A scary idea, Bell thought when he read this. But Chaum's suggestion for achieving electronic privacy inspired Bell in unexpected ways. Chaum suggested a method for using as identifiers not real names or social security numbers, but "digital pseudonyms." Such pseudonyms would make it much harder to connect various bits and pieces of information back to the same actual individual, just like in the good old days when cash payments were the norm.

"A few months ago, I had a truly and quite literally 'revolutionary' idea," Bell wrote in April 1995, still inspired by Chaum's ideas, "and I jokingly called it 'Assassination Politics.'"[73] Except Bell wasn't joking. *Assassination Politics* was the title of a sincere ten-part essay that was to become highly influential. To Bell, Chaum had it the wrong way around: The Amsterdam cryptographer was asking how the freedoms of ordinary life could be reproduced on the internet. For Bell, the more exciting question was, again, the opposite: "How can we translate the freedom afforded by the Internet to ordinary life?"

Bell's basic concerns seemed innocent enough: keeping the government from banning encryption and digital cash, the new technologies of freedom. But Bell reversed the cypherpunks' classic motivation: he didn't just want to ensure that well-established brick-and-mortar freedoms would apply on the new internet; he wanted to translate new internet freedom to the established social norms of ordinary life. Chaum suggested shaping cyberspace in the image of the real world; Bell suggested shaping the real world in the image of cyberspace. This reversal enabled a revolutionary outlook.

Bell suggested a market for assassinations. The MIT graduate had the idea to set up a legal organization that would announce a cash prize. The winner of that cash prize would be whoever correctly "predicted" the death of a specific person. The person on the hit list, Bell clarified, would be a violator of rights already, "usually either government employees, officeholders, or appointees."[74]

The fact that the suggested system was targeting the government made all the difference: Bell didn't claim that a person who hires a hit man is not guilty of murder. Obviously, the victim could be innocent. But he was not dealing with innocents. He wouldn't even initiate the use of force. The government, after all, held a monopoly on violence, and taxing citizens as well as enforcing the law represented a use of force already. By "taking a paycheck of stolen tax dollars" and being tied to police agents, a government employee had already violated the "non-aggression principle." Therefore, "any acts against him"—the government employee—"are not the initiation of force under libertarian principles."[75] Killing government employees, Bell argued, was a legitimate form of self-defense. Crypto was, indeed, like guns.

Strong crypto made legitimate hits possible—at least in Bell's mind. At the heart of his suggested assassination politics was a wish hit list administered by an organization. The list was to be made public. It had two columns: one with a government employee's name, and one with money pledged for that person's "predicted" death. (Bell always put that "predicted" in quotation marks, for somebody had to make the prediction come true). Anybody holding a grudge against a particular politician or government agent could bet a small, or large, amount of money on that person's life. "If only 0.1% of the population, or one person in a thousand, was willing to pay $1 to see some government slimeball dead, that would be, in effect, a $250,000 bounty on his head," Bell explained.[76] The bounty, once large enough, would be a market-driven incentive for assassins.

Bell's suggested system worked something like this: "Guessers" would create a file with their "guess"—that is, the politician or bureaucrat's name and the time stamp of his assassination. The person making the "prediction" would then encrypt that file with a private key, so that nobody else could read or edit this information without having access to the private key. Next, the "guesser" would put the sealed envelope and some digital cash in a second envelope, which would be encrypted with the organization's public key, so that only the prize-giving organization could open the envelope. The money was needed to avoid a large number of random guesses, Bell reasoned.

Once the hit had been made, the victim's death would become publicly known through press reports. The winner could then send an encrypted envelope to the organization that contained two things: a private key and a public key. The organization could use the private key to verify that the winner had correctly predicted the hit by opening the previously submitted bet with this private key. But how did the winner get paid? The second element in the envelope was a public key to which only the winner possessed the private key. The public key would effectively be used to "transfer" the prize cash to the winner— by publishing it online, so that everybody could see it. The winner, and only the winner, could then download the encrypted cash and unlock it with another private key. All this would be entirely untraceable. "Perfect anonymity, perfect secrecy, and perfect security," Bell boasted.[77]

Bell's vision was truly revolutionary. Nothing, he believed, would stay the same:

> Just how would this change politics in America? It would take far less time to answer, "What would remain the same?" No longer would we be electing people who will turn around and tax us to death, regulate us to death, or for that matter sent [sic] hired thugs to kill us when we oppose their wishes.[78]

Bell, predictably, became a divisive figure. Then–IRS inspector Jeff Gordon compared him to terrorist Timothy McVeigh, who bombed a federal building in downtown Oklahoma City in April 1995, killing 168 people. On the other end of the spectrum was John Young, an architect and cypherpunk who would found the first whistle-blowing portal in 1996: Cryptome. Young nominated Bell for a Chrysler Design Award for creating an "Information Design for Governmental Accountability."[79]

Crypto anarchy, some successes notwithstanding, seemed to meander toward the fringe. But curiously, the ideology lacked a proper book-length treatment. There was graffiti, as well as a deluge of rambling e-mails, magazine stories, interviews, and the messy and disorganized *Cyphernomicon*. The cypherpunks wrote code, but not books. May, the movement's self-styled essayist, had tried and failed.[80]

Then, in 1997, Simon & Schuster, one of the big New York publishing houses, published *The Sovereign Individual*. It was a strange book, full of apocalyptic yet optimistic predictions. The two authors, inspired by the political philosophy of cypherpunk, left out the jargon and the arcane crypto discussions, yet kept the boldness: cyberspace was about to kill the nation-state, they argued.

Lord William Rees-Mogg was a prominent, albeit sharply controversial, figure in British public life. From 1967 to 1981, the owlish Rees-Mogg was editor of the *Times*, chairman of the Arts Council of Great Britain, and the BBC's vice-chairman. In 1988 he was made a life peer in the House of Lords, as Baron Rees-Mogg of Hinton Blewett in the County of Avon. His coauthor was James Dale Davidson, a conservative American financial commentator and founder of the National Taxpayers Union, an advocacy group.

"As ever more economic activity is drawn into cyberspace, the value of the state's monopoly power within borders will shrink," Rees-Mogg and Davidson predicted. "Bandwidth is destined to trump the territorial state."[81] To back up their futurology, the two pundits called on the acid-dropping former cattle rancher from Wyoming, John Perry Barlow. He had it right, they said: "Anti-sovereign and unregulatable, the Internet calls into question the very idea of a nation-state."[82]

Echoing May and the cypherpunks, they argued that the state's threats of coercion would simply be ineffective online, shielded by strong crypto. "The virtual reality of cyberspace," they wrote, "will be as far beyond the reach of bullies as imagination can take it."[83] The advantage of large-scale violence, of police or military force, would be far lower than it had been at any time since the French Revolution. Individuals would no longer need, or tolerate, sovereign states above them. The age of violence was over. The individual would now become the sovereign, effectively taking over from the state. Soon, most of the world's commerce would be absorbed "into cyberspace," a novel realm where the governments of old would have "no more dominion" than they exercised over the bottom of the sea or indeed the solar system's outer planets. "In cyberspace, the threats of physical violence that have been the alpha and omega of politics since time immemorial will vanish."[84]

One big reason for this coming revolution was digital money. "Cybercash" would slash the state's ability to control its citizens. In the near future, any commercial transaction would happen over the "World Wide Web," paid for in untraceable digital cash. Taxation would become difficult, if not impossible, thus cutting the state back to size, if not destroying it entirely. As the two conservative authors put it in a twisted reference to Lennon and McCartney: "Cyberspace is the ultimate off-shore jurisdiction. An economy with no taxes. Bermuda in the sky with diamonds."[85]

The authors didn't use the phrase "sovereign individual" as mere slogan. "One bizarre genius" could achieve the same impact in cyberwar as a nation-state, they argued confidently. The Pentagon was no more powerful than some teenage whiz kid. Technology had truly leveled the playing field in future confrontations: "The meek and the mighty will meet on equal terms."[86] The consequences were profound: "Nation-states will have to be reconfigured to reduce their vulnerability to computer viruses, logic bombs, infected wires, and trapdoor programs that could be monitored by the U.S. National Security Agency, or some teenage hackers," Rees-Mogg and Davidson predicted.[87]

An advance Kirkus review called the best-selling book "astonishing" and "penetrating." The *Vancouver Sun* thought it was "must reading," Toronto's *Financial Post* described it as "sobering." Reviews appeared in the *Guardian* and the *Wall Street Journal*.[88] Predictably, the cypherpunks loved it. "*The Sovereign Individual* discusses many of the issues discussed on Cypherpunks," one list member wrote.[89]

"Strongly suggested for any cpunk," added Jim Choate, who had written one of the first remailers.[90] Some were skeptical. The book lacked precision and was full of bold overstatements. And although the two authors never referred to crypto anarchy or the c-punks, the publication gave wider currency to an emerging political philosophy. But the book's success was short-lived—sharing this fate with the utopian ideology from which it sprang.

Inspired by Vinge's story and the cypherpunk list, a few entrepreneurs took the idea of the sovereign individual rather literally in those enthusiastic years before the crash of the New Economy. One of them was Ryan Lackey. Digital cash had fascinated Lackey since he was fifteen years old. He had already started an e-money start-up on Anguilla, a loosely regulated island, but he had run into trouble with the ruling family. An avid c-punk, he had hosted the list archives on an MIT server when he was a student in Boston. Lackey even looked like a textbook cypherpunk: head shaved bald, pale skin from spending too much time in front of screens, rimmed glasses with a dark black frame, and usually dressed all in black. Privacy and internet freedom, as he saw it, were under siege: laws everywhere, particularly in the United States, were getting more and more restrictive and authoritarian.

Several crypto anarchists had long been looking for an offshore jurisdiction to run the automats of freedom: remailers, racks of servers dishing out encryption, and machines minting digital cash. Anguilla initially seemed a good option, as did Tonga. Then, in June 1999, Lackey and Sean Hastings discovered a curious place in a book, *How to Start Your Own Country*, by Erwin Strauss.[91] That place was the Principality of Sealand, a tiny artificial island on a World War II anti-aircraft platform in the North Sea known as Roughs Tower. The rust-covered 550-square-meter platform sat on two giant hollow pontoons, 60 feet above the brown-green waves of the harsh North Sea, 7½ miles off Felixstowe, on the Suffolk coast.

On September 2, 1967, Roy Bates, a retired army major and World

War II veteran, had declared the rig independent from Britain, bestowing the title "princess" on his wife (it was her birthday). Britain refused to accept the principality's sovereignty, as did the United States, the UN, and all other international organizations. Hastings and Lackey, however, had perfect timing. In 1999, Prince Roy was battling Alzheimer's and his health was deteriorating. The "royal" family considered leaving Sealand, making it available for other uses. In November that year, Hastings visited the platform for the first time. He already had experience in offshore financing and online gambling in Anguilla, where he had met Lackey. After inspecting Sealand that November, the entrepreneurs were inspired. They decided to move forward.

"The biggest inspiration was Vernor Vinge, *True Names*," recalled Lackey.[92] The vision was to have individuals acting in the Other Plane, able to "live on hardware and transact stuff on their own and not have to be under any government," Lackey recounted later in Palo Alto. These sovereign individuals were to be "the first class citizens," he said. The principality was the weakest country they could find in terms of jurisdictional challenges. The deserted antiaircraft platform in the North Sea had simply no legal system, no police, and no law enforcement. Sealand seemed like the perfect place to start for would-be sovereign individuals.

HavenCo Ltd. was to be the world's first real data haven, "physically secure against any legal actions," the business plan promised. The idea was to combine the best of the "first world"—high-quality infrastructure—with the best of the "third world": hosting data and running businesses "free of unnecessary regulation and taxation."[93] "Sealand is located less than three milliseconds (by light over fiber) from London," the business plan promised, in language indeed reminiscent of *True Names*.

Lackey, then twenty years old, became the chief engineer and moved to the barren platform. He spent the better part of two years on the rig. He was there mostly on his own, maintaining HavenCo's operations, with others coming just when journalists were visiting by boat. The media loved the story of Sealand.

No doubt, the start-up was a hot idea. It was so hot that *Wired* magazine had commissioned a cover story on it even before HavenCo had done anything. The magazine's reporter, Simson Garfinkel, was literally on Lackey's heels, in the same boat, when HavenCo's chief engineer visited Sealand for the very first time, in January 2000. Lackey recalls that the *Wired* reporter was "very credulous."

"The Ultimate Offshore Startup," was the magazine's title story in July 2000. "Meet the high-seas adventurers on a multibillion-dollar quest to build a fat-pipe data haven that answers to nobody." On *Wired*'s cover was the rusty antiaircraft platform with helicopter landing pad. Assembled on the platform was a fictional team of nine people (there were only four). The entire rig was not just sticking out of the ocean; it was gargantuan, reaching through the clouds and the atmosphere all the way into space, above the entire round, blue planet, *Whole Earth Catalog* style. The magazine presented the North Sea rig as a veritable Bermuda in the skies with diamonds. What a public launch.

After the *Wired* story came out, managing media requests and frequent visits of journalists took up more time than getting actual business operations up and running. Meanwhile, other things didn't go so well. Lackey had hired Winstar Communications to run a fiber-optic line from London to the shore. Winstar was one of the poster children of the late-1990s internet boom. But the overexposed company, with a revenue of $445 million in 1999, went belly-up in 2001. The fiber connection to the coast was never built, let alone the "fat pipe" to Sealand that *Wired* had announced so credulously. There was only a wireless link.

HavenCo's "crazy plan," Lackey said, was to lay a repeaterless fiber cable from London to the Netherlands via Sealand. That cable also remained a fat pipe dream. The ambitious internet start-up had to make do with a data transfer speed of just 10 megabits per second. That meant that downloading just one regular 1.5-GB movie would have taken nearly twenty minutes, clogging the pipe. "Low bandwidth killed the economics," Lackey recalled.[94]

Life on the c-punk rig was rough. Initially, the small HavenCo team had a nice spot in the edgy and hipster docklands in London. Living in East London was much more pleasant than on a deserted antiaircraft rig in the wintry sea. But money was in short supply, and living on Sealand was the cheapest option. Lackey moved there for good, now dining out of a pantry filled with canned food. Sean Hastings and his wife, Jo, brought a dog to Sealand from the Netherlands. But the black lab annoyed Lackey.

Then there was Colin, an English janitor in his sixties. Lackey didn't get along with him either. To avoid running into Colin in the claustrophobic 5,000 square feet of interior space, Lackey lived on San Francisco time, even though he was off the coast of England—sleeping during the day, awake at night. He would practically hide in his room, or in the data center. "It was pretty boring," he recalled. He would stay on the rusty rig for five to six months in one stretch, getting used to the permanent smell of diesel fuel. But he was busy tinkering with computer gear that he had smuggled in from the United States. "It really didn't matter too much," Lackey says. He was living in the Other Plane.[95]

The rough life reflected HavenCo's business situation. The company had five sturdy gray relay racks with blue plugs at the top, with space for forty-five servers. But it managed to put in and rent out only a dozen machines. The company never successfully raised sufficient seed money, not even in the bullish market of the New Economy before the crash. And the budget quickly ran thin. One of HavenCo's main investors, Avi Friedman, was worried about the Y2K problem, so he withdrew about $2 million in cash, in $100 bills, and kept the cash at home. He doled out $1,500 at a time, to make minimum payments.

Lackey started using his own credit cards, spending ever more money that he didn't have.

Businesses did not flock to the data haven as expected. By the summer of 2000, two of the three founders had jumped ship and left the start-up. A year later, Lackey managed to keep the company afloat with about ten customers, primarily casinos. A true cypherpunk at heart, Lackey ran a Mixmaster Type II anonymous remailer on the rig. That felt like the right thing to do, but it didn't make the company any money. HavenCo's business plan foresaw $25 million in profits in year three; Lackey ended up losing $220,000 after three years. One day in early 2001, Lackey was standing on the platform overlooking the wide North Sea horizon, when his phone rang. It was Google, offering him an engineering job. But he still believed the cypherpunk data haven could take off. He turned the offer down.

By the end of the decade, crypto anarchy had a mixed record of success. On the one hand, many of the cypherpunk projects had flopped: the list was in demise, and many of the projects that the activists had promoted with such youthful optimism—remailers, PGP, message pools, digital cash, offshore hosting—remained on the fringe, or they had failed outright. The cypherpunks were looking for individual sovereignty, a Bermuda in the sky with diamonds; what they found was a lone geek on a rusty rig in the North Sea without cash. Yet the ideology of crypto anarchy would become spectacularly successful, even without a nonfiction best seller to spread the gospel.

Like many other libertarians, May was fascinated by Friedrich Nietzsche's philosophy; he even called his cat Nietzsche. "Crypto is not going to enable the bottom 90 percent," May was sure. "Crypto enables the Übermensch," he believed. Just like a generation before him, some tech pioneers believed that cyborg technology would enable the superhuman. May ended up making a "huge amount" of money through the list, he says, by investing in promising ideas and companies that he learned about by running the list.

"This may smack of elitism," May realized, "but I have very little faith in democracy."[96] Instead, the anarchists put all their faith in technology. The power of large primes trumped the power of large institutions. Math decided, not man. Even under the most adverse conditions, technology was trustworthy—even if laws, even if society, and even if corrupt governments could not be trusted. May's vision was nothing less than one of an automated political order.

Crypto anarchy embodied the unshakable cybernetic faith in the machine. It combined Wiener's hubristic vision of the rise of the machines with Brand's unflinching belief that computers and networked communities would make the world a better place. A direct line connects the techno-utopianism of Timothy Leary to the techno-utopianism of Timothy May, cyberpunk to cypherpunk. Leary felt empowered by the personal computer. For May, just one ingredient was missing: the power of prime numbers. "Cryptography provides for 'personal empowerment,'" he wrote in 1999.[97]

The cypherpunks had not a trace of doubt that crypto itself was libertarian, that increasing its use would steadily increase degrees of freedom available to the individual. "This is just an inevitable consequence of technology," May said.[98] To the disciples, servers running remailers and encryption services were libertarian automata, subversive political machines. Whatever their input, their output was freedom.

Sometime in 1999, May looked back on the momentous changes of the previous two decades:

> The full-blown, immersive virtual reality of *True Names* may still be far off, but the technologies of cryptography, digital signatures, remailers, message pools, and data havens make many of the most important aspects of *True Names* realizable today, now, on the Net. Mr. Slippery is already here and, as Vernor [Vinge] predicted, the Feds are already trying to track him down.[99]

May need not have spoken in oblique science fiction metaphors. He didn't know it, but he was right. As he wrote these lines, the Feds were actually busy tracking down a real-world Mr. Slippery in the vast networks of the US military establishment. That Mr. Slippery was doing exactly what May himself had predicted a few years earlier: stealing vast amounts of commercial and military secrets, encrypting them, relaying scrambled versions of these files on machines in third countries, and then exfiltrating them to machines that seemed beyond the reach of the government. And, as in Vinge's story, the Feds were unable to stop the data theft. That metaphorical Mr. Slippery was not a freedom-loving American citizen. The FBI, after a long and painstaking investigation, was able to determine the culprit in the Other Plane: an intelligence agency of one of the most resourceful rivals of the United States.

8. WAR

"A CONFLICT IN WHICH HARDWARE BATTLES HARDWARE will inevitably catch human software in between—*cybernetic war*." This was how *Omni* magazine opened a feature article on the changing nature of war, in 1979. The cover boasted the magazine's sleek stylized logo, in postmodern neon green, hovering above the glowing silhouette of the blurred face of an embryo—a giant, transparent human embryo floating in outer space, in the dark among the stars, invoking the psychedelic final scenes of Kubrick's *2001: A Space Odyssey*.

The glossy cyberpunk monthly blended science and science fiction, in the escapist aesthetic that was so typical of the decade—lusciously futuristic, spiritually space-obsessed, in surreal bright colors. "Cybernetic War," the lead story that May, was no exception: a nonfiction story written by a science fiction author, Jonathan Post. Post confidently predicted the distant future of two decades away. By 1999, there will be roughly a billion computers in the world, Post foresaw, either on Earth or circling the planet in outer space, "almost all of which will be smaller than a large book," and connected to each other. One-third would be deployed in business and science, one-third in people's homes, and one-third in weapons and military equipment.

The soldiers of the future, "of course," will be computers as well. "Welcome to World War III, the Cybernetic War created by machines for machines."[1]

The arsenals of cybernetic war were stocked with an array of fancy weaponry: cruise missiles, smart bombs, sophisticated intercontinental missiles with multiple warheads, and tools such as robotic pattern recognition, code, game theory, cryptography, and simulation. In 1979 these terms were still somewhat vague and undefined, all sounding equally futuristic.

When Post's article on the future of war came out, its analysis was at the cutting edge of technology. The US military was still reeling from defeat in Vietnam, a decidedly low-tech war. The Apple II had been released less than two years earlier, in June 1977. "What kind of man owns his own computer?" asked an Apple advertisement in *Omni* just after Post's article: "Rather revolutionary, the idea of owning your own computer."[2] E-mail didn't exist yet. Usenet, one of the world's first computer network communication systems, had not been set up yet. CompuServe began offering a dial-up online information service to customers only four months later, in September 1979.[3] Predicting ubiquitous networked computers the size of a book was a daring move.

Omni's editors illustrated the visionary weaponry of the futuristic present with a picture of a NORAD console, a map of the United States glimmering in the darkness in the background, awaiting an attack of swarms of enemy missiles, an operator about to press a bright-orange button: "War rooms of World War III pulse with cathode-ray tubes," said the caption.[4]

Omni envisioned cybernetic war with four features: precision, speed, automation, and espionage. Cybernetic weapons were *precise*. War-fighting robots wouldn't be as cute as R2-D2 or C-3PO, the friendly servicing and translation robots from *Star Wars*. The first movie of the legendary series of films had been released in May 1977. Post didn't think the film articulated the present, let alone the future, very well: "Today's genuine robots are sleek subsonic assassins," he wrote. "They are better known by the name of cruise missiles."[5] The refined robotic weapons, as Post described cruise missiles, were small, fast, and light, and they packed an atomic sting.

> Thanks to the new science of robotics and artificial intelligence, cruise missiles strike within centimeters of their intended destinations, having searched with almost animal cunning from thousands of kilometers away.[6]

This was the time when the military's global positioning system—GPS—was deployed, with ten so-called Block I first-generation satellites being launched into orbit in the years after 1978.

Cybernetic war wouldn't be just a precise affair; it would also be *fast*: "Electrons flow through computer circuitry at awesome speeds." The speed of computation would be limited only by the speed of light, about 186,000 miles per second. "Compared to this, jets, bullets, and missiles are sluggish indeed." The author expected "computer warriors" to develop weapons that could strike at the speed of light.[7]

Yet another new feature of future war followed from this drastically increased speed of new weaponry: cybernetic war would be *automated*. Firepower could now be delivered so frighteningly fast that reaction at "human speed" became impracticable, Post believed, echoing a statement that air defense engineers had made for more than three decades. "The computer application that automates the entire battle is known as C3, communications, command, and control," *Omni* clarified. Ever-faster automation meant that machines were becoming more significant, and that humans were becoming less significant. Hardware and information would be more decisive in future wars than were soldiers and civilians: "The computer calls the shots."[8]

Finally, a good deal of cybernetic war would be about high-tech *spying*. Computers had brought a revolution to the shadowy art of cryptography. Post had read about public-key encryption and was mesmerized by this new development. "Anyone with a computer is now capable of communicating with anyone else with a computer by means of absolutely unbreakable code," he predicted for 1999.[9] Security was now intimately tied to the computer—because the most critical information was now stored on computers. This was true for governments, for private businesses, for intelligence agencies, and for the military and its terrifying array of new weapons and equipment that could now bring death faster, more precisely, and without human intervention. The rise of the machines was nigh.

All these new battle systems had one thing in common, Post believed: "All depend on one superweapon, the indefatigable computer, which, paradoxically, is also the best hope for human liberation."[10] This was a most remarkable statement. *Omni* beautifully captured the defining paradox of cyberpunk and applied it to war. The brightest minds of the budding Bay Area technology avant-garde agreed with their fellow counterculture intellectuals: they were sure the computerized future was unstoppable—yet they remained unsure how the machines would benefit the forces of good and evil. Their future vision was oscillating between dystopia and utopia.

That tension made cyberpunk fizz with excitement—and it was that tension between dystopia and utopia that would come to define the Pentagon's interest in "cybernetic war" at the same time. Because the new weapons could be used as tools for oppression, or liberation. Because technology could be used by state authorities, or by those resisting authority, by the army or the insurgency—by the empire or the rebels, in the language of *Star Wars*. Yes, the Department of Defense was pouring money into the development of fancy weaponry that no liberation movement could dream of ever matching. Yet, viewed up close, change was afoot: technology was empowering the strong and the weak. The rise of the personal computer, the spread of networks, and the discovery of public-key encryption were—paradoxi-cally—both strengthening *and* threatening hierarchy and authority. That tension, Post believed, would define the future of conflict:

> The gap between the rich and the poor will grow, but the poor will be more aware of this and more capable of action. Resentment of computerized-police-containment actions will mount. Sabotage of computer-managed production and distribution systems will provoke increased robotic security. Many will see the conflict as Man versus Machine.[11]

Defense intellectuals slowly began to discern an offensive and a defensive logic in what Post called "cybernetic war" in 1979. This development took some time. But over the next two decades, the military potential of computers would be exaggerated in both direc-tions: Washington's leading thinkers saw networked systems as an unprecedented strength in battle, and they saw interconnected control systems as an unprecedented vulnerability in battle. America, depend-ing on the perspective, was either the strongest or the weakest nation.

There was even a second similarity to cyberpunk: escapism. The 1970s had been a depressing time for the US armed forces. The world's mightiest military had just been humiliated by Vietcong jungle fight-ers in the rice paddies and forest trails of Southeast Asia. As a result, the institution turned on itself: the generals doubted their judgments, and the returning soldiers brought back with them corroding habits of drug consumption, alcoholism, a lack of discipline, and generally low morale. Depression, suicides, and stress were rife. To lead this bat-tered army was taxing. Technology and modernization offered a way out. Many senior officers soon projected their hopes onto a networked and automated future of intense yet short wars in which the military and the civilian sphere were clearly delineated. And shiny machines wouldn't doubt, do drugs, drink, disobey, or drift into depression.

In 1973, the Pentagon created the influential US Army Training and Doctrine Command, known simply as TRADOC. TRADOC's first commander was General William DePuy, veteran of the Battle of the Bulge in World War II and a famously tough leader. Standing on a windswept hill in Israel's Golan Heights in 1974, DePuy had seen the future of war. The barren landscape was littered with wrecked Syrian tanks and armored personnel carriers. In the blazingly quick Yom Kippur War, the Israelis had destroyed a much larger force, thanks to superior weapons and superior tactics—and in the process they showed the world what future war would look like. DePuy focused his reforms on firepower and active defense.

The lesson from Israel was that wars could be won nearly immediately. What really mattered was winning the first battle. DePuy enshrined some of these ideas in a famous doctrinal document, Field Manual 100-5. The manual's number is iconic in military circles. FM 100-5 was highly controversial and would evolve significantly between its first version in 1976 and its third in 1986.

One particular reformer had an outsize impact on TRADOC's evolving theory of war: John Boyd. Boyd was an aviator and former fighter pilot. He flew twenty-two combat sorties in the Korean War in an F-86 Sabre against Russian MiG-15s. Like the cybernetic pioneer Norbert Wiener, Boyd was fascinated by how the man-machine entity of pilot and plane behaved under stress: he wanted to explain how the American planes sustained a kill ratio of 10:1 in favor of the Sabre, despite a lower ceiling, a wider turn radius, and a slower maximum speed than the Russian machines had.[12]

Like Wiener, Boyd tried to predict when the design of the aircraft would enable evasive action. And like the cyberneticist, Boyd would develop a theory of feedback loops to come to terms with the problem. He even read Wiener, von Neumann, and the then famous plastic surgeon Maxwell Maltz.[13]

Boyd condensed his theory in the cybernetics-inspired OODA loop in an infamous presentation that he gave hundreds of times in the wider Department of Defense. OODA stood for "observe, orient, decide, act." The concept was simple and intuitive: it could be applied on the level of a Sabre pilot reacting to a MiG's maneuvers (the American machines had several technical features that enabled the pilot to accelerate its OODA loop—for instance, the bubble canopy to "observe" better and faster).

But the theory wasn't limited to a cockpit. Boyd's loop could be applied for an entire brigade or division. The model of this feedback loop made it into the 1982 version of the influential FM 100-5 doctrine: "To maintain the initiative, the attackers must see opportunities, analyze course of action, decide what to do, and act faster than the enemy—*repeatedly*," the document made clear.[14] Now, entire armies were cycling through a feedback loop. The new doctrine became known as AirLand Battle.

AirLand Battle emphasized firepower, superior command and control, the ability to react to changes on the battlefield faster than the adversary could. The winner of a battle was whoever could observe faster, orient faster, decide faster, and act faster. One way of achieving this goal was to use smart weapons, especially precision-guided munitions. Weapon systems could be guided into their precise targets by radio, by infrared, or by laser, and they could be guided from the ground, from the air, or even by satellite. Some of these state-of-the-art missiles were used for the first time in the early 1970s, during the Vietnam War. Their potential was obvious immediately.

TRADOC's Field Manual 100-5 now promoted a nonlinear view of battle, over an enlarged geographical area, unifying air and ground—not just firepower, but agile command and maneuver, as well as electronic warfare. Speed was key, tactical flexibility was key, and airpower was key. It was in the context of AirLand Battle and precision-guided munitions that the notion of "cyberwar" emerged.

AirLand Battle, again, sounded more exciting in *Omni*. In January 1987, the magazine published a landmark text on the future of war. Giant robots were ripping into the burning city under a sky red as blood with thick black clouds of smoke. The robots' claws were huge turning wheels, twenty times as big as skyscrapers, churning into Manhattan like chainsaws into fragile plywood models. A painting by Paul Lehr depicting this scene illustrated the article that first used the word "cyberwar." Lehr was known for his iconic illustrations of machinery and dream landscapes for many science fiction books of the 1980s. "CYBERWARS," blared the headline in large capitals, and the teaser read, "The Marines, and everyone else, are looking for a few good machines."[15]

The author again wrote in the confident style of a magazine that knew the future already: soon robotic warriors would clash in cyberwars. Flying robots would stay aloft for days, scanning terrain, eavesdropping on radio chatter, and feeding data to analysts "safely ensconced in a bunker 100 miles away." Kamikaze robot weapons would circle the skies waiting for a radar signal, and then swoop down on the enemy's gun positions. There would be automated tanks, autonomous minesweepers, generals watching the spectacle remotely from closed-circuit TV screens, and battlefields littered with "carcasses of crippled machines."[16] Mercenary replicants, "cloned humanoids with silicon intelligence" and reactive full-body armor, would be controlled by "brain waves" and tough enough to stop a slug from a .50 caliber gun.

The Pentagon was planning to equip its armed forces with "weapons that have brains of their own," *Omni* reported. The goal for the distant future was to put intelligent systems on the battlefield. Yet "independent cybernetic warriors," as *Omni* called them, were still a long way off. Already in the near future, however, the armed forces would "combine the best of both humans and machines."[17]

That future was indeed near. The first such cyberwar happened less than four years later. The 1986 version of Field Manual 100-5 codified the doctrine that the US Army took to the Iraqi desert in the Persian Gulf War of 1990–91.[18] The United States launched Operation Desert Storm with more than 520,000 troops and vastly superior military hardware. After a hundred hours the operation was over. The operational art on display in this high-tech war mesmerized observers everywhere: TV showed footage from the "bomb's eye view," recorded from the cones of autonomous cruise missiles and precision-guided munitions as they dove surgically into Saddam Hussein's bunkers and weapon factories.

The vast operation saw America's full force of naval, ground, and air power assembled against Iraq, with fast-moving ground maneuvers, smart weapons, precision-guided munitions, deception, and even the use of computer network attack and psychological operations to undermine the morale of the Iraqi Republican Guard. America's armed forces had indeed combined the best of both humans and machines—and thereby achieved one of the fastest victories in US military history, just as *Omni* had foreseen in 1979.

As soon as the battle in the Persian Gulf was over, another battle in the Pentagon started: how to interpret the results and extract the right lessons. One powerful group of officials saw a clear lesson in Iraq and Saudi Arabia; go in big, or not at all. It was the Powell Doctrine of "overwhelming force" that had brought victory, they argued. The cautious and superbly influential chairman of the Joint Chiefs of Staff, Colin Powell, had advocated going to war with Iraq only when the sheer mass of troops made victory practically predictable.

Another faction of military planners placed their bet on a technology-enabled Revolution in Military Affairs, or RMA. Victory in future wars would be guaranteed not by bulk and mass, but by a lean and agile high-tech strike force. This great procurement debate gave rise to a new concept of armed conflict: "The two sides are engaged in a long-term struggle that will redefine the relationship of human to machine in matters of war," the *Bulletin of the Atomic Scientists* reported in a long article on defense planning in September 1992.[19]

"The leading military concept of the new era might be called 'cyberwar,'" the *Bulletin* observed. In this new cyberwar, "robots do much of the killing and destroying without direct instructions from human operators." These weapons would be "autonomous," in one of the phrases then preferred by weapon designers. An example was the navy's Tomahawk cruise missile. Cyberwar meant crewless tanks, cruise missiles that behave like kamikaze robots, head-to-toe battle gear with microclimate control and hazard protection for the infantry, as well as "anti-missile satellites."[20]

Some thought this characterization was too simple. The following year, two Rand Corporation analysts—John Arquilla and David Ronfeldt—published an influential paper, "Cyberwar Is Coming!" Autonomous weapons weren't enough, they argued. The respected think-tank veterans injected a fresh and controversial idea into the Washington defense establishment. Arquilla and Ronfeldt also disagreed with the prevailing view within the Joint Staff, then still headed by Colin Powell: that overwhelming force was necessary to win the next war. Instead, they put knowledge center stage, not masses of tanks and troops. Rand suggested that attacking communication systems could overwhelm the enemy; the two analysts advocated messing with the enemy's mind, not its military machinery.

"Cyberwar," the authors wrote, "means trying to know everything about an adversary while keeping the adversary from knowing much about oneself."[21]

Computers and information technology would fundamentally change the "postmodern battlefield" on all levels, from the strategic down to the tactical level. But all sorts of communication systems mattered at war, not just technical systems. An army at war, no matter how large, needed to know itself, the Rand analysts wrote, reciting a century-old military mantra: the army needs to know who it is, where it is, what it can do when and where, why it is fighting, and which threats to counter first. Information was now becoming as valuable as capital and labor had been in the industrial age. Cyberwar was a knowledge competition.

Rand understood that information was power—not power in some other plane, but in the here and now. At times of war, this wasn't just an adage. It meant something concrete: information was organizational power. Superior command, control, communication, and intelligence had always been critical in war. New technologies would make it even more critical in future war. The prefix "cyber," so in vogue in the early 1990s, meant just that—"to steer or govern"—Arquilla and Ronfeldt pointed out. They stressed that "the prefix was introduced by Norbert Wiener in the 1940s."[22] Ronfeldt especially had been infatuated with cybernetics since the 1970s. He had even suggested, unsuccessfully, an entire set of neologisms inspired by Wiener's writings.[23] Cybernetics, as Arquilla and Ronfeldt saw it, captured how information bestowed organizational prowess:

> Cyberwar is about organization as much as technology. It implies new man-machine interfaces that amplify man's capabilities, not a separation of man and machine.[24]

In late 1992, cyberspace was still a novel concept, fresh out of the pages of the *Whole Earth Review*. Virtual reality had been embraced by the West Coast tech avant-garde, artists, and postmodern theorists, but not by national-security pundits inside the Beltway. Arquilla and Ronfeldt didn't think that the idea of a separate virtual space was so useful. They referred to it only once, and later explained that their view of conflict was much broader than simply "safety and security in cyberspace."[25]

To stress their critique of this narrow technology-centric view, the Rand authors pointed out that advanced information technology wasn't even necessary for cyberwar: their analysis was inspired by the Mongols of the thirteenth century, who had superior knowledge of the battlefield. Even the examples from the recent Persian Gulf War didn't include computer code: the Apache helicopter attack against Iraqi air defense systems at the beginning of the 1991 Gulf War was a "very important" example of a "cyber element" at play.[26]

Soon, in familiar fashion, the cybernetic myth asserted itself.

In late 1994, Colonel Mike Tanksley headed the army's information warfare center at Fort Belvoir, in northern Virginia. Working from a secure vault in the secretive Intelligence and Security Command, Tanksley had a simple vision of how to win a war before it even started. First a computer virus would be inserted into the telephone-switching stations of the enemy nation, thus causing widespread failure of the phone system. The next step would be to activate preimplanted "logic bombs," incapacitating the electronic routers that controlled the enemy's rail lines and military convoys. A third measure, simultaneously applied, would spoof the enemy army's command-and-control infrastructure, so that officers would obey phony orders that came in on their radios. As the confused military forces would disintegrate, the US Air Force would blanket the population with psychological operations, replacing state-sponsored TV with US propaganda that would turn the people against their aggressive rulers. Meanwhile, when the despot under attack would log into his Swiss bank account, the millions in cash that he had hoarded away would be zeroed out. All, in the army's vision, without firing a single shot.

"You can stop a war before it starts" said Tanksley in December 1994 at a defense industry meeting. "We think we have a paradigm shift here."[27]

The Pentagon and the wider defense and security establishment were soon seized by tech euphoria. The skeptics lost out, and cyberspace turned out to be at least as seductive in Washington as it had been in San Francisco five years earlier. Just as cyberspace cowboys and anarchists had arisen in the Bay Area, cyberspace warriors arose inside the Washington Beltway.

"One day national leaders will fight out virtual wars before they decide to go to war at all," Lieutenant General Jay Garner, head of the US Army's Space and Strategic Defense Command, told *Time* magazine in the summer of 1995 for a cover story on "cyberwar."[28] Information technology would revolutionize the battlefield. "This is America's gift to warfare," boasted Admiral William Owens,[29] then the vice chairman of the Joint Chiefs of Staff. Just a few months later, Owens published an influential paper about the emerging "system-of-systems," as he called tomorrow's networked armed forces. Despite shrinking post–Cold War budgets, the admiral argued, "electronic and computational technologies" would soon enable a qualitative jump in the military's ability to use force effectively and win wars faster than ever before. In August the air force published an equally bullish white paper, *Cornerstones of Information Warfare*.

The Pentagon was looking at a Revolution in Military Affairs, Owens was convinced: "The RMA is inevitable." That revolution was driven by better intelligence, better command and control, and better precision strike capabilities. With the right technology in place, Owens expected, the US sensor-to-shooter cycle would outpace that of the adversary: "This will give our forces a great fighting advantage."[30] America's military now had "information superiority," as a famous visionary document published by the chairman of the Joint Chiefs of Staff put it in 1996.[31] The fog of war, the nation's highest-ranking officer was sure, could not be entirely eliminated—but the battlefield could be made considerably more transparent.

"Cybernetic warfare—a form of information warfare involving

operations to disrupt, deny, corrupt, or destroy information resident in computers and computer networks."[32] This is how Patrick Hughes, the director of the Defense Intelligence Agency, introduced the future of war to the Senate Intelligence Committee on a Wednesday morning, January 28, 1998. Hughes was sitting next to CIA director George Tenet and the director of the FBI, who announced plans to open the FBI's new national infrastructure protection center, an office that would soon play a key role in discovering a surprising new dimension of human-machine interaction. Hughes even suggested an abbreviation: CYW.

This form of covert warfare, the DIA director believed, will get ever more important as technology enables new forms of attack. "Cybernetic warfare defies traditional rules of time and distance, speed and tempo," he told the senators.[33] The general didn't know it yet, but he was right. That future was already happening as he spoke—the mighty US military, for well more than a year had already been at the receiving end of a massive attack that defied traditional rules of time, distance, and speed. But that harsh reality would catch up with the Pentagon only four months later, in entirely unexpected ways.

The opposing view began to emerge in early 1991, just as the Persian Gulf War was at its most intense. As air force planners were getting ready to take out Iraq's command-and-control infrastructure, one eccentric IT entrepreneur in Tennessee had a vision while taking a shower.[34] By the time the hot water ran out, he knew what the future threat to the United States would look like: "For a motivated individual or organization, an assault on our information processing capabilities would be," Winn Schwartau wrote in *Computerworld* that January, "an electronic Pearl Harbor."[35]

Without data processing, it was inconceivable that society as we know it could function. Attacks to cripple society, he wrote, could be launched remotely, by a Trojan horse or "crystal virus" that could be inserted into their targets in various ways, including by radio frequency interception. Or the devastating blow could come by aerial assault against cellular towers or satellites.

A staffer in the US House of Representatives took notice, and one committee invited Schwartau to brief Congress as a witness. In June 1991, Schwartau repeated his dire and alarmist warning to the House Committee on Science, Space, and Technology: "Government and commercial computer systems are so poorly protected today that they can essentially be considered defenseless," he told the committee, "an electronic Pearl Harbor waiting to occur."[36] Later that year, Schwartau self-published a novel, *Terminal Compromise*, articulating his fears of the coming electronic doom.[37]

Such fears had been germinating for years already. Computer hackers had become a signature phenomenon of the 1980s, with several high-profile cases making national news.[38] In late 1988 the Morris worm emerged, one of the first computer worms on the still nascent internet, and the first to get mainstream media attention. One book in particular influenced the threat perception: Clifford Stoll's 1989 *The Cuckoo's Egg* chronicled how a German hacker breached Lawrence Berkeley National Laboratory and then sold stolen files to the KGB, Russia's spy agency.[39] In 1991 the Michelangelo virus caused a major scare.

Then, in February 1993, an Islamic extremist detonated a huge truck bomb underneath the North Tower of the World Trade Center in Manhattan. The tower didn't collapse, but the incident highlighted how vulnerable America's infrastructure was. Futurists Heidi and Alvin Toffler, in a widely read 1993 book, *War and Anti-war*, again repeated Schwartau's warning.[40] A single Islamic extremist with a computer and a modem, they argued, could cause immense damage to entire armies 10,000 miles away. Hackers weren't just pranksters, and technology empowered the weak.

The Pentagon became concerned—and decided to put these theories to a test. The Defense Information Systems Agency in Arlington, DISA, organized a team of in-house hackers. This red team staged a drawn-out mock attack against Pentagon machines that reportedly went on from 1993 to early 1995. Their success rate was phenomenal. The ethical white-hat hackers were able to seize control of 7,860 of the 8,932 systems they attacked. That was a success rate of 88 percent.[41]

Worse, for the most part the administrators and users in the Pentagon did not even notice that their machines were "owned." Only about 390 users detected the intrusion, and only about 20 of those reported the breach so that something could be done. It dawned on senior officials in the Department of Defense that they were sitting on a time bomb. Anonymous sources in the Pentagon acknowledged that they had no way to protect themselves "from cyberspace attacks."[42] The threat perception inched up.

Roger Molander, a nuclear protest leader–turned–security analyst, ran a series of crisis simulations for the Rand Corporation during the 1990s. Molander's "The Day After . . ." projects always gamed a hypothetical crisis in the spirit of arms control. In 1995, it was time to simulate a major crisis caused by computer attacks: "The Day After . . . in Cyberspace." Rand held the exercise six times, from January to June 1995. Some of the think tank's most innovative minds, including Arquilla and Ronfeldt, trained their offensive sights on the United States and its allies for simulation purposes.

The scenario was set five years into the future, in the spring of the year 2000. The Clipper chip's successor for a government-mandated encryption solution had failed, Rand assumed for the war game. The internet had become a virtual battlefield for autonomous bodiless robots made just out of software. In the Rand-simulated crisis, cyberwar was truly coming: hackers caused a catastrophic flow malfunction at the largest refinery in Saudi Arabia; crashed a high-speed train in Maryland, killing sixty; took CNN off-line; downed an airliner into a residential area in Chicago; grounded all types of a specific Airbus model; sabotaged the London and New York stock exchanges; messed with US command-and-control systems on ships and aircraft; triggered a revolution in Saudi Arabia; and brought about a host of other calamities. Computers, the Rand analysts were sure, would become veritable weapons.[43]

But computers weren't like other military weapon systems. Everybody could buy one, and everybody could write code, so everybody could attack—not just states. "It's the great equalizer," Alvin Toffler told *Time* magazine after the exercise. "You don't have to be big and rich to apply the kind of judo you need in information warfare."[44] National-security officials in Washington understood that hackers were not alone. America's adversaries would take advantage of this new attack vector. "We're more vulnerable than any nation on earth," added then NSA director Vice Admiral John McConnell.[45]

Soon, elected politicians became concerned as well. Sam Nunn, a senator from Georgia, had called John Deutch, director of the CIA, to testify before the Committee on Governmental Affairs to learn about this highly classified yet little-understood threat. That hearing took place in June 1996. "There are some who believe we are going to have an electronic Pearl Harbor, so to speak," Nunn told Deutch. "Do you think we're going to need that kind of real awakening, or are we fully alerted to this danger now?"

"I think that we are fully alerted to it now," Deutch responded, although that was an overstatement. "I don't know whether we will face an electronic Pearl Harbor, but we will have, I'm sure, some very unpleasant circumstances." Deutch fumbled for a good metaphor to explain to the lawmakers what was at stake. "The electron," he warned, "is the ultimate precision-guided weapon."[46]

Naturally, such language resonated with military officers, who had become keen to get their hands on precision-guided weapons, including the ultimate kind. The Joint Staff had the idea to test and simulate an electronic Pearl Harbor. Not just in a tabletop exercise at a think tank, but closer to the real thing. So the Pentagon's top brass turned to the actual Pearl Harbor—more specifically, to Hickam Air Force Base, in Pearl Harbor, Hawaii, the home of Pacific Command.

Every year, the Joint Staff holds so-called no-notice interoperability exercises, designed to stress-test the Pentagon's planning prowess in a crisis. Nineteen ninety-seven was the first time that such a no-warning exercise focused on what was then called "information operations." The exercise was dubbed Eligible Receiver. The target: US Pacific Command and other Department of Defense computer networks, including the commands for space and special operations. Pacific Command would be in charge of leading more than a hundred thousand troops if a military confrontation in the Pacific became serious. Already in 1997, PACOM at Pearl Harbor had the most advanced computer security of all commands, to be ready for a high-tech contingency that involved China. So the unannounced simulation might as well have been a real attack to paralyze forward-deployed units in the Pacific.

The exercise sounded like a great idea. The air force signed up, the navy signed up, the marine corps was in, but the army said no. The generals of the ground force knew that their networks were vulnerable and opted out. The air force, by contrast, had a potent intrusion detection system in place. The NSA formed the red team.[47]

The NSA red team was physically in one large room in the Friendship Annex, at FANX III (pronounced "FAN-ex"), near the Baltimore/Washington International Airport in Maryland, about 10 miles north of the NSA's main headquarters at Fort Meade. Operational security was tight: the red team needed special access to get into its operations center, packed with computers. The exercise was so intense that the NSA needed to make sure the red team ate and slept properly, because work on the mock attack was so exciting. Before the exercise kicked off, Kenneth Minihan, the NSA director and an air force general, came out to brief his hackers: "We're shaping history," he told his team at FANX III.[48]

The NSA team had no privileged intelligence about the systems it was supposed to bring down, but it did simple reconnaissance for six months. The team also wasn't allowed to use the spy agency's advanced gadgetry, but only publicly available techniques. "We put two people on every machine, one keyboard artist and one who was writing everything down," one of the planners recalled. "What we were doing was pretty big."[49] Very few people at Fort Meade knew what was going on. The red team even communicated through encrypted phones.

From June 9 to 13, the red team went on the offensive. Eighteen NSA geeks posed as surrogate agents for North Korea and Iran, cooperating in the mock cyber attack.[50] The NSA red-team members covered their digital traces with well-known techniques, and they didn't try to log into their target computers directly. The NSA hackers used dial-up modems and foreign internet service providers to enter machines abroad. Then, after multiple hops to mask their identity, they attempted to penetrate critical US networks. Some users of critical military systems, it turned out, simply used the word "password" as their password.

The NSA also tricked and socially engineered PACOM staff to open dodgy e-mail attachments laced with malware. The Joint Staff using another, more tested method, sent very special forces out to Hickam Field to "dumpster-dive" for discarded printouts with log-in information or other useful details. One day the red team at FANX III received an e-mail from PACOM: "Don't use MILNET," it said. "The network has been compromised." The officers at Pacific Command sending the e-mail of course did not realize that they had actually sent a warning to the ones attacking them. "Oh shit, this is not good," the red team thought at first.[51]

But then somebody had the idea to trick Pacific Command: the red team changed the body of the message and spoofed the e-mail's sender information: "We've corrected the problem," the government-sponsored hackers told PACOM, pretending to be PACOM's own technical team reporting that its systems were good to go again. The trick worked. When the exercise ended, at NSA director Minihan's suggestion, the red team took control of the PACOM commander's briefing computer.

"It's a very, very difficult security environment when you go through different hosts and different countries and then pop up on the doorstep of Keesler Air Force Base, and then go from there into CINCPAC," one official explained later.[52] Keesler is in Mississippi, and CINCPAC is shorthand for the command-and-control computers used by the commander in chief of Pacific Command.

The successful attack against military networks was bad enough. Suddenly it dawned on the Joint Staff that "information superiority," announced just a few months earlier, could quickly turn into inferiority. To make matters worse, reports started to surface that the Eligible Receiver red team could have shut down the power grid. Deputy Defense Secretary John Hamre boasted to a Fortune 500 forum of chief information officers in Aspen, Colorado, "We didn't really let them take down the power system in the country, but we made them prove that they knew how to do it."[53]

The NSA red team in FANX III did not attack any power systems, and it did not have the legal backing to touch any "boxes" (jargon for machines) that were outside the Department of Defense domain.[54] The civilian attacks were only simulated. The NSA did not actually attack the country's energy infrastructure.[55] "But the referees were shown the attacks and shown the structure of the power-grid control, and they agreed, yeah, this attack would have shut down the power grid," said one unnamed official later.[56] It remains unclear how much of this was realistic.

The experience was a wake-up call. "Frankly it scared the hell out of a lot of folks, because the implications of what this team had been able to do were pretty far-reaching,"[57] recalled John Campbell, then a general officer at the Joint Staff. The president was briefed on the worrying consequences of the exercise. "Eligible Receiver, I think, has succeeded beyond its planners' wildest dreams in elevating the awareness of threats to our computer systems," said Kenneth Bacon, then the official spokesperson for the Department of Defense.[58]

The Pentagon wanted to harness the electron as the ultimate precision-guided weapon. But by trying to extend information superiority into cyberspace, the Joint Staff had come around to Winn Schwartau's vision in his Tennessee shower: an electronic Pearl Harbor was waiting to happen.

"I think everybody has to realize that we are now entering a period where we have to worry about defending the homeland again," Hamre said in early 1998. "As computers are becoming interconnected, it is now possible for people to come in and disrupt our lives through computer connections."[59]

Meanwhile, an actual world crisis had slowly escalated. In breach of UN resolutions, Iraq had allegedly blocked Australian-led inspectors from visiting its stockpiles of weapons of mass destruction. Saddam Hussein accused the Americans of spying, and even threatened violence. But the White House was determined not to show weakness. At the end of January 1998, Bill Clinton increased the pressure on Iraq by indicating that the United States would be prepared to use military force. The White House dispatched twenty-two hundred marines and three aircraft carrier groups to the Gulf.

At the same time, the air force's computer intrusion detection system flagged an attempted breach at Andrews Air Force Base. Unknown intruders got into the air force's systems by exploiting a known vulnerability in Sun Microsystems' Solaris 2.4 and 2.6 Unix operating systems. But this wasn't a game; it was the real thing. The operation was code-named Solar Sunrise, inspired by the vulnerability in the Sun operating system.

When the Department of Defense and the FBI started looking into the incident, they found what Hamre described as "the most organized and systematic attack the Pentagon has seen to date."[60] A frantic investigation revealed that a number of military bases had been compromised, and whoever breached the Pentagon systems had come in via Emirnet, a service provider in the United Arab Emirates and one of the few electronic gateways into Iraq. The intruders, it seemed to the military officials, targeted logistics and communication systems that were at the heart of the military buildup in Iraq. "If you take one part of that machine, and disable it, you got a real problem trying to make a deployment operation take place," said Campbell, then the official on the Joint Staff responsible for information operations.[61]

Solar Sunrise was serious—or so it seemed, at least for a few weeks. The intruders had browsed through files and copied some of them, including log-in and password information. But they had not deleted or altered any data. Yet the unnervingly successful Eligible Receiver red-team exercise was fresh in the minds of the Pentagon leadership. It was not impossible that Saddam Hussein was preparing to plunge American cities into a power blackout. Hamre even briefed President Bill Clinton, telling him that the breaches could be the opening shots of an authentic cyberwar.[62]

But then the FBI caught up with the intruders. The hackers, it turned out, were three harmless teenagers: two sixteen-year-old Americans, with the screen names Makaveli and Stimpy, were sophomores at a high school north of San Francisco, and their Israeli online mentor was an eighteen-year-old hacker named Ehud Tenenbaum, a.k.a. The Analyzer. Little did the FBI know that one of the world's mightiest spy agencies was already inside the government's networks.

The vast spy campaign started ominously. On October 7, 1996, at eight-thirty at night, clandestine intruders breached the computer network at the picturesque Colorado School of Mines, a small engineering university on the outskirts of Golden, Jefferson County. As if to mock the utopians in California, the foreign intruders launched their breakin from a mysterious address at "@cyberspace.org." They rummaged into the robotics lab in the modernist Brown Building and hacked a machine called Baby_Doe.[63] The prowlers used a so-called "root kit" on Baby_Doe's Sun OS4 system. From their interim base in the Rockies, the digital spies then started probing NASA, the National Oceanic and Atmospheric Administration (NOAA), various navy and air force machines, and a long list of other computers. They hacked for the entire night. The operation looked like reconnaissance to the US Navy.[64] But were they dealing with an individual "Mr. Slippery" or actually a foreign spy agency?

Then, two months later, in December 1996, the intruders broke into NAVSEA Indian Head, in Maryland, the largest of the navy's gargantuan system commands, which swallows around one quarter of the entire naval budget. They got in through a known cgibin/phf vulnerability. NAVSEA logged 24 attempts to steal passwords. The spies had sometimes attempted breaches from a machine at the University of Toronto. But they actually came from ppp63.cityline.ru, in Moscow. This, however, was only the beginning. Over 1997, six different naval commands—including the Naval Space Command and the Naval Research Laboratory—detected intrusion attempts that originated from Moscow; four of the commands explicitly identified cityline.ru as the origin of the attacks.[65]

The navy wasn't alone. The Department of Energy discovered 324 intrusion attempts from overseas between October 1997 and June 1998. Sometimes the uninvited guests successfully gained "complete access and total control to create, view, modify or execute any and all information stored on the system," according to one leaked document.[66] There were even more breaches. But no US government agency was able to connect the dots. Not yet.

Then, on June 1, 1998, Wright-Patterson Air Force Base began detecting intrusions. Both the Institute of Technology and the Research Lab had been breached via connections from the computer lab at the engineering department at the University of Cincinnati, taking advantage of the university's fast 10-megabit connection.[67] DISA started investigating. The staff in Arlington, Virginia, worked grueling hours, feverishly trying to trace down the intruders.

DISA's digital sleuths soon discovered, to their surprise, that some isolated attacks seemed to come from the United Kingdom, one of America's closest allies. The machine in question, it turned out, was in London and belonged to a not-for-profit organization not far from the Wimbledon train station, the Institute of Personnel and Development. The site in South West London, DISA suspected, had been hacked and was now being used as a "hop" site to exfiltrate military secrets to a third country.

Such a relay site is a compromised computer that is used as a bridgehead to launch further attacks. The more such compromised sites are chained together to form the intruder's infrastructure, the more difficult it is to trace the attack back to its origin. A hop site works a bit like a connecting cargo airport, where new transportation documents are issued and old paperwork discarded, making it harder to find out where the cargo actually came from.

The staff at DISA operated like an incident response team. So they did what incident response teams do: they contacted the technicians in Wimbledon directly, asking them for help. The system administrators in London e-mailed a number of FTP transfer log files over to Arlington. FTP, short for "file transfer protocol," is widely used to move heavy loads of files around on computer networks. The administrator at the compromised UK site used the log-in "IT" and the password "IT," making it easy to hack into the HP-UX system he was running. And the London log files contained an ominous secret: DISA had followed one thin lead across the Atlantic to Britain, but now the log files showed an entire bundle of leads going back across the Atlantic to other compromised machines in the United States that the Pentagon didn't even know about. Worse, the outbound connection from London was connecting directly to machines in Moscow.

This was big. Both the Pentagon's lawyers and the FBI were brought in. "You did what? You contacted a foreign site?," the surprised legal experts asked the enterprising DISA tinkerer. They didn't appreciate that DISA had taken the initiative to contact system administrators in a foreign country. "That's a legal issue, we've got to pull Scotland Yard in."

The Metropolitan Police in London soon discovered that the server in Wimbledon, an HP 9000 make, was running outside the nonprofit institute's firewall, because it was being used as a test host at the time, dubbed "HRTest." That made the machine easy prey. The Met convinced the institute to leave the machine up and running so that they could monitor the activity. "We were watching military stuff coming down the pipe," one of the investigators remembered. GCHQ, the UK's electronic spy agency, also took note of the curious case, and code-named it "Astonish."

One of the London police force's whiz kids, nicknamed Clever Trevor, designed a so-called sandbox, a software security mechanism, that they connected to the server "to suck the information out of that machine without leaving a trace," one of the investigators recalled.[68] The FBI sent two experts to London to help set up the software. The Met knew that the intruders were very good, and they didn't want to tip off the Russians. So, Met officers saved the forensic data to magneto-optical discs, driving to the Wimbledon site several times a week to change the encased discs and then reporting what they found to FBI headquarters by phone. The Met passed the disks on to the FBI's rep at the US embassy in London. For security reasons, the bureau flew the files to Washington.

DISA had hit gold. Whoever had hacked the system in Britain had also hacked an entire range of systems stateside. The Pentagon investigators started matching forensic data to actual victim sites; then they contacted various IT departments in the United States and told them that their networks had been compromised. Some of the intrusions were known and had already been under FBI investigation. The various intrusions could now be grouped into something larger. The special agents working the case sat down to come up with a code name. The breaches were happening at night in the United States. And the entire situation was confusing, like a labyrinth. "Moonlight Maze," they thought, was an appropriate name for what seemed to be the first large state-on-state cyber attack. But bringing all this to the attention of law enforcement wasn't enough. Counterintelligence was looped in.

In late July, the intruders made a mistake. The address used by the Russian pilferers, the FBI noted, was 25.m9-3.dialup.orc.ru; they were entering via the South Carolina Research Authority, a research organization to promote high-tech development. The attackers also compromised Wright State University in Dayton, Ohio, and Advanced Technology Institute, a defense contractor in Charleston, South Carolina.[69] By early August, the FBI suspected that Russians had breached a series of systems at the Ohio base. "The intrusions into these U.S. systems appears [*sic*] to be originating from a dialup connection to four Internet Service Providers (ISPs) located in Russia," the FBI wrote. The investigators were intrigued by the address of the dial-up connection, "which suggests a local (i.e. Russian) point of origin," the FBI noted.[70] The bureau brought in Kevin Mandia, then one of the country's finest digital forensic experts, formerly with the Air Force Office of Special Investigations. Mandia feverishly started working the case, entering the Hoover building on a Wednesday and walking out sleep-deprived on Friday. Among other things, he found language artifacts in the coding, for example the Russian technical term for "child" process, "дочерний."

The Feds also started hunting for spies among faculty and students at some of the compromised universities, but their interviews didn't produce any good leads. By November 1998, the FBI had traced the Russian intrusions to more than a dozen colleges and universities in the United States, including some of the finest: Harvard University, Duke University, the California Institute of Technology, and the University of Texas at Austin.

The scale of the theft came as a shock. If piled up, the heap of files would be "equivalent to a stack of printed copier paper three times the height of the Washington Monument," as one senior Air Force intelligence officer put it. It was a drastic comparison: the famous stone column that dominates Washington is 555 feet high, taller than the Great Pyramid of Giza. The Russians were hauling away a pyramid-load of sensitive military research files.

The list of targets was extensive and went well beyond the known breaches at Department of Defense sites; various sites of the Department of Energy were affected too. NASA was hit badly; Inspector General Roberta Gross told reporters at the time that the breaches at the nation's space agency were "massive, really very massive" and "very, very surreptitious."[71] The Environmental Protection Agency was plundered. NOAA lost large amounts of data. Most of the nation's prized research labs were breached, including the national laboratories of Los Alamos, Sandia, Lawrence Livermore, and Brookhaven.

Some of the breached universities were targets; others were mainly way points. For example, the FBI discovered that the mysterious intruders had routed some attacks through McGill University and the University of Toronto in Canada. The foreign spies reckoned, probably correctly, that high-volume data flows between two universities wouldn't raise suspicions. To get into these unsuspicious hop sites, the intruders hacked systems in the United States, the United Kingdom, Canada, Brazil, and Germany. The evidence was "primarily transnational," the bureau noted, and the number of victims "ever expanding."[72]

The large-scale theft of sensitive and defense-related information was shocking enough to come to the attention of the White House. And the president's advisers were worried: "Such a thing has never happened before," one senior White House official later told the *Times* of London. "It's very real and very alarming."[73]

The attack's timing was perfect. Just when the curtain was slowly being pulled back on Moonlight Maze, both the Pentagon and the FBI were establishing new organizations to deal with the emerging threat of network breaches. Thus, America's largest ever digital forensic investigation got under way.

The case nearly overwhelmed the FBI. The public face of the investigation was Michael Vatis, head of the FBI's newly founded National Infrastructure Protection Center and previously a senior Department of Justice official. But the actual work was done by the Moonlight Maze Coordination Group, a team headed by Doris Gardner, an FBI special agent with fast-improving computer forensic skills, and Michael Dorsey, who had been sent over from the Pentagon. The task force had a staff of forty dedicated officers, recruited from two dozen of the nation's most potent security and intelligence agencies, including army, navy, and air force special investigation services. At busy times, as many as a hundred people were roped into the investigation. The group was housed in the Strategic Information and Operations Center, on the fifth floor of the FBI's sprawling modernist headquarters building in Washington, DC, at 935 Pennsylvania Avenue. The SIOC (pronounced "SYE-ahk"), the bureau's vast global watch and communications center, occupied 40,000 square feet of space.

The task force had a legal team, an analytical team, a technical team, and an investigative team.[74] The FBI issued orders to preserve fragile and perishable forensic evidence, and sent out special agents across the country, at night, on weekends. In some cases the forensic information was so perishable that agents were dispatched immediately after getting the lead. In one instance the FBI agent in question could not even go home to get a second set of clothes before taking a flight; in this case the intruders had buffered data in the RAM of a breached machine, and the FBI had to hurry up to extract and save the memory before it would disappear without a trace.

The obscure intrusions came at a fortunate moment. After the eye-opening Eligible Receiver exercise, the Pentagon leadership was primed for cyberwar. William Cohen, then the secretary of defense, decided that a new organization was needed. The military came up with an unwieldy name, the Joint Task Force for Computer Network Defense, or JTF-CND. Just before New Year's Eve, on December 30, 1998, the Pentagon announced that the new unit had achieved initial operating capacity.

The first JTF-CND commander was Campbell, who had already been involved in Eligible Receiver. The major general, nicknamed "Soup," was an energetic and experienced F-15 and F-16 fighter pilot with a gravelly voice. Now he led a team of six people, himself included, out of a temporary trailer in a parking lot behind DISA's then headquarters at 701 South Courthouse Road in Arlington, Virginia. One week after moving into the trailer, JTF-CND was pulled into Moonlight Maze.

Days later, on the morning of January 22, 1999, the president raised the cyber security stakes. Clinton gave a major speech at the National Academy of Sciences, five blocks from the White House. It was a carefully chosen spot. He spoke about keeping America secure for the twenty-first century. He mentioned the nearly devastating World Trade Center bombing of 1993, the Oklahoma City bombing of 1995, the Khobar Towers attack in Saudi Arabia, and Pan Am flight 103.

"The enemies of peace realize they cannot defeat us with traditional military means," the president said. "So they are working on two new forms of assault . . . : cyber attacks on our critical computer systems, and attacks with weapons of mass destruction."[75]

By now the federal government was coming to terms with the fact that it had been at the receiving end of the biggest and most sophisticated computer network attack against the United States in history, Moonlight Maze. The American public knew nothing about this evolving threat yet. And Clinton limited his remarks to a warning. "We must be ready," he said, "ready if our adversaries try to use computers to disable power grids, banking, communications and transportation networks, police, fire and health services—or military assets."[76]

Meanwhile, the Pentagon and the FBI still weren't ready, and they knew it. Intelligence officers and federal agents alike were surprised by how precisely and efficiently their mysterious adversary was going after its victims. The intruders knew exactly what they wanted. They would breach a system, go straight to root, do a full directory listing, save the list as a file, then leave, all in a few minutes. After a week they would come back, zoom down to very specific files, four levels, down to a subdirectory, and take five files. The hackers even knew the accounts of specific individuals who worked on military contracts at universities. The Pentagon investigators were dumbfounded: How could the intruders possibly know all this detailed information ahead of time?

One day an officer from the Defense Intelligence Agency with a thick white binder walked into the trailer at the temporary offices of the Pentagon's joint task force. The binder had the answer: a restricted but unclassified catalog of technologies that the Department of Defense sees as essential to maintaining superior military clout, the *Military Critical Technologies List*, or MCTL.

The document was a sprawling list of technologies with futuristic-sounding names, certainly by the standards of 1998: biological defense, chemical systems technology, ballistic missiles, cruise missiles, unmanned systems, electrothermal guns, high-performance computing, subsurface and deep submergence vehicles, vetronics, and even human performance enhancement, blood substitutes, and artificial skin. The MCTL was the perfect target list—prepared by the target itself: the Pentagon kept the list up to date and cataloged key partners with their contact details in government, industry, and academia, down to individual names and phone numbers. Indeed, the Pentagon decreed in January 1995 that the MCTL should be used as a "technical reference and guide for intelligence collection"[77]—but of course the Pentagon officials had in mind that the document would be used as a guide for intelligence collection *by* the United States, not *against* the United States.

Yet the mysterious intruders were, in fact, using it against the United States. Part III of this remarkable document outlined the most critical emerging technologies. The remote thieves rummaged through all the army and government systems that were working on technologies listed in the MCTL. "This document served as the intruders' 'roadmap' through all Army and government systems," one intelligence officer recalled.[78] He suspected that the intruders had stolen the document from the National Technical Information Service in Springfield, Virginia. "I remember the 'aha' moment when we found it and put the dots together," another investigator recalled.[79]

These dots pointed to a range of companies and universities with research relationships to the US Army's Major Shared Resource Centers, supercomputers that were designed to do high-performance computing for an entire community of developers and engineers in academia and the private sector. Whoever got so much money and worked on defense-related research with supercomputers, the mysterious intruders reckoned, had something worthwhile to hide. "That was genius and beyond the work of a simple hacking group," said one of the key forensic analysts at DISA.[80]

Despite all these details, it remained unclear who was behind the attack. The US investigators suspected the Russian government. But that was not more than a suspicion; an even more sophisticated adversary could have used the Russian sites as way points, just like the HRtest server in Wimbledon. That suspicion wasn't good enough. They needed more.

To the FBI's surprise, the attackers used another proxy site in the Rocky Mountains: the computer of the Jefferson County Public Library, in Lakewood, Colorado, a fifteen-minute drive west of Denver. The library's main administrative building resembles a large modernist fortress, with tall and rounded windowless walls. But it did not keep the foreign spies out. The county calls itself the gateway to the Rockies. Now it was the gateway to the Department of Defense. The spies checked out America's defense secrets via an innocent-looking library computer in Colorado.

But using the library machine was risky. Soon after the Jefferson County server was discovered, the Moonlight Maze task force decided to keep it up and running, not to shut it down. Now the Feds were in a position to spy on the spies, in real time. The FBI requested a so-called trap-and-trace warrant to monitor the incoming traffic and to download the log files from breached sites all across the United States. "This allowed the Task Force to conduct assessments of the data being taken," one investigator recalled.[81]

The method was highly sensitive. The FBI intercepted everything that happened on the library's computer. That included the searches and e-mails of ordinary citizens in the leafy Jefferson County who were using the library's public internet terminals. The FBI was careful not to retain or pass on such personal information to the Department of Defense. "Regular library patron searches and their information were not retained," added one Pentagon intelligence officer.[82] But the Jefferson County site allowed the counterintelligence specialists to trace the intrusions to a large number of additional sites. "The methodology was very, very subtle," said Michael Dorsey, who co-led the FBI's Moonlight Maze task force, "very sophisticated."[83]

The targets were revealing, but they were also wide-ranging. The investigating officers suspected that the haul included details on missile guidance systems, atmospheric data, oceanic data, cockpit designs, and even helmet-mounted displays. The irony was bittersweet: the first nation-state attackers in virtual space targeted the very Air Force Research Lab that had invented the idea of "virtual space" two decades earlier.

Later, the attackers also breached the Environmental Protection Agency. At the EPA the intruders went after files on computational fluid dynamics, data-intensive research on air and liquid flows. The thieves also stole information on pesticides. The government's environmental agency had detailed information about new chemical compounds on its servers because these needed to be evaluated for government approval for wider use in the field. Some members on the Moonlight Maze task force suspected that the intruders tried to weaponize this information. But the intruders also took information on water purification and fertilization, which was harder to explain. Some investigators suspected that Russia could have an interest in this kind of information. Lake Baikal in southern Siberia was polluted, and the country had a range of other pressing environmental problems.

The intruders operated on European business hours, the FBI and the Met noticed. They were active on Christmas Day 1998, but dormant on January 7–8, 1999—"both weekdays and Orthodox Christmas in Russia," a classified FBI memo noted.[84] That was a good piece of evidence, but the Americans needed more. The work on the case was so intense that the Moonlight Maze task force printed T-shirts with the operation's logo on the chest (a moon casting shadows on a blue-and-yellow maze) and "Byte Back!" printed on the back. And biting back they tried.

The FBI was working with army and navy operators to see whether setting up a "honeypot" would be doable and desirable "to assist attributing the intrusions."[85] Not only, however, was the FBI's bench too thin, as the bureau openly admitted, but it wasn't playing aggressive enough. Law enforcement officials, naturally, were concerned about the law. Some of the military officers in the Pentagon, naturally, were concerned about the enemy. They suggested hacking the hackers.

But the bureau took the conservative approach and preferred tracking the intruders only "passively," without using more aggressive techniques to go after them. Hacking back, they were concerned, would put them in a legal gray area. Worse, the Department of Justice was concerned that if the intruders were indeed state sponsored—if the Russian government attacked them—then crippling their capabilities could be construed as an "act of war."[86]

"Wait a minute," Dorsey said, recalling the discussion between the agencies. "Nobody has got absolute attribution; we don't know specifically who is doing this and where exactly it is coming from."[87]

The government's own coordination group was concerned as a result of the alarmist statements from US senior officials. "If you end up taking out a power grid, or if you end up taking out some critical services," the FBI's task force leadership responded to the Department of Defense, "de-facto we may have just committed an act of war against another country."[88] But others in the Pentagon and the NSA disagreed. Later in the investigation, the Pentagon upped its game. "We started doing fun things on our side too," one army operator working on the joint task force recalled.[89]

Adobe Acrobat is software commonly used to create and read files. It has a way to generate branded versions of files. These files can contain unique character sets with logos and templates. The standard Acrobat Reader can't handle such a branded file; specially equipped reader software is required. JTF-CND exploited this feature: "Why don't we put some real juicy named PDF files right in front of where we knew the Moonlight Maze people would come?"[90]

The idea was that the intruders would steal the juicy file, take it back home via the hop sites in Colorado and London, download it to their own system (perhaps in Moscow), and open it there. When they did that, they would get an error message prompting them to download the army's own reader. The hope was that they would go back and do that. The army reader was rigged so that, when installed and opened on a remote machine, it would beacon back to the Pentagon by sending a unique so-called DNS lookup. It was like sending a secret code back home that would hopefully reveal the thieves' location and, ideally, their identity. "We would watch for that one packet to ask a DNS query that no one else on the planet would ask," one of the Pentagon tinkerers recalled.[91]

It worked. Hacking back provided an additional piece of evidence: an IP address from a specific country that the intruders used to open the file. But even that wasn't good enough. JTF-CND and the NSA understood the limitations of this method. They had used it themselves before. The intruders could have opened the file on a compromised computer in another country, or tricked the machine to send a misleading identifier.

For the Department of Justice and the FBI, Moonlight Maze was "one of the highest priority cases" for a year.[92] In the ten months after the operation was discovered, the Russian intruders compromised an estimated 27 target systems, and more than 270 additional systems in order to breach their targets by exploiting established trust relationships. Something needed to be done.

On the morning of February 25, the FBI and members of the Pentagon task force gave a classified briefing to the House and Senate intelligence committees about the suspected large-scale intrusions from Russia. The members were alarmed by what they heard. "We are at war right now," Hamre told them in a closed seventy-five-minute session. "We are in a cyberwar."[93]

"Without compromising the investigation, what is going on?" one of the elected politicians demanded to know.[94] Curt Weldon, a congressman from Pennsylvania, felt that the public should know. He told the investigators behind closed doors that "the 'electronic Pearl Harbor' of which Hamre spoke last year has gone from if to when and the when is today."[95] Less than a week later, the story appeared in *Defense Week*. "It's got its own name," Weldon told a reporter. "It is a coordinated, organized effort. And it's serious."[96] The story quickly became prime-time news, and the news coverage had an immediate effect on the investigation.

"In light of the press coverage, the consensus among the participating agencies was that we had no real choice but to go directly to Moscow with a request for assistance," the FBI noted later.[97]

The intruders were unfazed. The Russians suspected that the machine in London had been watched, and immediately stopped using it. "The publicity stopped it in its tracks," one of the Met investigators watching the Wimbledon site recalled.[98] Then the intruders reviewed their operational security—and continued hacking. "In spite of the ABC story on 3/4/1999, intrusions continued," the FBI wrote in a memo on April 15.[99] One day after the story, the Russian spies broke into Lawrence Berkeley National Laboratory and into Argonne National Laboratory, both passing through the central hop point in Jefferson County, Colorado. The Feds found it "significant" that the spies apparently didn't shun the light.

But the FBI was about to get lucky. In early 1999, a Russian-language news site defamed Boris Yeltsin's daughter. Libel was illegal in Russia, so the Ministry of the Interior, or MVD, investigated. The MVD found that the site was hosted in California, so it needed the help of US law enforcement authorities to identify the culprits. When the Russian federal police contacted US federal law enforcement, the FBI agents working the computer intrusions had a shrewd idea: do the Russians a favor, and then ask them for a favor in return.

So in late March, the Moonlight Maze task force hosted a small Russian delegation from Moscow's MVD in Washington, wined and dined them in a fancy restaurant, drank vodka with them, and offered them the log files that they wanted. But they also presented them with five intrusion incidents related to the breaches at the various US military sites. The FBI formally requested assistance from its Russian counterparts. Vitaliy Degtyarev, from the Russian Ministry of Internal Affairs, led the small group and "pledged the aggressive investigative support" of his ministry.[100]

The delegation flew back to Moscow with souvenir pictures of a night out with FBI special agents. But the FBI had been careful not to give Degtyarev any documents relating to Moonlight Maze. "We didn't want them to have time to investigate without us being present in Moscow," one of the lead FBI agents recalled.[101]

Not even two weeks later, it was the FBI's turn to go to Moscow. On the afternoon of April 2, a group of seven went to Dulles International Airport in Virginia to catch the 5:10 p.m. Delta flight to Zurich. The delegation included two officers from the Air Force Office of Special Investigations, experts from the Department of Defense, a NASA computer security specialist, and two female FBI special agents working the case. They arrived in Moscow at 3:00 p.m. the following day. The FBI's high-powered coordination center in Washington was deploying a dedicated team to support the delegation on the ground in Moscow, working sixteen-hour shifts.

Degtyarev and another member of the MVD, a general, welcomed the US investigators with ample quantities of vodka. "So what brings you to Moscow?" the general asked over dinner, with a thick Russian accent. "We're getting hacked," the head of the Moonlight Maze task force responded. "We need your help."[102]

After dinner, the evening took a Russian turn. The group of about a dozen drank seventeen bottles of vodka, toasting liberally to America, to Russia, to mutual friendship, and later in the evening, to women. As the tanked general raised his glass for the ladies, he hugged the female FBI agents—and stuck his tongue in the ear of one of them. She was dumbfounded. The agent rushed to the ladies' room with her FBI colleague.

"He stuck his tongue into my ear," she said, enraged. "Did he stick his tongue into your ear as well?"

"No," the other agent responded, incredulous. The two women considered their options. They quickly agreed that they did not want to cause an international diplomatic incident.

"Just wipe it out," the FBI agent told her colleague.[103]

The next day, for reasons that remain mysterious, the MVD and the excited general cooperated with the US delegation. The Americans also had convincing evidence that seemed to implicate three internet service providers, including one of the country's oldest and biggest in Moscow, Cityline. Early the following morning, still hung over, the general arranged for one of his officers, Dmitry Chepchugov, to take the forensic investigators from Washington to the ISP's offices. The eight-day fact-finding trip started out promising. Cityline, it turned out, also served the government.

Perhaps the most revealing incident happened a day later. When the delegation was picked up by its MVD handlers in the morning, the van headed on a sightseeing tour across and around Moscow that the Americans had not signed up for, instead of back to the ISP. Chepchugov promised that the forensic investigation would be taken care of, and that there was no reason to worry. The FBI would receive a detailed letter with the results before returning to the United States.

On the third day of sightseeing, some in the delegation lost patience and pressed their handlers harder: would they be able to continue the investigation?

Their efforts were to no avail. The promised letter never arrived. The delegation's main contacts in the ministry went quiet. The special investigators and counterintelligence specialists flew back to Washington disappointed, with pictures of drinking feasts and group shots in front the colorful onion domes of St. Basil's Cathedral in Red Square next to the Kremlin.

Yet the Moscow trip had unexpected consequences. The embassy's FBI liaison officer in Moscow had an established working relationship with the computer crime unit in the MVD, and he soon reported back to the Moonlight Maze task force that the general with the loose tongue had simply disappeared. "As far as we know, he ceased to exist," said Bob Gourley, the leading intelligence officer working the case for the Department of Defense.[104] He had lost his job and wasn't seen again.

A closely guarded internal report revealed that the task force suspected a communication glitch between different parts of the Russian administration—a problem only too familiar to people who worked for the federal government in Washington. The Moonlight Maze operation was probably such a closely kept secret in Moscow that the MVD's own computer crime unit did not know about it. Russian law enforcement most likely mistook an intelligence operation for a criminal case. The fact that the FBI showed up at the Russian government's own ISP on the first day of the fact-finding trip seemed to have alerted somebody to the error. The agency in charge then switched into damage limitation mode, sending the American officers sightseeing for a few days—and removing the source of the problem. "The trip to Moscow was extremely important," said Dorsey, the Pentagon's co-leader on Moonlight Maze.[105]

Soon the FBI task force was shut down. The government stopped treating the issue as a criminal case, so law enforcement was not in charge anymore. The officers in the SIOC cleared their desks and packed their bags. This was now a counterintelligence matter, and the Pentagon took over. Meanwhile, Moonlight Maze continued. "After the trip to Moscow their tradecraft got even better," said Gourley.[106] The summer heat in DC and along the National Mall was scorching—with the Washington Monument as a grim reminder that the pile of stolen documents was still growing.

Over at the Pentagon, senior officials understood that cyberwar was here. The official ribbon-cutting ceremony at JTF-CND happened after the Moscow trip, on August 11, 1999, after the unit had moved into its new premises. John Hamre, the deputy secretary of defense, presided. On the lectern was the unit's freshly minted coat of arms, a bald eagle hovering over an abstract grid that looked a bit like a globe in low resolution against a blue background.

"The Department of Defense has been at cyberwar for the last half year," Hamre said, referring to Moonlight Maze, which in fact started around two-and-a-half years earlier. He was standing next to rows of gray computer monitors and large images projected against the walls in the windowless command center. "Several times I've testified and talked on Capitol Hill about the future electronic Pearl Harbor that might happen to the United States," he told the crowd of standing officers. "I've used that expression not to talk about surprise attacks. The most important message about Pearl Harbor was the way in which we had actually prepared well in advance for the war that came." As far as Hamre was concerned, computer network attacks were a game changer. "Cyberspace isn't just for geeks. It's for warriors now."[107]

By August 1999, the Department of Defense was so concerned about being paralyzed by the coming cyberwar that it took a drastic step: the Pentagon's computer security gurus ordered all of its employees, civilian and military, en masse, to change their log-in passwords.[108] This was the first time this had ever been done across the entire organization.

Several weeks later, on October 6, 1999, the mighty Senate Judiciary Committee held a highly sensitive hearing. The senators had invited some of the government's most eminent experts on the protection of the nation's fragile infrastructure, from the Pentagon, the FBI, and the General Accounting Office. They met in the Dirksen Building, just off the Capitol, promptly at 10:00 a.m. Jon Kyl, a senator from Arizona, opened the hearing. The United States, he said, was the world's only remaining superpower. America's armed forces would be technologically superior to those of any other country. So naturally, Kyl reasoned, future adversaries were "looking hard for an Achilles heel."[109]

"According to the National Security Agency," Kyl said, "over 100 countries are working on information warfare techniques." One recent case, he continued, illustrated that danger: Moonlight Maze. Just two weeks earlier *Newsweek* had publicly revealed the code name.

"Can you say anything on the record about that particular ongoing event and can you identify it by its code name?" asked Kyl. "The article called it Moonlight Maze, and that is, in fact, our name for an investigation that we have been conducting for over a year," the FBI's Michael Vatis responded. But the FBI lawyer felt uncomfortable speaking about an ongoing investigation in public. "About the furthest I can go is to say that the intrusions appear to originate in Russia."[110]

Senator Dianne Feinstein pressed on: "Has there been any penetration of classified systems?" Vatis held his ground: "I should not get into that area in this setting."[111] He neither confirmed nor denied that the Russian intruders had breached classified networks.[112]

The White House upped the ante. In early January 2000, the president was supposed to give a short speech on computer security on the South Lawn. The draft for the speech contained a particularly alarming line: any person on any computer could "double-click on a mouse, hack into a computer system, and potentially paralyze an entire company, or city, or government."

This was an extreme statement and an extraordinary claim. One politically minded White House staffer reviewing the document spotted it. He scribbled in the margin, "Do we really want to say that a single person can potentially paralyze a government?" The statement carried two risks. One was that there was no evidence, let alone a precedent, that one individual alone could take down the government by cyber attack. Y2K had just passed without incident, as Bill Clinton would point out in this very speech. But even if it could not be done, having the president of the United States announce such a monumental vulnerability would certainly inspire potential attackers to try. "I'll defer to Dick et al on this," the staffer added.[113]

Richard Clarke, then in charge of infrastructure protection and counterterrorism in the National Security Council, left the statement in the speech. The next morning at 9:30, Clinton stepped out on the South Lawn and delivered the announced statement on "cyberterrorism." Cyberspace, the president stressed, empowered people to create knowledge as well as to create havoc, and these powers rested in the same hands:

> We live in an age when one person sitting at one computer, can come up with an idea, travel through cyberspace and take humanity to new heights. Yet, someone can sit at the same computer, hack into a computer system and potentially paralyze a company, a city or a government.[114]

Cybernetic attacks at first appeared to be a shiny new weapon that the mighty US armed forces could hurl at its enemies, as America's gift to warfare. What looked at first like a military opportunity revealed itself more and more as a prime threat. The machines, it seemed, had turned against their American creators.

"Instead of using cruise missiles and bombs, as NATO did in Serbia, in cyberwar you use cyberattack, but they could have the same effect," Clarke told Steve Kroft, the host of *60 Minutes*, CBS's flagship news program. It was 7:00 on the East Coast on a Sunday evening, with millions of concerned Americans watching. Next cut: Cheyenne Mountain, Colorado, one of the US military's most prized sanctuaries. Forty years ago, Kroft explained, the Pentagon started blasting away 700,000 tons of granite. The goal: to bury America's top nuclear command deep inside a mountain to protect it from a Soviet nuclear strike that would crush most life on Earth. The CBS footage showed one of the heavy steel doors at Cheyenne Mountain closing slowly, like a force of nature. "These steel doors can withstand a 30-megaton blast, but in a cyberwar they won't protect the military's computers," a reporter bragged.[115]

CBS disclosed that even the Pentagon's inner sanctum, the nation's most secure and classified computer system, had been invaded by a virus in the year 2000. The system had no links to the internet, so the breach happened through a disk on a laptop that should never have been connected to the classified system. The reporter then turned to interview Herbert Browne, an admiral and number two at US Space Command, the military command in charge not just of space but also of cyberspace. Browne commanded thermonuclear warheads as well as computer attack capabilities.

"How many countries have the capability, do you think, to wage cyberwarfare against the United States?" Kroft asked. "Virtually any country that has a computer has an opportunity to enter into—into cyberspace and—and be disruptive," said the admiral.

While Kroft and the CBS crew were filming, Moonlight Maze was ongoing. CBS's reporters and editors had no idea how close they came to the actual Colorado beachhead in the live unfolding spy drama. The Lakewood public library was just a ninety-minute drive north of Cheyenne Mountain.

"I think what's going on right now throughout the world is cyberwar reconnaissance," Clarke said, confident of his analysis. Several countries were scanning each other's networks, probing for vulnerabilities, trying to find things to knock out when the virtual bombs start hissing down. "Think of it as prewar reconnaissance," Clarke told CBS.

Meanwhile, the Russian intelligence operators became more determined not to be caught again. The Moscow-based hackers started improving their operational security. They encrypted files before smuggling them out of their victims' networks so that a filter at the exit could not spot keywords in cleartext. They started moving more stealthily on their victims' networks. Later the Russian spies showed even more impressive ingenuity. Intelligence analysts at the NSA and GCHQ suspected that the Moonlight Maze intruders began to hijack satellite downstream links to cover their tracks. Satellites beam data down to an extended area on Earth, a bit like ones and zeroes raining down from the sky. A receiver catching this information coming down from above is easy to hide. Western intelligence analysts believed the Russians were concealing stolen data in legitimate satellite broadcasts. The nightly Moonlight Maze would extend even into outer space.

The result was a difficult situation for the intelligence officers working in the Pentagon and the NSA; the political stakes and the pitch of the debate were rising, but the intrusions became harder to trace after the Moscow trip.

The intelligence analysts tried everything to move the investigation forward, including methods that had an air of desperation—for example, using satellite imagery of buildings to find keyboard operators. JTF-CND studied high-resolution images of buildings they suspected to house computing facilities, comparing current overhead imagery with historical pictures. They hoped to find things like satellite dishes, or trenches that had been dug to lay cables around a specific building, or other changes in layout. "I left no stone unturned," said one intelligence officer.[116]

At some point later in the investigation, intelligence analysts came up with the hypothesis that shrewd Iranian operators could have used Russian servers as proxies. This hypothesis was discussed at senior intelligence levels, but then dismissed. JTF-CND was fraught, unable to stop the bleeding. Moonlight Maze didn't really end, although the Department of Defense and the NSA stopped using that name, since becoming public had compromised it. The breaches continued to be tracked under a different code name. Perhaps most distressing was one surprising aspect: there were few concrete consequences, despite all the alarmism. The Pentagon was at war, yet it wasn't clear what the damage was. Even worse, that kind of war seemed to be a permanent condition.

In August 1999, Hamre had called the massive Russian attack America's first "cyberwar"—but two years later, with the same intrusion campaign still active, the first cyberwar strangely moved back into the future, not into the past or present. The Pentagon task force had an FBI "Wanted" poster of Osama bin Laden up in the office. They called him "UBL" already before that fateful September day in 2001. Next to UBL were illustrations of the Chinese, the Russians, and the Iranians. The new task force had been in charge of computer network defense for a little more than two years, and from the president down, the country expected the next terrorist attack to be different.

The task force had changed its name from CND, "computer network defense," to CNO, "computer network operations," to include offense as well. The team was convinced that the next major terrorist offense was not going to be an airplane hijacking or a car bomb, as had been common throughout the 1990s. This was the twenty-first century. "It was going to be a cyber attack," said Marcus Sachs, one of the army operators on the task force. He had worked on computer network defense and offense in the Pentagon since early 1999.

When the next major terrorist attack came, it was even more innovative than expected. And at the same time, it was oddly old-fashioned, reminiscent of World War II kamikaze pilots, men who had steered their machines precisely into their targets. The world was shocked as Manhattan and the Pentagon reeled in smoking rubble. That day, the members of JTF-CNO came to work in battle gear, not in the coat-

and-tie uniforms that had been customary until 9/10. They expected to fight that day, in cyberspace. But it turned out differently. "We've spent all these years getting ready for ultimate cyberwar, and the little fuckers fly airplanes into buildings," Sachs recalled, incredulous.[117]

Meanwhile, the cybernetic myth retained its fervid force and would soon grow even stronger. Moonlight Maze was the first known state-on-state cyber attack in history, literally monumental in scale, and it was duly portrayed as the first "cyberwar" and as the long-awaited electronic Pearl Harbor, back in 1998 and early 1999. The pioneering Russian campaign would indeed accurately foreshadow the future of the most common and most costly computer network breaches over the next two decades, down to the level of tactics, techniques, and procedures; it was advanced, persistent, clandestine, and reactive. Identifying start, end, costs, and perpetrators required time and hard work. But when the long-anticipated future emerged in 1998, it wasn't good enough, yet again; it was too pale and unbefitting. Reality once more underdelivered and disappointed the visionaries, with all those unseemly details spoiling the brave prospect of what the future of networked machines had in stock. Thankfully, Norbert Wiener's devout descendants had already found a solution to that problem.

9. FALL OF THE MACHINES

CYBERNETICS STARTED AT WAR—AND EVENTUALLY CAME back to war. First came the *promise*, then the *rise*, and finally the *fall* of the machines.

Initial technical design proposals of how humans would interact with computers were always grand, ambitious, bold, even monumental. Often the US defense establishment initially sponsored these grand visions with lavish grants. But translating the early blueprints into technical reality proved hard again and again, and it took far longer than anticipated. Hopes were dashed. Projects were canceled. Products failed. Companies disappeared. Expectations fell.

The fate of cybernetics as a new engineering and scientific discipline began the pattern in the 1950s. The primitive computers and control systems of the time couldn't deliver the goods to turn design visions into practice. The story continued in the 1960s with Sputnik-shocked aeronautic engineers implanting osmotic pumps in rat "cyborgs," testing to modify the human body for life in outer space and the deep sea. By the early 1970s, bold promises of "cybernated" economies—a merger of cybernetics and automation—were revealed as fanciful. In the 1980s, virtual reality was the next big thing, with technologies portending access to the new frontier, "cyberspace."

After the promise came the rise of the machines—or, more precisely, the rise of cybernetic myths high and wide beyond their original purpose and application. Myths work as conceptual aids, reducing complexity, condensing narratives, and making novel yet unknown technologies approachable, either in a utopian or a dystopian way. Cybernetics immediately enthralled not only engineers, but a wide range of scientists, entrepreneurs, scholars, artists, and science fiction writers in the 1950s. Even charlatans and self-help gurus discovered the power of purpose-driven gadgetry.

Political activists in the 1960s took up the theme of automation, first as a negative vision and then as a positive one. Back-to-the-land communards of the 1970s embraced cybernetics as a way to achieve spiritual oneness with nature, reimagining networked computers as tools of liberation and self-discovery, not as military targeting devices. By the late 1980s, psychedelically enhanced hippies were trying to homestead the new frontier of cyberspace, and postmodernists recruited the cyborg to help them blur boundaries and subvert the established order of things in an improved virtual reality. In the 1990s, crypto activists plotted to upend the established political order. And by the turn of the millennium, the Pentagon was fixated on cyberspace as a "new domain of war," next to air, land, sea, and space. Through all its history, the cybernetic trajectory has simultaneously promised and exaggerated the rise and fall of the machines, utopia and dystopia. The technologies that could support pioneering designs would not appear in time.

Nevertheless, technological progress marched on. Decades after Wiener and his acolytes dreamt of autonomous machines and decades after NASA flirted with putting cyborgs into outer space, microprocessing and networking technology have radically transformed lives, work, play—even human bodies. Often the cybernetic origin remains invisible: the new discipline's founding challenge—guiding antiaircraft fire—has long been overcome. Computer-controlled pacemakers and insulin pumps are taken for granted, brakes and engines of internet-connected cars are run by software, and flying airliners have become ever more automated. Social media form a connection with communities, sometimes even independent from geographical location. Cryptography is widely used, often without users noticing it. Robots don't just toil in factories. Computers are an integral element of intelligence collection and even military operations, with battle-ready drones circling the skies over the world's most contested war zones, assassinating tiny humans below when remotely commandeered to do so. The machines have risen as widely used artifacts.

But now the machines can fall. The artifacts can be hacked. Control can be taken away from machines, or indeed from humans who use machines as tools. Systems can be accessed remotely to abuse, surveil, steal, damage, or delete data—or to deny service and even to cause physical damage and kill people as a result of tinkering with machines, through their controls and communications. Meanwhile, the cybernetic myth is still at work: promising and again exaggerating the fall. The dystopian fears of mass surveillance and cyberwar have the same historical ancestor. This cybernetic pattern—promise, rise, and fall—is one sweeping but coherent motion that can be understood only by taking into account its deep historic legacy. But it carries with it formidable risks and dangers.

Appreciating these risks requires taking a large step back.

Our tools acquired their mythical power in three stages. First came thinking about the machine in human terms, personifying the artificial as if made of flesh and blood. This anthropomorphic longing didn't start with the antiaircraft problem during World War II, or with Norbert Wiener's attempt to model pilot-plane interaction under enemy fire. The desire to enhance muscles by mechanical means, of course, is older than civilization. Already primitive preagriculture tools—for instance, a club or a lever—extended a user's physical strength. As these tools became industrial machines and as machines became computers, ambitions rose.

In various cultures, the idea emerged that humans could create, in their own image, artificial beings with independent will—a belief captured ingeniously in Stanley Kubrick's science fiction classic *2001: A Space Odyssey*, and especially in the film's famous four-million-year-spanning match cut: an ape throws a bone tool into the air, the camera follows the rotating bone against blue sky with scattered clouds—[cut]—a bone-shaped sentient spaceship glides through deep black outer space.

The idea of creating mechanical life predates the invention of computers by many centuries. Early Jewish folklore has the golem, a shapeless clay figure that is molded and brought to life, most famously in a story that involves Judah Loew ben Bezalel, a late-sixteenth-century rabbi of Prague. In Greek mythology, Hephaestus, the divine blacksmith, crafts automata out of bronze. His most prominent artificial creation is probably the giant Talos, built to protect Europa in Crete. Hephaestus also made the chain that tied Prometheus to a rock as he is tormented by an eagle for stealing fire from the Gods. The story of Prometheus was the inspiration for Mary Shelley's 1818 novel *Frankenstein*. The book's monster, brought to life by alchemy, is perhaps the most influential literary expression of human hubris. A more contemporary rendering of the humanoid machine is Karel Čapek's 1921 play *R.U.R.* ("Rossum's Universal Robots.")

The mechanical creation myth changed drastically midcentury. The invention of the digital computer in the late 1940s took this already potent narrative to a new level. Suddenly, machines would actually think, calculate, and make decisions on their own. Not just in fables and in mythical narratives, but for real, cast in iron, wired up with copper, run by transistors. Sperry gun predictors and radio-emitting variable-time fuses showed a degree of autonomy never seen before, and the first digital computer mesmerized its contemporaries. Cybernetics, the general theory of machines, emerged just in time to explain what was going on: self-regulating machines weren't powered by Jewish magic, Greek gods, or English alchemy; they were powered by negative feedback.

Wiener and the early cyberneticists thus replaced magic with science. They gave a scientific language and respectability to the idea of autonomous future machines endowed with their own will and even with the ability to self-reproduce. This vision inspired an entire generation of engineers and inventors. The goal became to create machines in the image of humans, or at least in the image of human faculties. Ross Ashby made an early start, with his homeostat clicking gently as it decided how best to overcome disturbances in the environment and find an elegant equilibrium. The machine, he believed, came to life in the English countryside.

Soon, engineers began envisioning and designing machines that could compete with humans and outperform them—first in strength, then in intelligence. John von Neumann and Norbert Wiener added the notion that machines could theoretically even build an improved version of themselves and thus evolve, and evolve faster than humans ever could. From here, it was only a small step to see that sentient, self-reproducing machines could not only outperform their own creators, but *outcreate* them. Once this moment was reached, Vernor Vinge later predicted, humankind would experience a "singularity," a point beyond which nobody could even imagine the future. Man, in short, could in theory not only create a superman; man could create a supergod. Cybernetics was indeed a dazzling new field of study.

The second phase of the myth was even more arresting. The rise of the machines continued apace when the logic of giving machines human traits was flipped over: if it was possible to think about the machine in human terms, then it was also possible to think about the human in mechanical terms. The goal wasn't any longer to create a machine in the image of humans, but to re-create humans in the machine's image—not to mechanically *eclipse* life, but to mechanically *enhance* life. And again this change in perspective applied both to the body and to the mind.

This shift happened around 1960: with his "cyborg," Manfred Clynes attempted to augment the flawed and fragile human body by adding machine parts and chemical tweaks, as did Ralph Mosher in his work on a range of exoskeletons at General Electric, where he added brushed-metal strength to brittle human bones. At the same time, Maxwell Maltz's "psycho-cybernetics" presented the human mind as a mechanism that could be reprogrammed and guided to a predefined goal, thus turning soggy gray goo into a whirring engine. Cybernetics was so convincing and seductive that it literally inspired cults; L. Ron Hubbard's infatuation with Wiener's new science produced only the first and most extreme example.

By the 1970s, the idea of cybernetically refining humans was extended from individual bodies and minds to the collective mind of entire communities. The first such attempt was Stewart Brand's *Whole Earth Catalog*, designed as a printed and mail-ordered cybernetic feedback loop for back-to-the-land communards, an analogue machine of loving grace to maintain the countercultural balance of the whole Earth. Gregory Bateson articulated the spiritual underpinnings in elegant and holistic prose that resonated with 1970s counterculture.

The Whole Earth approach was successful beyond anybody's expectations, and a dozen years later went digital with the WELL, the Whole Earth 'Lectronic Link. Man-machine interaction was no longer one man to one machine, but many to many—linked not by mail but by modem. An entire electronic subculture began to emerge. "We are the robots," as the German electronic music pioneers Kraftwerk put it so stylishly in 1978.

The third phase followed seamlessly. By the 1980s, computers and networks had given rise to the idea of a separate cybernetic space, a space inside the machines. This virtual space was, again, first developed in a secret air force project and then dubbed "cyberspace" by the *Whole Earth Review*, the hippie magazine and heir to Brand's catalog. Initially, utopia dominated, with optimism in abundance. Virtual space became the realm of the mind. It was an uncharted land of inspiration and creativity, a new frontier that could be colonized and explored and claimed and built anew, avoiding the mistakes of the original frontier. This space, as the new pioneers saw it, was unregulated, lawless, and free. There were no more borders, and everybody was equal, free to choose a virtual identity. And this was a good thing, in the eyes of the electronic pioneers, with hardly any downsides.

Cyberspace was beyond the grasp of the federal government and its dreaded law enforcement and intelligence agencies, at least in the perspective of the early internet activists. Securing cyberspace meant securing the machines that carried the new vast virtual planes of the open range, and maintaining the law of the open range on the electronic frontier. And such perfect freedom was possible by running software that enabled strong cryptography, protecting anonymity and thereby cutting the proverbial electronic barbed wire fences.

Once such a system was set up, the most hardened cyberspace anarchists believed, a freer and better political order would emerge. That order would emerge automatically, enabled and driven and protected by impartial machines and pure math, not by biased bureaucrats and partisan politicians. The countercultural cyberspace cowboys were deeply disappointed by politics and power, and counted on code and crypto instead. The machines—or more precisely, the space inside the machines, cyberspace—were the salvation, a screen to project humankind's noblest hopes and dreams of a truly free society.

Counterculture and intellectuals were not the only ones exhilarated by cybernetics and the new technical possibilities of networked machines. America's military establishment was equally excited.

Network-centric warfare meant that technologically superior armies could now dominate the battlefield, and win wars before wars even started, crippling an adversary's communication infrastructure while maintaining one's own—blinding and paralyzing the enemy while retaining a high-resolution overview and agility to maneuver. The Pentagon called this "information superiority."

But hope was soon joined by fear, even in America's mighty armed forces and its spy agencies. As the internet branched out, more and more people and more and more control systems were plugged into the global network. The infrastructure of entire countries became vulnerable to remote tampering—the specter of an "electronic Pearl Harbor" raised its head. The electricity grid, chemical plants, air traffic control, stock exchanges, banks—all these critical systems were sitting ducks for skilled hackers, the White House feared, no ballistic prediction required. In the next crisis, cyberspace would be the new battlespace, and the digital open range would quickly turn into a digital shooting range.

The history of cybernetics holds several warnings. The new field's scientific popularity peaked around 1969.[1] By the 1980s, cybernetics had lost a great deal of academic respectability; it had "died of dry rot," as *Wired*'s Kevin Kelly memorably put it.[2] Today, the most serious scholars working on cybernetics are historians of science and cultural scholars; engineers and computer scientists abandoned it in droves, renaming their projects and their degrees.[3] "Cybernetics," nearly seventy years after the word was coined, sounds oddly old-fashioned and out-of-touch.

Yet the myth has lost none of its formidable force. For technology, as ever, exudes certainty: Flick a switch and the engine will stop. Push a button and the light will go out. Run a calculation and the computer will display the result. It is with such mechanized certainty that futurists have, again and again, predicted the future. Cybernetics, as Ross Ashby put it so eloquently, was a theory of all machines, includ-

ing those machines that had not been built yet. Cybernetics was thus
equipped with uncanny powers. It professed to master the mechanics
of the future. It claimed to control the wiring that linked yesterday
to tomorrow. On closer examination, cybernetics thus carried several
patterns through the decades—patterns that are also warnings.

One pattern is spiritual. Not always, but often, the machine has
become a godhead, an idol. Norbert Wiener, mesmerized by his own
invention, wanted to understand magic and religion through his
theory of the machine—but he achieved the opposite: the spiritual
encroached on the mechanical, and Wiener himself came to see the
machine in mythical terms. First, humans were seen as functioning
like machines, then the community became a machine, and finally
the spiritual itself became cybernetic. Science created a totem. The
machine became the avatar.

When he published *God and Golem*, Wiener sensed what he had
done, although he never expressed it as clearly as he could have. His
fascination with the story of the jinni, or "The Monkey's Paw," had
a straightforward reason: the stubborn scientist realized that he had
unleashed a jinni himself: cybernetics. The new discipline was meant
to be a general theory of machines, but too often it tended toward a
theology of machines: for Alice Mary Hilton, for Manfred Clynes,
for young Jaron Lanier, for Timothy Leary, for John Perry Barlow, for
Tim May, or for John Hamre. Often it was—and still is—faith, not
facts, that dominated the discussion. Cyberculture turned cybercult.

A second powerful pattern is contradiction. The history of cybernetic ideas has a perennial tension built in. The machines were always
a positive and a negative force at the same time, utopian and dystopian
at once, although most of the time optimism dominated. Automated
factories would free workers from undignified drudgery, yet deprive
them of their dignity. Robotics would take away labor and create
more labor. Computers were dumb and could be hacked by teenagers,
yet they could outsmart humans. Autonomous machines could seize

control, and as personal computers they would provide more control. More networked computers would lead into a "dossier society" of ubiquitous surveillance—and enable anonymity and a freer and better political order. Networked information systems would make nations more vulnerable and more fragile than ever, and networked command-and-control systems would make their armies more dominant and more lethal than ever. Machines would be future society's hard-charging overlords and its soft underbelly. The myth hides these contradictions and makes them acceptable.

A third pattern is that technology, again and again, has outperformed the myth. Historical myths hail from the past and thus cannot be overwhelmed by reality. But technology myths hail from the future and therefore can come into conflict with reality. Even bold predictions about the future a quarter century out will face a moment of truth a quarter century later. This has occurred many times in the seven decades since World War II.

Whenever actual technical developments catch up with the mythical vision of technology, one of two things tends to happen. The first is that work on the myth slowly stops, it doesn't get repeated anymore, and eventually recedes. This is what happened with Manfred Clynes's original idea of the cyborg, and with John von Neumann's vision of the self-reproducing machine. The second possibility is that the myth simply shifts its shape and escapes again into the future. This is what happened with Alice Mary Hilton's dreams of automation in the early 1960s, with Jaron Lanier's cyberspace in the early 1990s, with Timothy May's crypto anarchy in the late 1990s, and, indeed, with John Hamre's cyberwar around the same time. First a free cyberspace arrived; then it was a vision of the future again. One day the Pentagon was at cyberwar; the next day, cyberwar had not happened yet.

Another pattern is an extraordinary appetite for new terms. New myths need new words, pointed out Roland Barthes, an influential mid-twentieth-century philosopher and literary theorist.[4] This is especially true for technology myths that carry a promise of a new future. Old words cannot convey the novel. And cybernetics, as the story here demonstrates, has spawned a range of new terms that were eagerly taken up by scientists, writers, activists, and officials.

The prefix "cyber-" has proved perennially attractive for the past seventy years, from cybernation to cyborgs, from cyberculture to cyberspace, from cyberpunk to cyberwar. Timothy Leary was an unrivaled master in the jazzy art of cybernetic wordsmithing, and surely the psychedelic guru would have a fabulous time with Pentagon and NSA slide decks dotted with similar artistry. Norbert Wiener himself disdained such "initial jargon" as a curse of modern times. The newspeak irked him. It often sounded "like a streetcar turning on rusty nails," he wrote to Alice Mary Hilton a year before his death.

But this perennial hunger for trendy words holds a warning: they are ephemeral and always have been. Many of these phrases have already receded into history. That even applies to Wiener's master discipline. Once, it was a vibrant new idea that "set bells jangling wildly in a dozen different sciences," in the words of *Time* magazine.[5] Then it went away within just a few years. Yes, its ideas may continue to be highly relevant, but the term was forgotten faster than many could have imagined.

The final pattern is irony. The thinking machines have risen, in breathtaking and revolutionary fashion. In the decades since World War II, during the second half of the twentieth century, a time that is still in living memory, humankind has experienced a faster pace of change in its communication behavior than ever before in human history. By the late 1970s, the three main ingredients were available: the computer, the internet, and public encryption. Over the next decade, input-output devices became ever sleeker and networking platforms more social and more competitive. But the machine's basic ingredients remained unchanged—as did the human tendency to think about the machine in perfection, not in limitation. If something goes wrong, it must be human error: a human flaw in the machine's operation, in its design, in its programming, or in its maintenance must have caused the problem, whatever it was. For our instrument can't make a mistake, even if it is created in our image; only we can make mistakes. And therein lies the ultimate irony.

Cybernetics, first and foremost Norbert Wiener, tried to disenchant the machine—but achieved the opposite, the enchantment of the machine. The science of negative feedback itself created a powerful positive feedback loop, driving our visions of the future into perennial overdrive, the opposite of equilibrium: the persistent expectation is that everything will change—fast, completely, and inevitably. Wiener died about fifteen years before engineers at Wright-Patterson Air Force Base started to talk about the virtual space inside the machines. Wiener probably would have both loved and hated the talk of cyberspace. He would have appreciated it because it carried the myth that he had created to its logical conclusion, by opening up a novel space inside the machines. But at the same time he would have disdained the idea and the jargon. The entire notion of a separate space, of cordoning off the virtual from the real, is getting a basic tenet of cybernetics wrong: the idea that information is part of reality, that input affects output and output affects input, that the line between system and environment is arbitrary.

This triple cybernetic progression from *man-machine* to *machine-man* to *machine-worlds* was indeed dangerous. But not in the way Wiener and his acolytes had foretold. The machines weren't about to take over; the myth took over, sending the debate into trembles, like Palomilla. That seductive power of the cybernetic mythos has increased over the decades, not decreased.

The futurists, of course, didn't always get the future wrong, but almost always they got the speed, the scale, and the shape wrong. They continue to do so. Yet we choose to remember history very selectively, favoring foresight over failure. It is thus useful, and unusually humbling, to remember Wiener's original challenge of 1940, when the German Luftwaffe sent its fighters across the channel to bomb London: he tried predicting how the cybernetic system that was the pilot and his aircraft would behave for the next twenty seconds under stress. He failed.

Acknowledgments

THIS BOOK IS THE MOST LABOR-INTENSIVE AND COLLABORATIVE project I have ever engaged in—by far. There is no way I could have done it without the help of a large number of people. W. W. Norton was just phenomenal—above all Brendan Curry, my deft editor; and Sophie Duvernoy. Philip Gwyn Jones at Scribe in London adroitly added a few final etchings to the manuscript. Stephanie Hiebert's hawkish copyediting and Michael Adrian's attention to detail were priceless. Catherine Clarke and George Lucas made it all possible.

Researching the Moonlight Maze campaign involved more than one year of persistent investigative work and more than three dozen interviews. I am deeply grateful for the confidence placed in me by so many, and I especially want to thank James Adams, Dave Bryan, John "Soup" Campbell, Paul Cox, Michael Dorsey, Michael Hayden, Jason Healey, Richard Kaplan, Richard Marshall, Marcus Sachs, Dion Stempfley, and a number of people who prefer not being named at all, including several law enforcement and intelligence sources in the United States and the United Kingdom. These anonymous sources were some of the most crucial.

I would also like to thank the FBI for favorably responding to FOIA requests by sending 286 pages of redacted yet helpful files, enabling me to date several events that the individuals I interviewed could not place precisely anymore, fifteen years after the fact. Sometimes I went back to sources many times to verify details. Errors are mine alone.

I spent the spring of 2014 researching parts of the book in San Francisco and the Bay Area, interviewing pioneers of 1980s electronic counterculture, as well as 1990s crypto activists. I am particularly grateful to John Gilmore, Kevin Kelly, Ryan Lackey, and notably Tim May for being so helpful and generous with their time. Barbara Hayward, Alice Mary Hilton's daughter, gave me access to unpublished correspondence and papers from the 1960s. I would also like to thank several people who helped answer arcane questions over the past two years on Twitter, particularly Daniel Bilar and Scott Carson.

For intellectual, archival, and logistical help, my thanks to Mick Ashby, Nora Bateson, Jeff Bauer, Kurt Baumgartner, Richard Bejtlich, Michael Benedikt, Stewart Brand, Ben Buchanan, Myles Crowley, Thomas Furness, Ken Goffman, Karl Grindal, Juan Andrés Guerrero-Saade, Phillip Guddemi, Ralph Langner, Jaron Lanier, Robert Lee, Charles Levinson, David Omand, Barry Schwartz, Wolfgang Seibel, Tim Stevens, Fred Turner, Vernor Vinge, Cameo Wood, Graeme Wood, Thomas Zimmerman, as well as the staff at MIT Libraries; the Museum of Innovation and Science in Schenectady, New York; and the British Library. At King's College London, I am grateful to Theo Farrell, Lawrence Freedman, Joe Maiolo, and Chris Mottershead for their unflinching support.

Finally, thank you, Annette—for everything.

Notes

PREFACE

1. Roland Barthes, *Mythologies* (Paris: Editions du Seuil, 1957).
2. Hans Blumenberg, *Arbeit am Mythos* (Frankfurt: Suhrkamp, 1984).

RISE OF THE MACHINES

1. André-Marie Ampère, a French physicist and mathematician, was the first to use the term "cybernétique" in an essay in 1834. Wiener did not know this in 1948. Ampère suggested an ambitious terminological classification of all human knowledge, as was common at the time. Like Wiener more than a century later, he got his inspiration from the Greek. His neologism was meant to refer to the "art of government in general." Unlike Wiener's, Ampère's attempt to coin an expression did not catch on, and remains an obscure reference in a long-neglected volume. In French: "Ce n'est donc qu'après toutes les sciences qui s'occupent de ces divers objets qu'on doit placer celle dont il est ici question et que je nomme *Cybernétique*, du mot κυβερνήτης [*sic*], qui, pris d'abord, dans une acception restreinte, pour l'art de gouverner un vaisseau, reçut de l'usage, chez les Grecs même, la signifi- cation, tout autrement étendue, de l'art *de gouverner en général*." See André-Marie Ampère, *Essai sur la philosophie des sciences* (Paris: Bachelier, 1843), 140–41.
2. Unless otherwise attributed, all quotes in the Introduction come from "In Man's Image," *Time* 52, no. 26 (December 27, 1948): 47.
3. Richard Brautigan, "All Watched Over by Machines of Loving Grace," in *The Pill versus the Springhill Mine Disaster* (San Francisco: Four Seasons Foundation, 1968).

1. CONTROL AND COMMUNICATION AT WAR

1. For an overview of the raids, see Richard Overy, *The Battle of Britain* (New York: Penguin, 2010).

2. Joseph Cerutti, "The Battle of Britain," *Chicago Tribune*, September 19, 1965, G34.

3. Ibid.

4. John Keegan, *The Second World War* (London: Pimlico, 1989), 78.

5. Frederick Arthur Pile, *Ack-Ack* (London: Harrap, 1949), 39.

6. Keegan, *Second World War*, 73.

7. Rexmond C. Cochrane, *The National Academy of Sciences: The First Hundred Years, 1863–1963* (Washington, DC: The Academy, 1978), 387.

8. Charles A. Lindbergh, *The Spirit of St. Louis* (New York: Scribner, 1953), 486.

9. Cochrane, *National Academy of Sciences*, 387.

10. Vannevar Bush, in classified draft memorandum, quoted in David A. Mindell, *Between Human and Machine: Feedback, Control, and Computing before Cybernetics* (Baltimore: Johns Hopkins University Press, 2002), 187.

11. Ibid.

12. For instance, Norbert Wiener, *Cybernetics; or, Control and Communication in the Animal and the Machine* (Cambridge, MA: MIT Press, 1948), 113.

13. Quoted in Mindell, *Between Human and Machine*, 350.

14. Alfred D. Crimi, *Crimi* (New York: Center for Migration Studies, 1988), 150.

15. Steven Gould Axelrod, Camille Roman, and Thomas J. Travisano, *The New Anthology of American Poetry*, vol. 3 (New Brunswick, NJ: Rutgers University Press, 2003), 96.

16. Crimi, *Crimi*, 152.

17. See ibid., 151.

18. William White, "Secrets of Radar Given to World," *New York Times*, August 15, 1945, 1.

19. *Radar Operator's Manual*, Radar Bulletin no. 3 (RADTHREE) (Washington, DC: United States Fleet, Navy Department, 1945), 3–10.

20. David Zimmerman, *Top Secret Exchange: The Tizard Mission and the Scientific War* (Montreal: McGill-Queen's Press, 1996), 90–91.

21. George Raynor Thompson and Dixie R. Harris, *The Signal Corps: The Outcome* (Washington, DC: US Army Center of Military History, 1991), 303.

22. H. A. H. Boot and J. T. Randall, "Historical Notes on the Cavity Magnetron," *IEEE Transactions on Electron Devices* 23, no. 7 (1976): 724–29.

23. Quoted in Zimmerman, *Top Secret Exchange*, 135.

24. Ivan A. Getting, interview by Frederik Nebeker, IEEE History Center, Hoboken, NJ, #077, June 11, 1991.

25. See "Tech's Radar Specialists Now Return to Peace Jobs," *Christian Science Monitor*, August 15, 1945, 2.

26. American Defense Preparedness Association, "Army Ordnance," *National*

Defense 12, no. 67 (1931): 38. Also quoted in Mindell, *Between Human and Machine*, 89.

27. Bernard Williams, *Computing with Electricity, 1935–1945* (Ann Arbor, MI: University Microfilms International, 1984), 204–5.

28. Glenn Zorpette, "Parkinson's Gun Director," *IEEE Spectrum*, April 1989, 43.

29. Clarett A. Lovell and David Parkinson, Range computer, US Patent 2,443,624, filed April 23, 1942, and issued June 22, 1948.

30. Quoted in Mindell, *Between Human and Machine*, 136.

31. The quote is from Project 2's final report, see ibid., 233; source on p. 375.

32. Several countries tried to develop the technology at the time. The Germans worked on no less than thirty different types of proximity fuses before and during the war, but none saw service. See James Phinney Baxter, *Scientists against Time* (Cambridge, MA: MIT Press, 1968), 222.

33. Lee McCardell, "Now It Can Be Told: How Hopkins Kept Secret Fuse Secret," *Baltimore Sun*, November 25, 1945, A1.

34. Quoted in ibid.

35. David A. Mindell, "Engineers, Psychologists, and Administrators: Control Systems Research in Wartime, 1940–45," *IEEE Control Systems* 15, no. 4 (August 1995): 91–99.

36. Pesi Masani, *Norbert Wiener, 1894–1964* (Basel: Burkhäuser, 1990), 182.

37. Peter Galison, "The Ontology of the Enemy: Norbert Wiener and the Cybernetic Vision," *Critical Inquiry* 21, no. 1 (Autumn 1994): 233.

38. See ibid., 234.

39. M. D. Fagen, Amos E. Joel, and G. E. Schindler, *A History of Engineering and Science in the Bell System: Communications Sciences (1925–1980)* (Indianapolis, IN: AT&T Bell Laboratories, 1984), 359.

40. Quoted in Galison, "Ontology of the Enemy," 228.

41. Masani, *Norbert Wiener*, 182.

42. Wiener, *Cybernetics*, 5.

43. Flo Conway and Jim Siegelman, *Dark Hero of the Information Age* (New York: Basic Books, 2005), 111. Conway and Siegelman's book quotes from several interviews with Bigelow, but it seems to be unreliable on some of the technical details.

44. Masani, *Norbert Wiener*, 188.

45. Quoted in Conway and Siegelman, *Dark Hero*, 114.

46. Wiener, cited in Galison, "Ontology of the Enemy," 236.

47. Masani, *Norbert Wiener*, 188.

48. Notebook entry from Stibitz, quoted in Galison, "Ontology of the Enemy," 243.

49. Ibid., 243.

50. Quoted in Mindell, *Between Human and Machine*, 281.

51. Quoted in Galison, "Ontology of the Enemy," 242.

52. Quoted in Mindell, *Between Human and Machine*, 281.

53. Ibid.

54. Quoted in Galison, "Ontology of the Enemy," 245. The full source is Norbert Wiener to Warren Weaver, January 28, 1943, Norbert Wiener Papers, collection MC-22, box 2, folder 64, Institute Archives and Special Collections, MIT Libraries, Cambridge, MA.

55. Wiener, *Cybernetics*, 15.

56. From Sperry company history, probably 1942, quoted in Mindell, *Between Human and Machine*, 69.

57. Ibid.

58. Preston R. Bassett, "Review of Ground Anti-aircraft Defense," *Sperryscope* 11, no. 7 (Autumn 1948): 16.

59. Ibid., 19.

60. Ibid.

61. Preston R. Bassett, "Sperry's Forty Years in the Progress of Science," *Sperryscope* 12, no. 2 (Summer 1950): 20–21.

62. John Sanderson, "The Sperry Corporation—A Financial Biography (Part II)," *Sperryscope* 10 (1952): 19.

63. Claus Pias and Heinz von Foerster, *Cybernetics: The Macy-Conferences 1946–1953* (Zurich: Diaphanes, 2003), 12.

64. Quoted in Mindell, *Between Human and Machine*, 104.

65. Frederick Pile, "Ack-Ack," *Anti-Aircraft Journal*, May–June 1950, 40 (serialized from book).

66. A good sound recording of the British World War II air-raid siren can be found in "Air Raid Sirens Followed by the All Clear," YouTube video, posted January 10, 2009, https://youtu.be/erMO3m0oLvs.

67. See Mindell, *Between Human and Machine*, 232.

68. Joe Maiolo, "Hitler's Secret Weapon," *BBC Knowledge*, November/December 2011, 59–63.

69. Pile, *Ack-Ack*, 14.

70. "Ford Makes U.S. 'Robombs,'" *Christian Science Monitor*, October 23, 1944, 1.

71. D. M. Dennison and H. R. Crane, "The V-T Fuze," *Michigan Technic* 64, no. 7 (May 1946): 24.

72. Pile, *Ack-Ack*, 14.

73. Ibid., 41.

74. Ibid., 326.

2. CYBERNETICS

1. Michael C. Quinn, "American V-2 Rocket Facilities," HAER no. NM-1B (Washington, DC: Historic American Engineering Record, National Park Service, Department of the Interior, 1986), sheet 1-6.

2. Malcolm Macdonald and Viorel Bedesco, *The International Handbook of Space Technology* (Heidelberg: Springer, 2014), 8.

3. Norbert Wiener, "A Scientist Rebels," *Atlantic* 179, no. 1 (January 1947): 46.

4. Norbert Wiener, *I Am a Mathematician* (Cambridge, MA: MIT Press, 1956), 308.

5. Ibid.

6. Ibid.

7. Wiesner, quoted in David Jerison, I. M. Singer, and Daniel W. Strook, eds., *The Legacy of Norbert Wiener: A Centennial Symposium in Honor of the 100th Anniversary of Norbert Wiener's Birth, October 8–14, 1994, Massachusetts Institute of Technology, Cambridge, Massachusetts* (Providence, RI: American Mathematical Society, 1997), 19.

8. Entropy is a fundamental and related concept in physics as well as in information theory. See James Gleick, *The Information* (New York: Pantheon, 2011), chap. 9.

9. Norbert Wiener, *The Human Use of Human Beings* (New York: Houghton Mifflin, 1954), 263.

10. Ibid., 33.

11. Ibid., 24.

12. Norbert Wiener, *Cybernetics* (Cambridge, MA: MIT Press, 1948), 43.

13. Norbert Wiener, *God and Golem, Inc.* (Cambridge, MA: MIT Press, 1963), 74.

14. Ibid., 76.

15. Ibid.

16. Wiener, *Cybernetics*, 157–58; Harry Davis, "An Interview with Norbert Wiener," *New York Times*, April 10, 1949, BR23.

17. William Laurence, "Cybernetics, a New Science, Seeks the Common Elements in Human and Machine," *New York Times*, December 19, 1948, E9.

18. John Pfeiffer, "The Stuff That Dreams Are Made on: Cybernetics," *New York Times*, January 23, 1949, BR27.

19. "The Thinking Machine," *Time*, January 24, 1949, 66.

20. Andrew Pickering, *The Cybernetic Brain* (Chicago: Chicago University Press, 2010), 92.

21. "Thinking Machine," *Time*, 66.

22. "Checkmate in 390,625," *Daily Mail*, December 13, 1948, 3.

23. W. Ross Ashby, *Design for a Brain* (London: Butler & Tanner, 1952), 103.

24. "The Clicking Brain Is Cleverer Than Man's," *Daily Herald*, December 13, 1948.

25. W. Ross Ashby, "Design for a Brain," *Electronic Engineering* 20 (December 1948): 379–83. The article was reprinted in the United Sates in March 1949 in the magazine *Radio-Electronics*, pp. 77–80.

26. Margaret Mead, "Cybernetics of Cybernetics," in *Purposive Systems*, ed. Heinz von Foerster (New York: Spartan Books, 1968), 1.

27. W. Ross Ashby, "Homeostasis," in *Cybernetics: Circular Causal and Feedback Mechanisms in Biological and Social Systems: Transactions of the Ninth Conference, March 20–21, 1952, New York, NY*, ed. Heinz von Foerster, Margaret Mead, and Hans Lukas Teuber (New York: Josiah Macy Jr. Foundation, 1953), 73.

28. Ibid., 74.

29. Ibid., 89.

30. Ibid.

31. Ibid.

32. Ibid., 97.

33. Ibid., 103.

34. Ibid., 95.

35. Ibid., 106.

36. Stefano Franchi, "Life, Death, and Resurrection of the Homeostat," in *The Search for a Theory of Cognition*, ed. Stefano Franchi and Francesco Bianchini (Amsterdam: Rodopi, 2011), 3.

37. William Grey Walter, *The Living Brain* (London: Duckworth, 1953), 123.

38. Wiener, *Human Use of Human Beings*, 34.

39. Ibid., 38.

40. Judging by the number of citations, Ashby's *Introduction to Cybernetics* (London: Chapman & Hall, 1956) is the most influential single work on cybernetics after Wiener's foundational 1948 book.

41. Ashby, *Introduction to Cybernetics*, 1.

42. Ashby, "Design for a Brain," 379.

43. W. Ross Ashby, "The Nervous System as Physical Machine," *Mind* 56, no. 221 (January 1947): 44. This article also reveals Ashby's admiration for Pavlov.

44. Ashby, *Design for a Brain*, 35.

45. Ibid., 39.

46. Ibid.

47. Ashby refers to the experiment but doesn't provide the reference to the article. It is R. W. Sperry, "Effect of Crossing Nerves to Antagonistic Limb Muscles in the Monkey," *Archives of Neurology and Psychiatry* 58 (October 1947): 452–73.

48. Ashby, *Introduction to Cybernetics*, 110.

49. Charles Stafford, "Man Called Future Slave of Machine," *Los Angeles Times*, October 23, 1961, H1.

50. See Paul L. Edwards, *The Closed World: Computers and the Politics of Discourse in Cold War America* (Cambridge, MA: MIT Press, 1997).

51. Ashby, *Introduction to Cybernetics*, 2.

52. Wiener, *God and Golem*, 71.

53. Ibid., 69.

54. "See Machines Bordering on Reproduction," *Chicago Daily Tribune*, June 4, 1961, A14.

55. Twenty-four years later, in 1983, a classic science fiction movie immortalized Wiener's nightmare: *WarGames*, with a screenplay by Lawrence Lasker and Walter Parkes, and directed by John Badham. By implying that the machines could deliberately turn on their creators, not merely accidentally, Wiener went even further than the film would go. Wiener, quoted in Roy Gibbons, "Machines That Think Called Peril to Man," *Chicago Daily Tribune*, December 28, 1959, 1.

56. Norbert Wiener, quoted in "Scientist Claims Electronic Brain Could Victimize Man," *Baltimore Sun*, December 28, 1959, 1.

57. Stafford, "Man Called Future Slave."

58. Wiener, quoted in Gibbons, "Machines That Think."

3. AUTOMATION

1. See "Push-Button War," *Newsweek*, August 27, 1945, 25.

2. Half a century later, commanders would have a similar intuition when weapons could be made from software: the machine removed first humans and then its own hardware. What remained were lines of code—a revolution so radical, some believed, that new rules would apply "in cyberspace." See Thomas Rid, *Cyber War Will Not Take Place* (Oxford: Oxford University Press, 2013).

3. Henry Arnold, "If War Comes Again," *New York Times*, November 18, 1945, 39.

4. Quoted in Michael S. Sherry, *Preparing for the Next War* (New Haven, CT: Yale University Press, 1977), 19.

5. Norbert Wiener, "Moral Reflections of a Mathematician," *Bulletin of the Atomic Scientists* 12, no. 2 (February 1956): 55.

6. George E. Valley, "How the SAGE Development Began," *Annals of the History of Computing* 7, no. 3 (1985): 197.

7. Ibid., 198.

8. "$61 Billions [*sic*] for a 2-Hour Warning against Sneak Attack," *US News and World Report*, September 6, 1957, 81.

9. Morton M. Astrahan and John F. Jacobs, "History of the Design of the SAGE Computer—The AN/FSQ-7," *Annals of the History of Computing* 5, no. 4 (October 1983): 343.

10. "IBM on Guard! The Semi-Automatic Ground Environment," 1956 educational film produced by IBM Corporation, Military Products Division (with the US Air Force), YouTube video, posted December 18, 2013, https://youtu.be/IzfxVd5nJUg.

11. Valley, "How the SAGE Development Began," 200.

12. Valley recounted this anecdote, including the subsequent details, in ibid., 202.

13. Kenneth Schaffel, *The Emerging Shield: The Air Force and the Evolution of Continental Air Defense 1945–1960* (Washington, DC: Office of Air Force History, 1991), 210.

14. E. F. O'Neill, *A History of Engineering and Science in the Bell System: Transmission Technology (1925–1975)* (Indianapolis, IN: AT&T Bell Laboratories, 1985), 739.

15. Astrahan and Jacobs, "History of the Design," 348.

16. O'Neill, *History of Engineering and Science*, 704–6.

17. M. D. Fagen, *A History of Engineering and Science in the Bell System: National Service in War and Peace (1925–1975)* (Indianapolis, IN: AT&T Bell Laboratories, 1978), 579.

18. "Sage Electronic Brain Teams with Radar in Pushbutton Air Defense System," *Electrical Engineering* 75, no. 3 (March 1956): 306. See also Astrahan and Jacobs, "History of the Design."

19. "Sage Electronic Brain Teams with Radar," *Electrical Engineering*, 306.

20. "IBM Sage Computer Ad, 1960", YouTube video, posted December 29, 2009, https://youtu.be/iCCL4INQcFo?t=2m18s.

21. Norbert Wiener, *Cybernetics* (Cambridge, MA: MIT Press, 1948), 27–28.

22. Norbert Wiener, *The Human Use of Human Beings* (New York: Houghton Mifflin, 1954), 220.

23. "Revival of R.U.R. with New Prologue," *New York Times*, May 7, 1950, 163.

24. Wiener, *Human Use of Human Beings*, 167.

25. Bertrand Russell, "Are Human Beings Necessary?" *Everybody's*, September 15, 1951, 13.

26. See David Standish, "Interview with Kurt Vonnegut Jr.," *Playboy* 20, no. 7 (July 1973): 57–60, 66, 68, 70, 72, 74, 214, 216.

27. Ibid.

28. Frank Riley, "The Cyber and Justice Holmes," *Worlds of If* 5, no. 1 (1955): 55.

29. Ibid., 57.

30. R. H. Macmillan, *Automation: Friend or Foe?* (Cambridge: Cambridge University Press, 1956), 61.

31. John Johnsrud, "Computer Marks Fifteenth Year," *New York Times*, November 2, 1961, 51.

32. David R. Francis, "Self-Producing Machines," *Christian Science Monitor*, June 2, 1961, 16.

33. Norbert Wiener, *God and Golem, Inc.* (Cambridge, MA: MIT Press, 1963), 4.

34. Ibid., 5.

35. Ibid., 10.

36. L. Landon Goodman, "Automation and Its Social and Economic Implications" (paper presented at the British Electrical Development Association annual conference, April 12, 1956), 1.

37. Alistair Cooke, "Big Brains," *Letter from America*, BBC Radio 4, January 21, 1962, 21:00.

38. Norbert Wiener, "Some Moral and Technical Consequences of Automation," *Science* 131, no. 3410 (May 6, 1960): 1358.

39. Astrahan and Jacobs, "History of the Design," 349.

40. John Diebold, *Automation* (New York: Von Nostrand, 1952), 154.

41. John Diebold, *Beyond Automation* (New York: McGraw Hill, 1964), 105.

42. Ibid., 106.

43. Peter F. Drucker, "The Promise of Automation," *Harper's Magazine* 210, no. 1259 (April 1, 1955): 41.

44. Diebold, *Beyond Automation*, 108.

45. "7-Radar Net to See Rocket 3,000 Mi. Away," *Chicago Tribune*, March 13, 1959, A5.

46. John F. Kennedy, "News Conference 24," February 14, 1962, 11.00 a.m., Accession Number JFKWHA-073, John F. Kennedy Presidential Library and Museum.

47. Donald N. Michael, *Cybernation: The Silent Conquest* (Santa Barbara, CA: Center for the Study of Democratic Institutions, 1962), 6.

48. Ibid.

49. Ibid., 9.

50. Ibid., 44.

51. See, for instance, "Rule Cyberneteira," *Washington Post*, February 12, 1962, A14.

52. "Report on Automation Predicts Job Losses and Social Unrest," *New York Times*, January 29, 1962, 1.

53. Alice Mary Hilton, *Logic, Computing Machines, and Automation* (Washington, DC: Spartan Books, 1963), xvi.

54. Ibid.

55. Norbert Wiener to Alice Mary Hilton, March 8, 1963, private archive of Barbara Hayward (AMH's daughter).

56. Alice Mary Hilton to "Mrs. Norbert Wiener," March 23, 1964, private archive of Barbara Hayward (AMH's daughter).

57. "Dr. Norbert Wiener Dead at 69; Known as Father of Automation," *New York Times*, March 19, 1964, 1.

58. Alice Mary Hilton to "Mrs. Norbert Wiener," March 23, 1964.

59. Quoted in Robert MacBride, *The Automated State* (New York: Chilton, 1967), 192.

60. Ibid., 168.

61. Ibid., 192.

62. Alice Mary Hilton, ed., *The Evolving Society* (New York: ICR Press, 1966), dedication.

63. Hilton's personal papers are very limited. Barbara Hayward (Alice Mary Hilton's daughter), interview by the author, March 28, 2014.

64. Hilton, *Evolving Society*, xii.

65. See "Cybernation—Automation—Computerization: 3 Big Words That Could Mean Your Job," *Baltimore Afro-American*, March 13, 1965, A3.

66. "Dr. King Explains How, Why He Came to New York," *New York Amsterdam News*, August 8, 1964. The quote is from Martin Luther King Jr., "Negros-Whites Together," *New York Amsterdam News*, August 15, 1964, 18.

67. James Boggs, "The Negro and Cybernation," in Hilton, *Evolving Society*, 172.

68. Hannah Arendt, "On the Human Condition," in Hilton, *Evolving Society*, 219.

69. "Cyberculture and Girls," *New Yorker*, July 4, 1964, 21–22.

70. Marshall McLuhan, "Cybernation and Culture," in *The Social Impact of Cybernetics*, ed. Charles R. Dechert (Notre Dame, IN: University of Notre Dame Press, 1966), 98–99.

71. John Diebold, "Goals to Match Our Means," in Dechert, *Social Impact of Cybernetics*, 4.

72. Curtis Gerald, *Computers and the Art of Computation* (Reading, MA: Addison-Wesley, 1972), 319.

73. "Automation: Jobs Change, Clamor Dies," *Chicago Tribune*, October 26, 1969, B8.

74. Gregory R. Woirol, *The Technological Unemployment and Structural Unemployment Debates* (Westport, CT: Greenwood, 1996), 78.

75. David Fouquet, "Automation Held Threat to US Value Code," *Washington Post*, May 12, 1964, A24.

76. Diebold, *Beyond Automation*, 10.

77. Fouquet, "Automation Held Threat."

78. Diebold, *Automation*, 170.

79. Diebold, *Beyond Automation*, 206.

80. Kahn left Rand before publishing *The Year 2000*.

81. Herman Kahn and Anthony Wiener, *The Year 2000: A Framework for Speculation on the Next Thirty-Three Years* (London: Macmillan, 1967), 350.

4. ORGANISMS

1. Pesi Masani, *Norbert Wiener, 1894–1964* (Basel: Burkhäuser, 1990), 225.

2. Paul E. Ceruzzi, *A History of Modern Computing* (Cambridge, MA: MIT Press, 2003), 21.

3. Masani, *Norbert Wiener*, 184.

4. John von Neumann to Norbert Wiener, November 29, 1946, in ibid., 243.

5. John von Neumann, *Theory of Self-Reproducing Automata*, ed. Arthur W. Burks (Urbana: University of Illinois Press, 1966), fifth lecture, 78.

6. Ibid., 79.

7. Ibid., 86.

8. Ibid., 87.

9. Edward F. Moore, "Artificial Living Plants," *Scientific American*, October 1956, 118–26.

10. Ibid., 118.

11. Ibid., 119.

12. Ibid., 121.

13. Ibid., 122.

14. Ibid., 126.

15. David R. Francis, "Self-Producing Machines," *Christian Science Monitor*, June 2, 1961, 16.

16. Ibid.

17. Norbert Wiener, "Some Moral and Technical Consequences of Automation," *Science* 131, no. 3410 (May 6, 1960): 1355.

18. Arthur C. Clarke, "Machina Ex Deux," *Playboy*, July 1961, 66.

19. Ibid.

20. Ibid., 70.

21. Ibid., 66.

22. Manfred Clynes to Norbert Wiener, November 13, 1961, Norbert Wiener Papers, MC 22, box 21, folder 305, Institute Archives and Special Collections, MIT Libraries, Cambridge, MA.

23. Alexis Madrigal, "The Man Who First Said 'Cyborg,' 50 Years Later," *Atlantic*, September 30, 2010.

24. See Daniel S. Halacy, *Cyborg: Evolution of the Superman* (New York: Harper & Row, 1965), 147.

25. Manfred E. Clynes and Nathan S. Kline, "Cyborgs and Space," *Astronautics*, September 1960, 27.

26. Halacy, *Cyborg*, 148.

27. Clynes and Kline, "Cyborgs and Space."

28. "Man Remade to Live in Space," *Life*, July 11, 1960, 77–78; "Man in Space," *Life*, October 2, 1964, 124.

29. "Man Remade to Live in Space," *Life*.

30. Quoted in Bob Ward, *Dr. Space: The Life of Wernher von Braun* (Annapolis, MD: Naval Institute Press, 2005), 156.

31. Albert Rosenfeld, *The Second Genesis: The Coming Control of Life* (Englewood Cliffs, NJ: Prentice-Hall, 1969), 276–77. The *Michigan Technic*, in December 1969, examined various life support systems and called the idea "a bit of science fiction." David Lucas, "Life Support Systems," *Michigan Technic*, December 1969, 11.

32. Robert W. Driscoll, *Engineering Man for Space: The Cyborg Study*, contract no. NASw-512 (Farmingdale, CT: United Aircraft Corporate Systems Center, 1963), republished in Chris Hables Gray, ed., *The Cyborg Handbook* (New York: Routledge, 1995), 76.

33. 456th Fighter Interceptor Squadron, "An Airfield without a Runway," https://web.archive.org/web/20110509065307/http://www.456fis.org/AN_AIR-FIELD_WITHOUT_A_RUNWAY.htm, cached on May 9, 2011.

34. "Discovery Channel: Nuclear Airplane—Part 1," Discovery Channel episode

"The Atomic Bomber" of the documentary series *Planes That Never Flew*, 2003, YouTube video, posted May 4, 2010, https://youtu.be/xb7uZQ1_n4w.

35. See James R. Berry, "I Was an 18-Foot Robot," *Popular Mechanics*, October 1965, 202.

36. Ralph S. Mosher, "Industrial Manipulators," *Scientific American* 211, no. 4 (October 1964): 88–96.

37. See Berry, "I Was an 18-Foot Robot," 202.

38. Ralph S. Mosher, *Applying Force Feedback Servomechanism Technology to Mobility Problems*, contract no. DAAE07-72-C-0109, Technical Report 11768 (LL 144) (Warren, MI: US Army Tank-Automotive Command, 1973), 26.

39. Ralph S. Mosher, *Handyman to Hardiman*, SAE Technical Paper 670088 (Warrendale, PA: Society of Automotive Engineers, 1967), 4.

40. Mosher, "Industrial Manipulators," 88–96.

41. Mosher, *Applying Force Feedback*, 7.

42. See final report in ibid.

43. A video of the experimental machine can be found at "Walking Machine in U.S.A. 1966," British Pathé, http://www.britishpathe.com/video/walking-machine-in-u-s-a.

44. Mosher, *Handyman to Hardiman*, 8.

45. Berry, "I Was an 18-Foot Robot," 118.

46. Quoted in Robert A. Freitas Jr., "The Birth of the Cyborg," in *Robotics*, ed. Marvin Minsky (New York: Omni Publications International, 1985), 151.

47. Ibid., 150.

48. Ralph S. Mosher, J. S. Fleszar, and P. F. Croshaw, "Test and Evaluation of the Limited-Motion Pedipulator," AD0637681 (Ft. Belvoir: Defense Technical Information Center, 1966), abstract.

49. See Berry, "I Was an 18-Foot Robot," 118.

50. A video of the cybernetic walking machine in action can be found at "GE Walking Truck—Cybernetic Anthropomorphous Machine (CAM) 1969," YouTube video, posted January 30, 2010, http://youtu.be/ZMGCFLEYakM.

51. Freitas, "Birth of the Cyborg," 152.

52. See "The Fabulous Walking Truck," *Popular Science* 194 (March 1969): 76–79.

53. "Huge 'Beetle' to Handle Nuclear Power," *Missiles and Rockets* 8, no. 15 (October 9, 1961): 39.

54. Ralph S. Mosher, "Manipulation and Viewing," *Transactions of the American Nuclear Society* 5, no. 2 (1962): 310.

55. Mosher, *Handyman to Hardiman*, 2.

56. Ibid., 5.

57. Ibid., 10.

58. Ibid., 10–11.

59. Bruce R. Fick and John B. Makinson, *Final Report on Hardiman I Prototype for Machine Augmentation of Human Strength and Endurance*, ONR contract no. N00014-66-C0051 (Philadelphia: General Electric, 1971).

60. Berry, "I Was an 18-Foot Robot," 66.

61. Freitas, "Birth of the Cyborg," 159.

62. Walter Troy Spencer, "Not Robots, They're Cyborgs," *New York Times*, December 14, 1969.

63. Halacy, *Cyborg*, 144.

64. Marvin Minsky, "Telepresence," *Omni* 2, no. 9 (June 1980): 50.

65. Charles P. Comeau and James S. Bryan, "Headsight Television System Provides Remote Surveillance," *Electronics* 34, no. 45 (November 10, 1961): 89.

66. Halacy, *Cyborg*, 11.

67. Ibid., 19.

68. David M. Rorvik, *As Man Becomes Machine: The Evolution of the Cyborg* (New York: Doubleday, 1970), 16.

69. Ibid., 13.

70. Gray, *Cyborg Handbook*, 36.

71. Ibid., 37.

72. William Aspray and Arthur Norberg, *An Interview with J. C. R. Licklider* (Cambridge, MA: Charles Babbage Institute, 1988), 13.

73. Chigusa Ishikawa Kita, "JCR Licklider's Vision for the IPTO," *IEEE Annals of the History of Computing* 25, no. 3 (2003): 66.

74. J. C. R. Licklider, "The Truly SAGE System, or, Toward a Man-Machine System for Thinking," August 20, 1957, Licklider Papers, box 6, folder "1957," MIT Libraries, Cambridge, MA. See also J. C. R. Licklider and Welden E. Clark, "On-line Man-Computer Communication," *American Federation of Information Processing Societies Proceedings* 21 (1962): 113.

75. Aspray and Norberg, *Interview with J. C. R. Licklider*, 38.

76. Ibid., 24.

77. Ibid., 24–25.

78. Quoted in Kita, "JCR Licklider's Vision," 68.

79. Ibid.

80. J. C. R. Licklider, "Man-Computer Symbiosis," *IRE Transactions on Human Factors in Electronics* 1 (March 1960): 5.

81. Kita, "JCR Licklider's Vision," 68.

82. Licklider, "Man-Computer Symbiosis," 5.

83. Licklider and Clark, "On-line Man-Computer Communication," 115.

84. Licklider, "Man-Computer Symbiosis," 11.

85. J. C. R. Licklider, "Topics for Discussion at the Forthcoming Meeting," Memorandum for Affiliates of the Intergalactic Computer Network, Advanced Research Projects Agency, Washington, DC, April 25, 1963.

86. Dan van der Vat, "Jack Good," *Guardian*, April 29, 2009.

87. Irving J. Good, "Speculations concerning the First Ultraintelligent Machine," *Advances in Computers* 6 (1965): 31–88.

88. John von Neumann discussed the effects of ever-accelerating technological progress with colleagues. In one such discussion, he allegedly said that humankind is approaching an essential "singularity" after which human affairs will be altered forever. See the recollection of Stanislaw Ulam: "Tribute to John von Neumann," *Bulletin of the American Mathematical Society* 64, no. 3 (1958): 5.

89. Vernor Vinge, "First Word," *Omni* 5, no. 1 (January 1983): 10. For Vinge's weak scientific output, see his Google Scholar profile, http://bit.ly/vinge-scholar+.

90. Jürgen Kraus, "Selbstreproduktion bei Programmen" (master's thesis, Universität Dortmund, Abteilung Informatik, 1980).

91. Ibid., 2.

92. Ibid., 154.

93. Ibid., 161.

94. Ibid., 160.

95. Ronald R. Kline, *The Cybernetics Moment* (Baltimore: Johns Hopkins University Press, 2015).

96. A search for "cyborgs" on Google Scholar will illustrate the extent to which critical philosophers have taken over the concept.

97. The commonly cited version is Donna Haraway, "A Cyborg Manifesto," in *Simians, Cyborgs, and Women* (London: Routledge, 1991), 149.

98. Ibid.

99. Ibid., 177.

100. Ibid., 151.

101. Ibid., 152.

102. Gray, *Cyborg Handbook*, xiii.

103. Haraway, in ibid., xix.

104. Ibid., 12.

105. N. Katherine Hayles, "The Life Cycle of Cyborgs: Writing the Posthuman," in ibid., 322.

106. Interview with Clynes, in Gray, *Cyborg Handbook*, 49.

107. W. R. Macauley and A. J. Gordo-Lopez, "From Cognitive Psychologies to Mythologies," in Gray, *Cyborg Handbook*, 442.

108. Cynthia J. Fuchs, "Death Is Irrelevant," in Gray, *Cyborg Handbook*, 291.

109. Clynes, quoted in Gray, *Cyborg Handbook*, 47.

5. CULTURE

1. Kevin Kelly, *Out of Control* (Reading, MA: Addison-Wesley, 1995).

2. L. Ron Hubbard, *Dianetics* (Los Angeles: Bridge Publications, 1950), 56.

3. See also Ronald R. Kline, *The Cybernetics Moment* (Baltimore: Johns Hopkins University Press, 2015), 91–93.

4. Hubbard, *Dianetics*, 23.

5. William Schlecht to Norbert Wiener, June 29, 1950, Norbert Wiener Papers, MC 22, box 8 ("Correspondence 1950"), folder 121, Institute Archives and Special Collections, MIT Libraries, Cambridge, MA.

6. Norbert Wiener to William Schlecht, July 8, 1950, Norbert Wiener Papers, MC 22, box 8 ("Correspondence 1950"), folder 121, Institute Archives and Special Collections, MIT Libraries, Cambridge, MA.

7. Norbert Wiener to Frederick Schuman, August 14, 1950, Norbert Wiener Papers, MC 22, box 8 ("Correspondence 1950"), folder 122, Institute Archives and Special Collections, MIT Libraries, Cambridge, MA.

8. Norbert Wiener to L. Ron Hubbard, July 8, 1950, Norbert Wiener Papers, MC 22, box 8 ("Correspondence 1950"), folder 121, Institute Archives and Special Collections, MIT Libraries, Cambridge, MA.

9. L. Ron Hubbard to Norbert Wiener, July 26, 1950, Norbert Wiener Papers, MC 22, box 8 ("Correspondence 1950"), folder 121, Institute Archives and Special Collections, MIT Libraries, Cambridge, MA.

10. L. Ron Hubbard to Claude Shannon, December 6, 1949, Claude Elwood Shannon Papers, box 1, MSS84831, Library of Congress, Washington, DC.

11. Norbert Wiener to William Schlecht, July 8, 1950.

12. Norbert Wiener, "Some Maxims for Biologists and Psychologists," *Dialectica* 4, no. 3 (September 15, 1950): 190.

13. Ibid., 191.

14. Ibid.

15. William Grey Walter, *The Living Brain* (London: Duckworth, 1953), 223.

16. Ibid.

17. Maxwell Maltz, *Psycho-Cybernetics* (New York: Pocket Books/Simon & Schuster, 1969), cover.

18. The figure of thirty million is provided by the book's publisher. See the back cover of the 2001 Penguin edition.

19. Maxwell Maltz, *Psycho-Cybernetics* (Englewood Cliffs, NJ: Prentice-Hall, 1960), viii.

20. Ibid., 18.

21. Ibid., 13.

22. Ibid., 28–29.

23. Quoted in Joseph Durso, "The Secret Weapon," *New York Times*, July 15, 1968, L37.

24. Maltz, *Psycho-Cybernetics* (1960), x.

25. Kelly, *Out of Control*, 379.

26. For a detailed version of this remarkable rise, see Fred Turner, *From Counterculture to Cyberculture* (Chicago: University of Chicago Press, 2006).

27. Richard Brautigan, "All Watched Over by Machines of Loving Grace," in *The Pill Versus the Springhill Mine Disaster* (San Francisco: Four Seasons Foundation, 1967), reproduced here with permission of Sarah Lazin Books.

28. Brand, quoted and interviewed by Fred Turner in *From Counterculture to Cyberculture*, 69.

29. Ibid.

30. Ibid.

31. Brand recounts the story in detail in Stewart Brand, "Photography Changes Our Relationship to Our Planet," Smithsonian Photography Initiative, http://web.archive.org/web/20080530221651/http://click.si.edu/Story.aspx?story=31, cached on May 30, 2008.

32. Quoted in Katherine Fulton, "How Stewart Brand Learns," *Los Angeles Times Magazine*, November 30, 1994, 40.

33. Quoted in Turner, *From Counterculture to Cyberculture*, 79.

34. *Whole Earth Catalog*, Fall 1968, 34.

35. Ibid., 35.

36. *The Last Whole Earth Catalog: Access to Tools* (Menlo Park, CA: Portola Institute, 1971), 316.

37. See the preface of every Whole Earth publication, all catalogs and supplements.

38. Michael Rossman, *On Learning and Social Change* (New York: Random House, 1972), 109.

39. Ibid., 203.

40. Ibid., 113.

41. Ibid., 260–61.

42. Ibid., 262.

43. Stewart Brand, "Both Sides of the Necessary Paradox," *Harper's* 247, no. 1482 (November 1973): 20.

44. For more details, see Bateson's short biography in Gregory Bateson, *Mind and Nature* (New York: Dutton, 1978), xiii.

45. Gregory Bateson, *Steps to an Ecology of Mind* (Northvale, NJ: Jason Aronson, 1972), xi.

46. Ibid., 481. Stewart Brand would later be fascinated by this quote. See Brand, "Both Sides of the Necessary Paradox."

47. Bateson, *Steps to an Ecology of Mind*, 321.

48. Ibid., 323.

49. Ibid., 324.

50. Ibid., 315.

51. Ibid., 316.

52. Ibid., 501.

53. For this detail I would like to thank Phillip Guddemi, one of Bateson's former students, who bought two Ashby books on Bateson's recommendation. Phillip Guddemi, "Gregory Bateson and Ross Ashby," e-mail to the author, August 27, 2015. The reading list was published in the winter 1974 issue of *Co-evolution Quarterly*, p. 28.

54. Bateson, *Steps to an Ecology of Mind*, 323.

55. Ibid., 467.

56. Ibid., 468.

57. Ibid., 469.

58. Stewart Brand, *Two Cybernetic Frontiers* (New York: Random House, 1974), 9.

59. See ibid., 24–25, where the story is reprinted.

60. Ibid., 29.

61. Ibid., 7.

62. Ibid., 48.

63. Ibid., 39.

64. Ibid., 49.

65. Stewart Brand, "SPACEWAR: Fanatic Life and Symbolic Death among the Computer Bums," *Rolling Stone*, December 7, 1972, 58.

66. Brand, *Two Cybernetic Frontiers*, 78.

67. Stewart Brand, "We Owe It All to the Hippies," *Time* 145, no. 12 (March 1, 1995): 54–56.

68. Ken Goffman, "Wake Up, It's 1984!" *High Frontiers* 1 (1984): 3.

69. Ibid.

70. Terence McKenna, "Phychopharmacognosticon," *High Frontiers* 4 (1988): 12.

71. Ibid., 11–12.

72. Quoted in Goffman, "Wake Up, It's 1984!" 23.

73. "Apple Macintosh Ad—Aired during the SuperBowl 1984," YouTube video, posted November 9, 2009, http://youtu.be/8UZV7PDt8Lw.

74. Timothy Leary, "Access Codes & Carnival Blasts," *High Frontiers* 1 (1984): 24.

75. Elizabeth Mehrin, "People in the News: Timothy Leary," *PC Magazine*, February 7, 1984, 62.

76. Timothy Leary, *Chaos and Cyber Culture* (Berkeley, CA: Ronin, 1994), 120.

77. Mehrin, "People in the News: Timothy Leary."

78. Katie Hafner, "The Epic Saga of the Well," *Wired* 5.5 (May 1997): 98–142.

79. Ibid.

80. Ibid.

81. For a video of one of these parties, see Susan Hedin, "WELL Party 1989," YouTube video, posted August 24, 2009, http://youtu.be/RhWbQMrqyEc.

82. DEC's VAX systems were still used for testing the ICBM program in 2012; the WELL replaced its VAX in 1988. Benj Edwards, "If It Ain't Broke, Don't Fix It: Ancient Computers in Use Today," *PC World*, February 19, 2012.

6. SPACE

1. See Megan Prelinger, *Inside the Machine: Art and Invention in the Electronic Age* (New York: Norton, 2015).

2. Thomas A. Furness, interview by the author, February 26, 2015.

3. Ibid.

4. Thomas A. Furness and Dean F. Kocian, "Putting Humans into Virtual Space," in *Aerospace Simulation II: Proceedings of the Conference on Aerospace Simulation II, 23–25 January, 1986, San Diego, California*, Simulation Series 16, no. 2 (San Diego: Society for Computer Simulation, 1986), 215.

5. *New Developments in Computer Technology Virtual Reality: Hearing before the Subcommittee on Science, Technology, and Space of the Committee on Commerce, Science, and Transportation, United States Senate, One Hundred Second Congress, First Session, May 8, 1991* (Washington, DC: Government Printing Office, 1992), 24.

6. Furness and Kocian, "Putting Humans into Virtual Space," 214.

7. At the time, Furness did not know about the work of Ivan Sutherland and Morton Heilig, two pioneers of innovative human-machine interfaces often cited in histories of virtual reality. Furness, interview, February 26, 2015.

8. "HITD601: Guest Lecture Tom Furness," YouTube video, posted March 1, 2012, http://youtu.be/JpmM3O4vLto?t=23m55s.

9. Thomas A. Furness, "Visually-Coupled Information Systems," in *Biocybernetic Applications for Military Systems: Conference Proceedings, Chicago, April 5–7, 1978*, ed. Frank E. Gomer (St. Louis, MO: McDonnell Douglas Corporation, 1980).

10. Furness, interview, February 26, 2015.

11. Ibid.; *New Developments in Computer Technology Virtual Reality*, 26.

12. "Top Gun and Beyond," *Nova* season 15, episode 11 (PBS, January 19, 1988).

13. Furness, interview, February 26, 2015.

14. Ibid.

15. David Underwood, "VCASS: Beauty Is in the Eye of the Beholder," *Rotor & Wing International* 20, no. 3 (February 1986): 72.

16. Furness, interview, February 26, 2015.

17. For an example of the experimental voice command and eye tracking in action, see *At the Edge—Aviation Documentary*, director Chris Haws (London: INCA, broadcast October 2, 1988), YouTube video, posted December 9, 2012, https://youtu.be/XloCadboCwU?t=40m0s.

18. Furness, interview, February 26, 2015.

19. *New Developments in Computer Technology Virtual Reality*, 26.

20. "HITD601: Guest Lecture Tom Furness."

21. Furness, interview, February 26, 2015.

22. Dean F. Kocian, *A Visually-Coupled Airborne Systems Simulator (VCASS): An Approach to Visual Simulation* (Wright-Patterson Air Force Base, OH: Aerospace Medical Research Laboratory, 1977), 5.

23. Furness, "Visually-Coupled Information Systems," 43.

24. *New Developments in Computer Technology Virtual Reality*, 26.

25. Furness, interview, February 26, 2015.

26. Dan Rather and David Martin, "US Air Force; High Tech Weaponry," *CBS Evening News*, March 27, 1984, 17:48:10–17:52:00 (-0500), YouTube video, posted March 30, 2015, https://youtu.be/9gScsC7VTFQ?t=3m28s.

27. Furness, interview, February 26, 2015.

28. Stanley W. Kandebo, "Navy to Evaluate Agile Eye Helmet-Mounted Display System," *Aviation Week & Space Technology* 128, no. 7 (August 15, 1988): 94.

29. Furness and Kocian, "Putting Humans into Virtual Space."

30. Thomas A. Furness, "Fantastic Voyage," *Popular Mechanics* 162, no. 12 (December 1986): 63.

31. Thomas A. Furness, "Harnessing Virtual Space," *Digest of Technical Papers* (SID International Symposium), 1988, 4–7.

32. Ibid.

33. Vernor Vinge, *True Names*, in *Binary Star #5*, ed. George R. R. Martin and Vernor Vinge (New York: Dell, 1981).

34. James Frenkel and Vernor Vinge, *True Names and the Opening of the Cyberspace Frontier* (New York: Tor, 2001).

35. Vinge, *True Names*, 250.

36. Ibid.

37. Dmitry (Dima) Adamsky, "The 1983 Nuclear Crisis," *Journal of Strategic Studies* 36, no. 1 (2013): 4–41.

38. One science fiction novel that is sometimes credited with articulating cyberspace as the "Metaverse" is Neal Stephenson, *Snow Crash* (New York: Bantam, 1992).

39. William Gibson, "Cyberpunk Era," *Whole Earth Review* 63 (Summer 1989): 80.

40. "William Gibson: Live from the NYPL," New York Public Library, April 19, 2013, 19:00, YouTube video, posted on July 6, 2013, http://youtu.be/ae3z7Oe3XF4.

41. Ibid.

42. William Gibson, "Burning Chrome," *Omni* 4, no. 10 (July 1982): 76.

43. William Gibson, *Neuromancer* (New York: Ace Books, 1984), 10–11.

44. Ibid., 69.

45. Larry McCaffery, "An Interview with William Gibson," *Mississippi Review* 16, no. 2/3 (1988): 224.

46. Ibid.

47. William Gibson, *Count Zero* (New York: Arbor House, 1986), 33.

48. For a review of the Commodore 64 version of *Moondust*, see "LGR—Moondust—Commodore 64 Game Review," YouTube video, posted June 30, 2009, http://youtu.be/DTk4SqKL-PA.

49. Lanier recounts the story in Jaron Lanier, "Virtually There," *Scientific American* 284, no. 4 (2001): 68.

50. See Thomas G. Zimmerman, Optical flex sensor, US Patent 4542291 A, filed September 29, 1982, and issued September 17, 1985.

51. "Brain Scan: The Virtual Curmudgeon," *Economist*, September 2, 2010.

52. Thomas Zimmerman, interview by the author, April 15, 2014.

53. See Adam Heilbrun's description in Heilbrun, "An Interview with Jaron Lanier," *Whole Earth Review* 64 (Fall 1989): 109.

54. Zimmerman, interview, April 15, 2014.

55. "An Interview with Mitch Altman (Inventor and Virtual Reality Pioneer from the 80's)," YouTube video, posted January 28, 2015, https://youtu.be/5TrRO_j_efg.

56. For a list of companies, see Rudy Rucker, R. U. Sirius, and Queen Mu, *Mondo 2000: User's Guide to the New Edge* (New York: Harper, 1992), 315.

57. "Interview with Mitch Altman."

58. "Virtual Reality from 1990, Jaron Lanier, Eye Phones," YouTube video, posted December 3, 2014, https://youtu.be/ACeoMNux_AU?t=29s.

59. Heilbrun, "Interview with Jaron Lanier," 109.

60. Ibid., 110.

61. Ibid., 114.

62. Ibid., 115.

63. Timothy Leary and Eric Gullichsen, "Artificial Reality Technology," *Reality Hackers* 5 (1988): 23. Thanks to Daniel Bilar for pointing out the Buddha allusion.

64. Andrew Pollack, "For Artificial Reality, Wear a Computer," *New York Times*, April 10, 1989, A1.

65. John Walker, *Through the Looking Glass: Beyond "User Interfaces"* (Sausalito, CA: Autodesk, Inc., 1988).

66. Walker referred to Stanley Kandebo's press article about the Agile Eye in his Autodesk memo, ibid.

67. Walker, *Through the Looking Glass.*

68. Rudy Rucker, *Seek!* (New York: Running Press, 1999), 91.

69. Stewart Brand and Kevin Kelly, "Cyberspace," *Whole Earth Review*, no. 63 (Summer 1989): 84.

70. Ibid., 84–87.

71. Pascal G. Zachary, "Artificial Reality: A Kind of Electronic LSD?" *Wall Street Journal*, January 23, 1990, A1.

72. Jaron Lanier, interview by the author, April 27, 2014.

73. Timothy Leary, *Chaos and Cyber Culture* (Berkeley, CA: Ronin, 1994), 40–41.

74. "Timothy Leary: From LSD to Virtual Reality" (lecture, Sonoma State University, October 19, 1992), YouTube video, posted April 14, 2015, http://youtu.be/7IxZkeE1wQc.

75. Leary, *Chaos and Cyber Culture*, 41.

76. Ibid., 37.

77. Ibid., vii.

78. Quoted in Richard Kadrey, "Cyberthon 1.0," *Whole Earth Review* 70 (Spring 1991): 57.

79. Quoted in Pollack, "For Artificial Reality, Wear a Computer."

80. Timothy Leary, statement in John Forbes, "Cyberspace—The New Explorers," Autodesk, [1989], YouTube video, posted July 28, 2013, http://youtu.be/yYiX42rqbbs?t=5m30s.

81. Ibid.

82. David Sheff, "The Virtual Realities of Timothy Leary," *Upside*, April 1990, 70.

83. Ibid.

84. John Perry Barlow, "Being in Nothingness," *Mondo 2000* 2 (Summer 1990): 38.

85. Quoted in ibid., 39.

86. Ibid., 36.

87. Ibid.

88. Ibid., 41.

89. Ibid.

90. Barlow, quoted in Fred Turner, *From Counterculture to Cyberculture* (Chicago: University of Chicago Press, 2006), 165.

91. Barlow, "Being in Nothingness," 38.

92. See F. Randall Farmer, "Lucasfilm's Habitat Promotional Video" (ca. 1986), YouTube video, posted May 17, 2008, http://youtu.be/VVpulhO3jyc.

93. Chip Morningstar and F. Randall Farmer, "The Lessons of Lucasfilm's Habitat," in *Cyberspace: First Steps*, ed. Michael Benedikt (Cambridge, MA: MIT Press, 1991), 273–77.

94. Vernor Vinge, interview by the author, April 18, 2014.

95. Morningstar and Farmer, "Lessons of Lucasfilm's Habitat," 275.

96. Ibid., 279.

97. Chip Morningstar and Randy Farmer, interview by the author, April 24, 2014.

98. Aaron Britt, "Avatar," *New York Times*, August 10, 2008, MM12.

99. Morningstar and Farmer, "Lessons of Lucasfilm's Habitat," 298.

100. Michael Benedikt, interview by the author, February 26, 2015.

101. Erik Josowitz to comp.society.futures, October 18, 1989, https://groups.google.com/forum/#!forum/comp.society.futures.

102. Michael Benedikt, ed., *Cyberspace: First Steps* (Cambridge, MA: MIT Press, 1991), iii.

103. Benedikt, interview, February 26, 2015.

104. Nicole Stenger, "Mind Is a Leaking Rainbow," in *Cyberspace: First Steps*, ed. Michael Benedikt (Cambridge, MA: MIT Press, 1991), 57.

105. Ibid., 51.

106. Ibid., 52.

107. Ibid.

108. Ibid., 54.

109. Ibid., 56.

110. Allucquère Rosanne Stone, "Will the Real Body Please Stand Up?" in *Cyberspace: First Steps*, ed. Michael Benedikt (Cambridge: MIT Press, 1991), 109.

111. Morningstar and Farmer, interview, April 24, 2014.

112. Ibid.

113. Howard Rheingold, "Teledildonics: Reach Out and Touch Someone," *Mondo 2000* 2 (Summer 1990): 52–54.

114. John Perry Barlow, Lee Felsenstein, and Clifford Stoll, "Is Computer Hacking a Crime?" *Harper's* 280, no. 1678 (March 1, 1990): 51–52.

115. Ibid., 53.

116. John Perry Barlow, "Crime and Puzzlement: In Advance of the Law on the Electronic Frontier," *Whole Earth Review* 68 (Fall 1990): 47.

117. Michael Alexander, "Secret Service Busts Alleged Crime Ring," *Computerworld*, May 14, 1990, 128.

118. Barlow et al., "Is Computer Hacking a Crime?," 53.

119. Barlow, "Crime and Puzzlement."

120. Ibid.

121. Ibid., 57.

122. Kevin Kelly, interview by the author, April 24, 2014.

123. Peggy Orenstein, "Get a Cyberlife," *Mother Jones*, May/June 1991, 62.

124. Ibid., 63.

125. Gregg Keizer, "Virtual Reality," *Compute!* 130 (June 1991): 30.

126. Kadrey, "Cyberthon 1.0," 54.

127. Quoted in Orenstein, "Get a Cyberlife," 64.

128. Quoted in Antonio Lopez, "Networking Meets Authentic Experimental Space," *Santa Fe New Mexican*, January 22, 1999, 46.

129. Keizer, "Virtual Reality," 30.

130. Quoted in Gregg Keizer, "Explorations," *Omni* 13, no. 4 (January 1991): 17.

131. Quoted in Jack Boulware, "Mondo 1995," *SF Weekly* 14, no. 35 (October 11, 1995): 51.

132. "Virtual Reality," *Cryptologic Quarterly* 12, no. 3–4 (Fall/Winter 1993): 47, DOCID 3929132.

133. "The Rather Petite Internet of 1995," *Royal Pingdom*, March 31, 2011.

134. John Perry Barlow, "A Declaration of the Independence of Cyberspace," February 8, 1996, https://projects.eff.org/~barlow/Declaration-Final.html.

135. Ibid.

7. ANARCHY

1. James Ellis, *The Story of Non-secret Encryption* (Cheltenham, UK: GCHQ/CESG, 1987), para. 4.

2. Walter Koenig, *Final Report on Project C-43: Continuation of Decoding Speech Codes*, NDRC contract no. OEMsr-435 (New York: Bell Telephone Laboratories, 1944).

3. James Ellis, *The Possibility of Secure Non-Secret Digital Encryption*, research report no. 3006 (Cheltenham, UK: GCHQ/CESG, 1970).

4. Quoted in Steven Levy, *Crypto* (New York: Penguin, 2000), 396.

5. Clifford Cocks, quoted in Simon Singh, *The Code Book* (London: Fourth Estate, 1999), 285.

6. Ellis, *Possibility*.

7. "You did more with it than we did," Ellis once told fellow cryptographer Whitfield Diffie, but he refused to say more. See the last paragraph of Steven Levy's *Crypto*.

8. Whitfield Diffie and Martin Hellman, "New Directions in Cryptography," *IEEE Transactions on Information Theory* 22, no. 6 (November 1976): 644–54.

9. For an excellent and more detailed description, see Levy, *Crypto*, 90–124.

10. Martin Gardner, "A New Kind of Cipher That Would Take Millions of Years to Break," *Scientific American* 237, no. 2 (August 1977): 120–24.

11. Singh mentions three thousand letters; Levy, seven thousand. Singh, *Code Book*, 278.

12. Deborah Shapley, "Intelligence Agency Chief Seeks 'Dialogue' with Academics," *Science* 202, no. 4366 (October 27, 1978), 408.

13. Bobby R. Inman, "The NSA Perspective on Telecommunications Protection in the Non-governmental Sector," *Signal* 33, no. 6 (March 1979): 7.

14. Ron L. Rivest, Adi Shamir, and Leonard Adleman, "A Method for Obtaining Digital Signatures and Public-Key Cryptosystems," *Communications of the ACM* 21, no. 2 (February 1978): 120.

15. David Chaum, "Security without Identification: Transaction Systems to Make Big Brother Obsolete," *Communications of the ACM* 28, no. 10 (October 1985): 1030.

16. Ibid.

17. David Chaum, "Numbers Can Be a Better Form of Cash Than Paper," in *Computer Security and Industrial Cryptography*, ed. Bart Preneel, René Govaerts, and Joos Vandewalle, Lecture Notes in Computer Science 741 (Berlin: Springer, 1993), 174, 177.

18. Levy, *Crypto*, 213.

19. Andy Greenberg, *This Machine Kills Secrets* (New York: Dutton, 2012), 55–56.

20. Ibid., 56.

21. Vernor Vinge, *True Names*, in *Binary Star #5*, ed. George R. R. Martin and Vernor Vinge (New York: Dell, 1981), 35.

22. Timothy May, interview by the author, April 17, 2014.

23. Ibid.

24. Timothy C. May, "True Nyms and Crypto Anarchy," in *True Names and the Opening of the Cyberspace Frontier*, ed. James Frenkel and Vernor Vinge (New York: Tor, 2001), 83.

25. Ibid.

26. Timothy C. May, "The Crypto Anarchist Manifesto," e-mail to cypherpunks@ toad.com, November 22, 1992. The document dates back to mid-1988.

27. May, interview, April 17, 2014.

28. Kevin Kelly, "Cypherpunks, e-Money, and the Technologies of Disconnection," *Whole Earth Review*, no. 79 (Summer 1993): 45.

29. Timothy C. May, "Announcement: 'Cyphernomicon' FAQ Available," e-mail to cypherpunks@toad.com, September 11, 1994.

30. Ibid.

31. Jude Milhon (as St. Jude), "The Cypherpunk Movement: Irresponsible Journalism," *Mondo 2000*, no. 8 (1992): 36–37.

32. See cover of *Wired* magazine 1.02, May/June 1993.

33. May, "Announcement."

34. Ibid.

35. Lackey archived the entire list and started hosting the archives online in 1995/1996. My research was possible largely thanks to his archive at cypherpunks .venona.com. Ryan Lackey, e-mail to the author, February 16, 2015.

36. John Gilmore, "Summary of Cypherpunks Discussion Volume and Participants," e-mail to Gene Grantham, April 13, 1999.

37. May, "True Nyms and Crypto Anarchy," 83.

38. Ibid., 61.

39. Kelly, "Cypherpunks, e-Money," 46.

40. May, "Announcement."

41. Ibid.

42. Kelly, "Cypherpunks, e-Money," 42.

43. Kevin Kelly, *Out of Control* (Reading, MA: Addison-Wesley, 1995), 178.

44. May, "True Nyms and Crypto Anarchy," 82.

45. James Bamford, *The Puzzle Palace* (New York: Houghton Mifflin, 1982), 47–56.

46. The find wasn't too remarkable. James Bamford had prominently reported on the wider Friedman papers in an influential book on the NSA: *The Puzzle Palace*, published in 1982. John Gilmore, interview by the author, April 7, 2014.

47. Gilmore, interview, April 7, 2014.

48. Ibid.

49. John Markoff, "In Retreat, U.S. Spy Agency Shrugs at Found Secret Data," *New York Times*, November 28, 1992.

50. Gilmore, interview, April 7, 2014.

51. John Perry Barlow, "Remarks," in *First International Symposium: "National*

Security & National Competitiveness: Open Source Solutions": Proceedings, vol. 2 (Reston, VA: Open Source Solutions, 1992), 182–83.

52. Timothy May, "'Stopping Crime' Necessarily Means Invasiveness," e-mail to cypherpunks@toad.com, October 17, 1996. See also May, "True Nyms and Crypto Anarchy," 81.

53. Quoted in Philip Elmer-DeWitt, "Who Should Keep the Keys?" *Time*, March 14, 1994, 90.

54. Matt Blaze, "Protocol Failure in the Escrowed Encryption Standard," in *Proceedings of the 2nd ACM Conference on Computer & Communications Security, November 02–04, 1994* (New York: Association for Computing Machinery), 59–67.

55. Tommy the Tourist, "Please Read: Philip Zimmerman Arrested," e-mail to cypherpunks@toad.com, April 1, 1994.

56. Stewart Brand, "Re: Philip Zimmerman Arrested [Not!]," e-mail to cypherpunks@toad.com, April 3, 1994.

57. Eric Hughes, "Philip Zimmerman Arrested [Not!]," e-mail to Stewart Brand, April 4, 1994.

58. Gilmore, interview, April 7, 2014.

59. Hughes, "Philip Zimmerman Arrested [Not!]."

60. "Introduction to BlackNet," e-mail to cypherpunks@toad.com, August 18, 1993.

61. Ibid.

62. Ibid.

63. May, "True Nyms and Crypto Anarchy," 57.

64. Al Billings, "BlackNet," e-mail to cypherpunks@toad.com, August 18, 1993.

65. Christian D. Odhner, "Re: BlackNet," e-mail to cypherpunks@toad.com, August 18, 1993.

66. Paul Leyland, "The BlackNet 384-Bit PGP Key Has Been BROKEN," posting to alt.security.pgp, June 26, 1995.

67. Timothy May, "Who Is L. Detweiler," e-mail to cypherpunks@toad.com, January 11, 1994.

68. Timothy May, "Untraceable Digital Cash, Information Markets, and Black-Net" (paper presented at the Seventh Conference on Computers, Freedom, and Privacy, Burlingame, CA, March 11–14, 1997).

69. Greenberg, *This Machine Kills Secrets*, 91.

70. The name of the agent was Jeffrey Diehl, who then worked on the FBI's computer analysis and response team in the greater Denver area.

71. May, interview, April 17, 2014.

72. David Chaum, "Achieving Electronic Privacy," *Scientific American*, August 1992, 96.

73. Jim Bell, "Assassination Politics," essay in 10 parts (New York: Cryptome, 1996).

74. Ibid., 1.

75. Ibid.

76. Ibid., 2.

77. Ibid.

78. Ibid.

79. Declan McCullagh, "Crypto-Convict Won't Recant," *Wired*, April 14, 2000.

80. May, interview, April 17, 2014.

81. James Dale Davidson and William Rees-Mogg, *The Sovereign Individual* (New York: Simon & Schuster, 1997), 178–79.

82. Ibid., 197.

83. Ibid., 15.

84. Ibid., 23.

85. Ibid., 24.

86. Ibid.

87. Ibid., 193.

88. *Kirkus Reviews*, December 15, 1996; Peter C. Newman, "Beyond Boom and Bust," *Vancouver Sun*, March 1, 1997, C5; John Chevreau, "Escape Route for Tax Prisoners," *Financial Post*, March 15, 1997, 65; Decca Aitkenhead, "The Sovereign Individual," *Guardian*, April 5, 1997, A3; John H. Fund, "Taking Future Stock," *Wall Street Journal*, March 31, 1997, 12.

89. Howard Campbell, "fighting the cybercensor," e-mail to cypherpunks@toad.com, January 26, 1997.

90. Jim Choate, e-mail to cypherpunks@ssz.com, August 4, 1997.

91. Erwin S. Strauss, *How to Start Your Own Country* (Port Townsend, WA: Loompanics Unlimited, 1984).

92. Ryan Lackey, interview by the author, April 27, 2014.

93. "HavenCo Ltd., Business Plan," [2002,] http://web.archive.org/web/20030425191342/http://www.seanhastings.com/havenco/bplan/ExecSummary.html, cached on April 25, 2003.

94. Lackey, interview, April 27, 2014.

95. Ibid.

96. May, "True Nyms and Crypto Anarchy," 73.

97. Ibid., 60.

98. May, interview, April 17, 2014.

99. May, "True Nyms and Crypto Anarchy," 35–36.

8. WAR

1. Jonathan Post, "Cybernetic War," *Omni* 2, no. 5 (May 1979): 45.

2. Apple, advertisement, *Omni* 2, no. 8 (August 1979): 29.

3. Dylan Tweney, "Sept. 24, 1979: First Online Service for Consumers Debuts," *Wired*, September 24, 2009.

4. Post, "Cybernetic War," 45.

5. Ibid.

6. Ibid., 46.

7. Ibid., 48.

8. Ibid.

9. Ibid., 49.

10. Ibid., 45.

11. Ibid., 104.

12. Frans Osinga, *Science, Strategy and War: The Strategic Theory of John Boyd* (London: Routledge, 2007), 21–22.

13. See ibid., 72, especially notes.

14. US Army, *Field Manual 100-5* (Washington, DC: Headquarters of the Army, 1982), 8-5.

15. Owen Davies, "Robotic Warriors Clash in Cyberwars," *Omni* 9, no. 4 (January 1987).

16. Ibid., 76.

17. Ibid., 78.

18. Henry G. Gole, *General William E. DePuy* (Lexington: University Press of Kentucky, 2008), 297.

19. Eric H. Arnett, "Hyperwar," *Bulletin of the Atomic Scientists* 48, no. 7 (September 1992): 15.

20. Ibid.

21. John Arquilla and David Ronfeldt, "Cyberwar Is Coming!" *Comparative Strategy* 12, no. 2 (1993): 146.

22. Ibid., 162.

23. See David Ronfeldt, "Cyberocracy Is Coming," *Information Society* 8 (1991): 243n1.

24. Arquilla and Ronfeldt, "Cyberwar Is Coming!" 147.

25. John Arquilla and David Ronfeldt, *In Athena's Camp* (Santa Monica, CA: Rand, 1997), 155.

26. Arquilla and Ronfeldt, "Cyberwar Is Coming!" 152.

27. Quoted in Douglas Waller and Mark Thompson, "Onward Cyber Soldiers," *Time* 146, no. 8 (August 21, 1995): 38–45.

28. Ibid.

29. Ibid.

30. William A. Owens, "The Emerging U.S. System-of-Systems," *Strategic Forum*, no. 63 (February 1996): 2.

31. John M. Shalikashvili, *Joint Vision 2010* (Washington, DC: Joint Chiefs of Staff, 1996), 1.

32. Select Comm. on Intelligence, US Senate, *Current and Projected National*

Security Threats to the United States (Washington, DC: US Government Printing Office, 1998), 60–68.

33. Ibid., 68.

34. Winn Schwartau, interview by the author, March 31, 2015.

35. Winn Schwartau, "Fighting Terminal Terrorism," *Computerworld*, January 28, 1991, 23.

36. *Computer Security: Hearing before the Subcommittee on Technology and Competitiveness of the Committee on Science, Space, and Technology, U.S. House of Representatives, One Hundred Second Congress, First Session, June 27, 1991*, no. 42 (Washington, DC: Government Printing Office, 1991), 10.

37. Winn Schwartau, *Terminal Compromise* (Old Hickory, TN: Interpact Press, 1991).

38. Steven Levy, *Hackers* (New York: Doubleday, 1984).

39. Clifford Stoll, *The Cuckoo's Egg* (New York: Doubleday, 1989).

40. Alvin Toffler and Heidi Toffler, *War and Anti-war: Survival at the Dawn of the 21st Century* (Boston: Little, Brown, 1993).

41. Bob Brewin and Elizabeth Sikorovsky, "Information Warfare: DISA Stings Uncover Computer Security Flaws," *Federal Computer Week* 9, no. 3 (1995): 1, 45.

42. Neil Munro, "The Pentagon's New Nightmare: An Electronic Pearl Harbor," *Washington Post*, July 16, 1995, C03.

43. Roger C. Molander et al., *The Day After . . . in Cyberspace*, report on a simulation run at the National Defense University, Washington, DC, on June 3, 1995 (Rand Corporation, 1996).

44. Waller and Thompson, "Onward Cyber Soldiers."

45. Ibid.

46. Tim Weiner, "Head of C.I.A. Plans Center to Protect Federal Computers," *New York Times*, June 26, 1996, B7.

47. The limitation of the target set to military targets alone had complicated legal reasons, since the mission was justified as an information assurance exercise. Anonymous NSA supervisor of the operation, interview by the author, November 19, 2014.

48. Ibid.

49. Ibid.

50. Ibid.

51. Ibid.

52. Bill Gertz, "'Infowar' Game Shut Down U.S. Power Grid, Disabled Pacific Command," *Washington Times*, April 16, 1998, A1.

53. John Hamre, remarks at the Fortune 500 CIO Forum, Aspen, Colorado, July 21, 1998.

54. Anonymous NSA supervisor, interview, November 19, 2014. See also Gertz, "'Infowar' Game Shut Down."

55. "Exercise ELIGIBLE RECEIVER 97," undated Joint Chiefs presentation, Department of Defense (declassified by Brigadier General Bruce Wright on December 11, 2006).

56. Gertz, "'Infowar' Game Shut Down."

57. Jason Healey, *Transcript: Lessons from Our Cyber Past—The First Cyber Cops*, Cyber Statecraft Initiative event, Atlantic Council, Washington, DC, May 16, 2012.

58. Laura Myers, "Security Team Finds Pentagon Computers Unsecured," Associated Press, April 16, 1998.

59. Bill Pietrucha, "US Government to Hack Its Own Computers," *Newsbytes*, March 11, 1998.

60. Quoted in William M. Arkin, "Sunrise, Sunset," *Washington Post*, special supplement at washingtonpost.com, March 29, 1999.

61. FBI, "Solar Sunrise: Dawn of a New Threat," 1999 (published by Kevin Poulsen and *Wired* magazine), YouTube video, posted September 22, 2008, https://youtu.be/bOr5CtqYnsA.

62. Bryan Burrough, "Invisible Enemies," *Vanity Fair*, June 2000.

63. David M. Larue, "CSM CC baby@doe hack," Colorado School of Mines, 6 November 1996.

64. National Infrastructure Protection Center, Memorandum, sender and recipient redacted, "IP ADDRESS (fwd)," 10 August 1998, Washington, DC.

65. Incident report A1010.1, email from [redacted]@rock.mines.edu to navcirt@fiwc.navy.mil, 10 October 1996, 13:22.

66. Jeff Gerth and James Risen, "1998 Report Told of Lab Breaches and China Threat," *New York Times*, 2 May 1999.

67. FBI, "Request for Computer Forensic Media Analysis," File no. 288-CI-68562, October 6, 1998.

68. Anonymous case investigator, interview by the author, January 21, 2015.

69. FBI, Letter, [subject and addressee redacted,] August 4, 1998. Traffic probably originated from aticorp.org; see https://web.archive.org/web/19981212022933/http://www.aticorp.org, cached on December 12, 1998.

70. Ibid.

71. Bob Drogin, "Russians Seem to Be Hacking into Pentagon," *Los Angeles Times*, October 7, 1999.

72. FBI, "RE: (U) 'MOONLIGHT MAZE'" (memo), April 15, 1999, 6.

73. Campbell, "Russian Hackers Steal."

74. Anonymous FBI source, interview by the author, November 20, 2014.

75. Bill Clinton, "Keeping America Secure for the 21st Century" (speech, National Academy of Sciences, Washington, DC, January 22, 1999).

76. Ibid.

77. *The Military Critical Technologies List, Part II: Weapons of Mass Destruction Technologies* (Washington, DC: Department of Defense, 1998), II-iii.

78. Anonymous JTF-CND source, e-mail conversation with the author, November 3, 2014.

79. Anonymous FBI source, interview, November 20, 2014.

80. Anonymous JTF-CND source, e-mail conversation with the author, November 3, 2014.

81. Anonymous senior investigator, interview by the author, November 6, 2014.

82. Anonymous JTF-CND source, e-mail conversation with the author, November 16, 2014.

83. Michael Dorsey, interview by the author, November 6, 2014.

84. FBI, "RE: (U) 'MOONLIGHT MAZE,'" 5.

85. Ibid., 6.

86. Vernon Loeb, "NSA Adviser Says Cyber-Assaults on Pentagon Persist with Few Clues," *Washington Post*, May 7, 2001, A02.

87. Dorsey, interview, November 6, 2014.

88. Ibid.

89. Anonymous former army operator, interview by the author, September 17, 2014.

90. Ibid.

91. Ibid.

92. FBI, "Request for Computer Forensic Media Analysis," File no. 288-CI-68562, October 6, 1998.

93. John Donnelly and Vince Crawley, "Hamre to Hill: 'We're in a Cyberwar,'" *Defense Week*, March 1, 1999, 1.

94. FBI, "RE: (U) 'MOONLIGHT MAZE,'" 5.

95. Ibid., 8.

96. Donnelly and Crawley, "Hamre to Hill," 1.

97. FBI, "RE: (U) 'MOONLIGHT MAZE,'" 10.

98. Anonymous Met source, interview by the author, January 21, 2015.

99. FBI, "RE: (U) 'MOONLIGHT MAZE,'" 10.

100. FBI Baltimore, Squad 14/MMOC, to London, Ottawa, "UNSUB(S); ARMY RESEARCH LAB—VICTIM; INTRUSIONS—INFO SYSTEMS; OO:BA" (memo), April 7, 1999, 1.

101. Anonymous FBI source, e-mail to the author, undisclosed date.

102. Anonymous, interview by the author, undisclosed date.

103. Ibid.

104. Gourley was not part of the Moscow delegation. Bob Gourley, interview by the author, September 18, 2014.

105. Dorsey, interview, November 6, 2014.

106. Bob Gourley, e-mail to the author, November 17, 2014.

107. Jim Garamone, "Hamre 'Cuts' Op Center Ribbon, Thanks Cyberwarriors," *American Forces Press Service*, August 24, 1999.

108. Gregory L. Vistica, "We're in the Middle of a Cyberwar," *Newsweek*, September 20, 1999, 52.

109. *Critical Information Infrastructure Protection: The Threat is Real, Hearing before the Subcommittee on Technology, Terrorism, and Government Information of the Committee of the Judiciary, United States Senate, One Hundred Sixth Congress, First Session on Examining the Protection Efforts Being Made against Foreign-Based Threats to United States Critical Computer Infrastructure, October 6, 1999*, serial no. J-106-53 (Washington, DC: Government Printing Office, 2001), 27.

110. *Critical Information Infrastructure Protection*, 27.

111. Ibid., 30.

112. Michael Vatis declined to be interviewed for this book.

113. Draft of "Remarks by the President on Cyberterrorism," *White House*, January 7, 2000, as of January 6, 2000, 6:00 p.m. EST.

114. "Remarks by the President on Cyberterrorism," White House, January 7, 2000.

115. Steve Kroft, "Cyber War," *60 Minutes* (CBS, April 9, 2000).

116. Gourley, interview, September 18, 2014.

117. Marcus Sachs, interview by the author, September 17, 2014.

9. FALL OF THE MACHINES

1. For a graphical illustration of the rise and fall of cybernetics, see Google's *n*-gram data for "cybernetics," https://goo.gl/qmHGQ3.

2. Kevin Kelly, *Out of Control* (Reading, MA: Addison-Wesley, 1995).

3. See, for instance, David A. Mindell, *Between Human and Machine: Feedback, Control, and Computing before Cybernetics* (Baltimore: Johns Hopkins University Press, 2002); David Tomas, "Feedback and Cybernetics: Reimagining the Body in the Age of the Cyborg," in *Cyberspace/Cyberbodies/Cyberpunk: Cultures of Technical Embodiment*, ed. Mike Featherstone and Roger Burrows (London: Sage, 1995), 21–43.

4. Roland Barthes, *Mythologies* (Paris: Editions du Seuil, 1957), 126.

5. "In Man's Image," *Time* 26, no. 52 (December 27, 1948): 47.

Illustration Credits

The publisher and author make grateful acknowledgment for permission to reproduce the following illustrations.

Frontispiece

Pencil sketch by Alfred Crimi. Courtesy of Northrop Grumman Corporation.

First Insert

Photo of B-17 with turret. US Air Force. Public domain.

Pencil sketch by Alfred Crimi. Courtesy of Northrop Grumman Corporation.

SCR-584 impression. US Army. Public domain.

Photo of V-1 cruise missile. Bundesarchiv, Bild 146-1973-029A-24A. Photo: Lysiak | 1944/1945.

Photo of SCR-584 panel. US Army. Public domain.

Photo of sea fort. Courtesy of the Royal Navy.

Photo of Norbert Wiener with senior army officers. Courtesy MIT Museum.

Photo of Norbert Wiener seated by Alfred Eisenstaedt/The LIFE Picture Collection/Getty Images.

Photo of Stewart Brand playing with the Earth Ball at the New Games. © Ted Streshinsky/CORBIS.

Photo of Stewart Brand holding an issue of *Whole Earth Catalog*. © Roger Ressmeyer/CORBIS.

Second Insert

Photo of Staff Sergeant Vernon Wells with Visually Coupled Airborne Systems Simulator (VCASS) helmet. Department of Defense. Public domain.

Computer-generated image projected inside the VCASS Helmet. Department of Defense. Public domain.

Photo of Timothy Leary. Frans Schellekens/Redferns/Getty Images.

Psychedelic Buddha illustration in *Reality Hackers*, Courtesy Ken Goffman

Photo of William Gibson in *Mondo 2000*. Courtesy Ken Goffman

Illustration for *Mondo 2000*. Courtesy Ken Goffman

Group photo at Cyberconf. Courtesy of Michael L Benedikt.

Photo of Nicole Stenger in VPL gear. Public domain.

Jaron Lanier. Photo: Kevin Kelly.

VPL suits and gloves. Photo: Kevin Kelly.

Full-body VPL suit. Photo: Kevin Kelly.

VPL diagram. Photo: Kevin Kelly.

Photo of VR machine at Whole Earth Institute's Cyberthon, *Mondo 2000*. Courtesy Ken Goffman.

Cyberspace illustration, *Mondo 2000*. Courtesy Ken Goffman.

Cybersex illustration, *Mondo 2000*. Courtesy Ken Goffman.

R.U. a Cyberpunk? *Mondo 2000*, 1993, Nr 10, p. 30. Courtesy Ken Goffman.

Photo of John Perry Barlow. *Mondo 2000*. Courtesy Ken Goffman.

John Gilmore. Photo: Kevin Kelly.

Timothy C. May. Photo: Kevin Kelly.

May, Gilmore, and Hughes. Photo: Larry Dyer, WIRED Magazine.

Ryan Lackey. Photo: Kim Gilmour.

HavenCo platform at Roughs Tower. Photo: Kim Gilmour.

HavenCo servers. Photo: Kim Gilmour.

Cyberwars illustration by Paul Lehr. Published in *Omni* magazine in January 1987.

"Electronic Pearl Harbor" illustration by Harry Whitver. Courtesy of Winn Schwartau, Interpact Inc., 1989.

Photo of aviator with night vision goggles. Department of Defense. Public domain.

John Hamre announcing the creation of the JTF-CND. Photo: John Kandrac. Department of Defense. Public domain.

Washington Monument. Photo: Ron Cogswell.

Moonlight Maze t-shirt. Photo: Thomas Rid.

Index